Group Theory and Chemistry

David M. Bishop
Department of Chemistry
University of Ottawa

Dover Publications
Garden City, New York

This Dover edition, first published in 1993, is an unabridged and corrected republication of the work first published by The Clarendon Press, Oxford, in 1973. A new section of *Answers To Selected Problems* has been added to this edition.

Library of Congress Cataloging-in-Publication Data

Bishop, David M.
Group theory and chemistry / David M. Bishop.
p. cm.
Originally published: Oxford : Clarendon Press, 1973.
Includes bibliographical references and index.
ISBN-13: 978-0-486-67355-4 (pbk.)
ISBN-10: 0-486-67355-3 (pbk.)
1. Group theory. 2. Chemistry, Physical and theoretical.
I. Title.
QD455.3.G75B57 1993
541.2′2′015122—dc20
92-39688
CIP

Manufactured in the United States of America
67355314 2023
www.doverpublications.com

TO MY TEACHERS

J. A. W., D. P. C., R. G. P.

Preface

THIS book is written for chemistry students who wish to understand how group theory is applied to chemical problems. Usually the major obstacle a chemist finds with the subject of this book is the mathematics which is involved; consequently, I have tried to spell out all the relevant mathematics in some detail in appendices to each chapter. The book can then be read either as an introduction, dealing with general concepts (ignoring the appendices), or as a fairly comprehensive description of the subject (including the appendices). The reader is recommended to use the book first without the appendices and then, having grasped the broad outlines, read it a second time with the appendices.

The subject material is suitable for a senior undergraduate course or for a first-year graduate course and could be covered in 15 lectures (without the appendices) or in 21 lectures (with the appendices).

The best advice about reading a book of this nature was probably that given by George Chrystal in the preface to his book *Algebra:*

> Every mathematical book that is worth reading must be read "backwards and forwards", if I may use the expression. I would modify Lagrange's advice a little and say, "Go on, but often return to strengthen your faith". When you come on a hard or dreary passage, pass it over, and come back to it after you have seen its importance or found the need for it further on.

Finally, a word of encouragement to those who are frightened by mathematics. The mathematics involved in actually applying, as opposed to deriving, group theoretical formulae is quite trivial. It involves little more than adding and multiplying. It is in fact possible to make the applications, by filling in the necessary formulae in a routine way, without even understanding where the formulae have come from. I do not, however, advocate this practice.

London D. M. B.
November 1972

Acknowledgements

I would like to thank Professor Victor Gold for the hospitality he extended to me while I was on sabbatical leave at King's College London, where the major part of this book was written. I also owe a particular debt of gratitude to Dr. P. W. Atkins and Dr. B. A. Morrow who read the final typescript in its entirety and to Professor A. D. Westland who read Chapter 12.

I acknowledge with thanks permission to reproduce the following figures:

Fig. 1-2.1 (Trustees of the British Museum (Natural History)),
Fig. 1-2.3 (National Monuments Record, Crown Copyright),
Fig. 1-2.5 (Victoria and Albert Museum, Crown Copyright),
Fig. 12-7.2 and Fig. 12-7.3 (B. N. Figgis, *Introduction to ligand fields*, Interscience Publishers).

I am similarly grateful to Dr. D. S. Schonland for permission to reproduce, in Appendix I, the character tables of his book *Molecular symmetry* (Van Nostrand Co. Ltd.).

Last, I would like to thank Mrs. M. R. Robertson for her immaculate typing.

Contents

List of symbols

R	symmetry element
$\boldsymbol{R}$	symmetry operation
$D(\boldsymbol{R})$	matrix representing $\boldsymbol{R}$
$D_{ij}(\boldsymbol{R})$	element in the ith row, jth column of $D(\boldsymbol{R})$
$\boldsymbol{O}_{\boldsymbol{R}}$	transformation operator corresponding to $\boldsymbol{R}$
A	matrix, elements are displayed between two pairs of vertical lines
A_{ij}	element in the ith row, jth column of A
$\mathscr{A}_{ij}$	cofactor of A_{ij} in $\det(A)$
$\det(A)$	determinant of A
$\mathrm{Trace}(A)$	trace of A
A^*	conjugate complex of A
$\tilde{A}$	transpose of A
$A^\dagger$	adjoint of A
A^{-1}	inverse of A
X	matrix
x_i	ith column of X
x_{ij}	element in the ith row, jth column of X
E	identity matrix
0	null matrix
$\mathscr{G}$	point group
g	order of a group (number of elements in a group)
g_i	number of elements in the ith class of a group
δ_{ij}	Kronecker delta (equals 0 if $i \neq j$, equals 1 if $i = j$)
x_1, x_2, x_3	Cartesian coordinates of a point
Γ	a representation of a point group
Γ^μ	the μth representation of a point group
$D^\mu(\boldsymbol{R})$	the matrix representing $\boldsymbol{R}$ in Γ^μ
$D^\mu_{ij}(\boldsymbol{R})$	the matrix element in the ith row and jth column of $D^\mu(\boldsymbol{R})$
n_μ	the dimension of Γ^μ or the order of $D^\mu(\boldsymbol{R})$
$\chi(\boldsymbol{R})$	the character of $\boldsymbol{R}$ in Γ
$\chi^\mu(\boldsymbol{R})$	the character of $\boldsymbol{R}$ in Γ^μ
$\boldsymbol{P}^\mu(\boldsymbol{R})$	the projection operator $\sum_{\boldsymbol{R}} \chi^\mu(\boldsymbol{R})^* \boldsymbol{O}_{\boldsymbol{R}}$
$\boldsymbol{P}^\mu_{ij}(\boldsymbol{R})$	the projection operator $\sum_{\boldsymbol{R}} D^\mu_{ij}(\boldsymbol{R})^* \boldsymbol{O}_{\boldsymbol{R}}$
a_μ	the number of times Γ^μ occurs in Γ
k	the number of classes in a group
$D^{\mathrm{reg}}(\boldsymbol{R})$	the matrix representing $\boldsymbol{R}$ in the regular representation Γ^{reg}
$\boldsymbol{C}_i$	any operation of the ith class of a point group
$\boldsymbol{R}^m_j$	the jth operation of the mth class of a point group
$\oplus$	symbol linking the irreducible components of a reducible representation
$\otimes$	symbol linking two representations in a direct product representation

E_v	vth energy level
ψ^v	a wavefunction associated with E_v
X	a set of coordinates for a number of particles
X_{nuc}	a set of coordinates for a number of nuclei
X_{el}	a set of coordinates for a number of electrons

(a)

(b)

FIG. 1-2.1. (a) Cymothoe aloatia; (b) primrose.

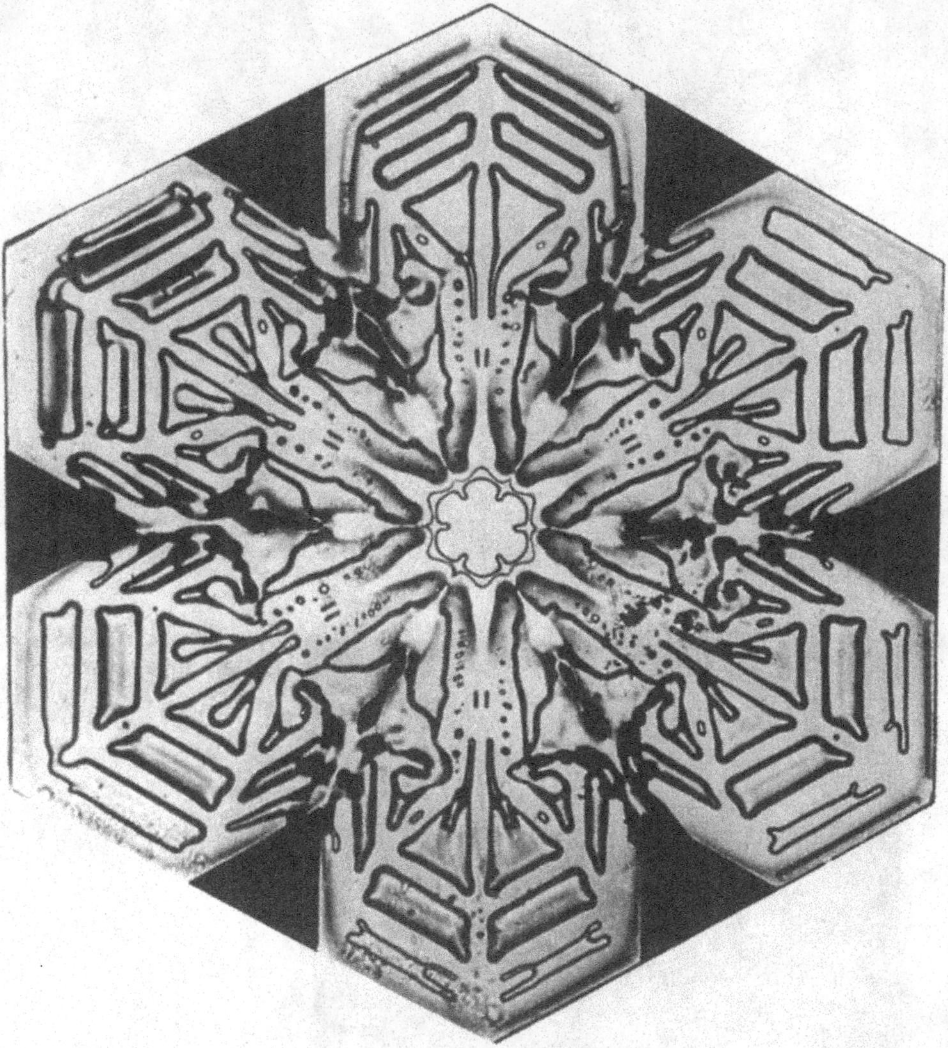

FIG. 1-2.1. (c) ice crystal.

FIG. 1-2.3. The octagonal ceiling in Ely Cathedral.

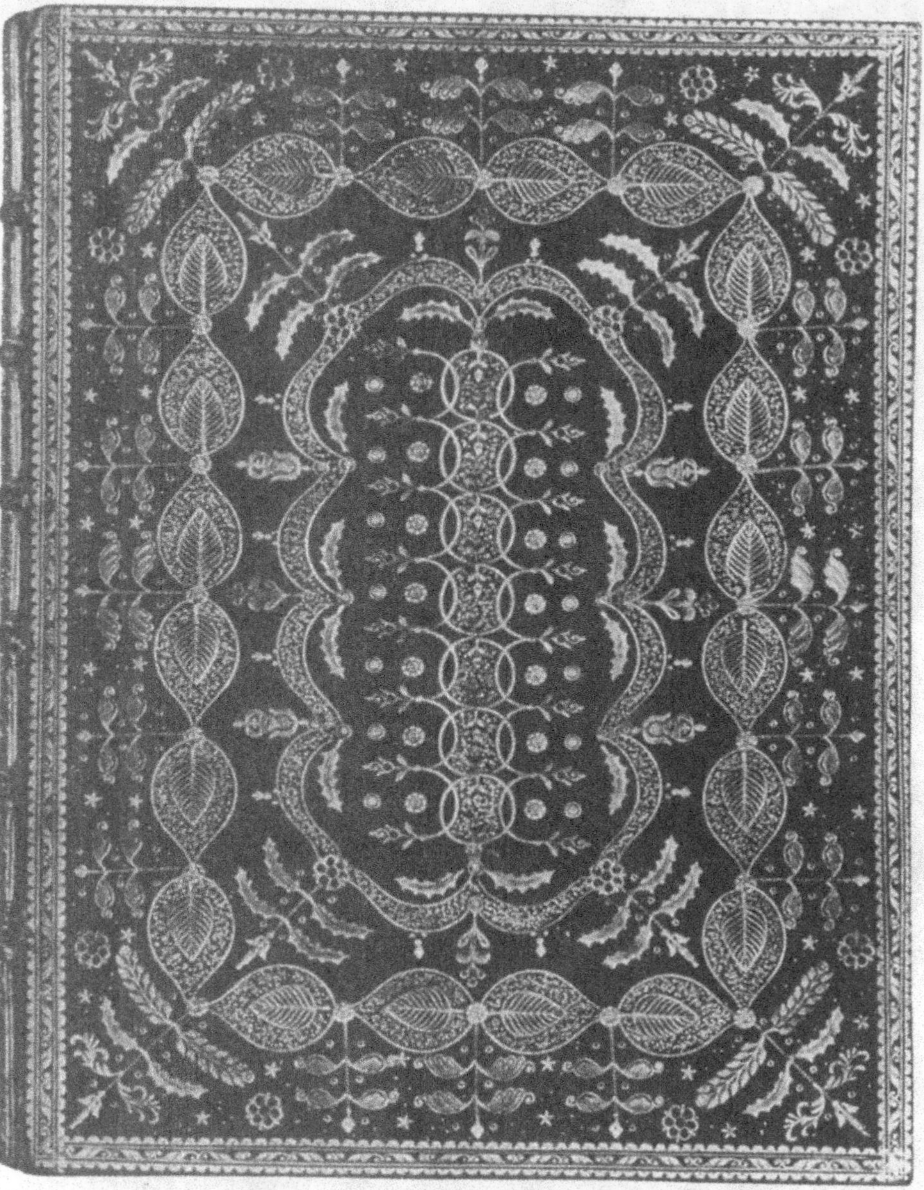

FIG. 1-2.5. An example of Scottish bookbinding, circa 1750.

1. Symmetry

1-1. Introduction

In everyday language we use the word *symmetry* in one of two ways and correspondingly the Oxford English Dictionary gives the following two definitions:

(1) Mutual relation of the parts of something in respect of magnitude and position; relative measurement and arrangement of parts; proportion.

(2) Due or just proportion; harmony of parts with each other and the whole; fitting, regular, or balanced arrangement and relation of parts or elements; the condition or quality of being well proportioned or well balanced.

The first definition of the word has a more scientific ring to it than the second, the second being related to some extent to the rather more nebulous concept of beauty, for example John Bulwer wrote in 1650: 'True and native beauty consists in the just composure and symetrie of the parts of the body'.† It is nonetheless interesting that when we go deeper into the scientific meaning of symmetry we find that the underlying mathematics involved has itself a beauty and elegance which could well be described by the second definition.

In this chapter we will first look at symmetry as it occurs in everyday life and then consider its specific role in chemistry. We will end the chapter by giving a historical sketch of the development of the mathematics which is used in making use of symmetry in chemistry.

1-2. Symmetry and everyday life

The ubiquitous role of symmetry in everyday life has been neatly summarized by James Newman in the following way:

> Symmetry establishes a ridiculous and wonderful cousinship between objects, phenomena, and theories outwardly unrelated: terrestial magnetism, women's veils, polarized light, natural selection, the theory of groups, invariants and transformations, the work habits of bees in the hive, the structure of space, vase designs, quantum physics, scarabs, flower petals,

† This quotation comes from a book with the extraordinary title, *Anthropometamorphosis: Man Transform'd; or the Artificial Changeling. Historically presented, in the mad and cruel Gallantry, foolish Bravery, ridiculous Beauty, filthy finenesse, and loathsome Loveliness of most Nations, fashioning and altering their Bodies from the Mould intended by Nature. With a Vindication of the Regular Beauty and Honesty of Nature. And an Appendix of the Pedigree of the English Gallant.*

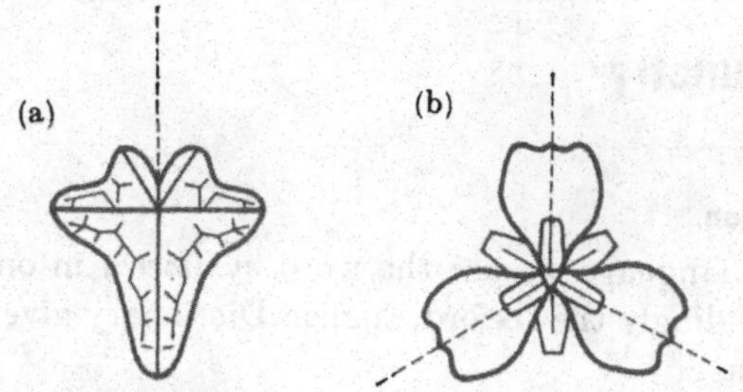

FIG. 1-2.2. (a) Ivy leaf; (b) iris. Dotted lines show planes of symmetry perpendicular to the page.

X-ray interference patterns, cell division in sea urchins, equilibrium positions in crystals, Romanesque cathedrals, snowflakes, music, the theory of relativity.†

In nature we find countless examples of symmetry and in Fig. 1-2.1 we show some rather beautiful examples from the animal, vegetable, and mineral kingdoms. Externally, most animals have bilateral symmetry that is to say they contain a single *plane of symmetry;* such a plane bisects every straight line joining a pair of corresponding points. This is the same thing as saying that the plane divides the object into two parts which are mirror images of each other. In Fig. 1-2.2 it is seen that the ivy leaf and iris have, perpendicular to the plane of the page, one and three planes of symmetry respectively. Actually, the most frequent number of planes of symmetry in flowers is five. Anyone interested in the predominance of bilateral symmetry in the animal world, with its corollary of left and right handedness, is recommended to read *The ambidextrous universe.*‡ In the iris we also notice that there is a three-fold *axis of symmetry*, that is, if we rotate the flower by $2\pi/3$ radians about the axis perpendicular to the page and running down the centre of the flower, then we cannot tell that it has been moved. Similarly, the ice crystal in Fig. 1-2.1 has a six-fold axis of symmetry: a $2\pi/6$ rotation leaves it apparently unmoved.

Because of its basic aesthetic appeal (regularity, pleasing proportions, periodicity, harmonious arrangement) symmetry has, since time immemorial, been used in art. Probably the first example a child experiences of the beauty of symmetry is in playing with a kaleidoscope. More erudite examples occur in: poetry, for example the *abccba* rhyming sequence in many poems; architecture, for example the octagonal ceiling in Ely Cathedral (see Fig. 1-2.3); music, perhaps the most astute use of symmetry in art is a two part piece of music which is sometimes

† *The world of mathematics*, vol. 1, p. 669, Allen and Unwin, London (1960).
‡ M. Gardner, *The ambidextrous universe*, Allen Lane, Penguin Press, London (1967).

FIG. 1-2.4. The score of 'Mozart's' Table Music for Two.

attributed to Mozart,† one part being simply the upside down version of the other; consequently a single copy of the score can be used by both players (see Fig. 1-2.4); and painting and design, see for example the specimen of Scottish bookbinding shown in Fig. 1-2.5.

† In Köchel–Einstein's *Thematic catalogue of Mozart's works* this piece is in the *Anhang* of doubtful and spurious pieces (Anh. 284dd) and Einstein states: 'These items are surely not by Mozart'.

One thing we notice is that all of these examples involve either a plane, an axis or a centre of symmetry, which in turn define a plane, a line or a point about which the object is symmetric.

1-3. Symmetry and chemistry

The involvement of symmetry in chemistry has a long history; in 540 B.C. the Society of Pythagoras held that earth had been produced from the regular hexahedron or cube, fire from the regular tetrahedron, air from the regular octahedron, water from the regular icosahedron, and the heavenly sphere from the regular dodecahedron. Today, the chemist intuitively uses symmetry every time he recognizes which atoms in a molecule are equivalent, for example in pyrene it is easy

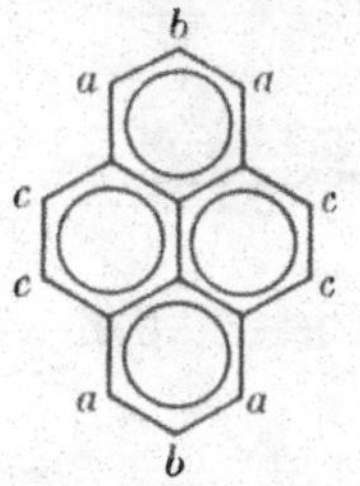

to see that there are three sets of equivalent hydrogen atoms. The appreciation of the number of equivalent atoms in a molecule leads to the possibility of determining the number of substituted molecules that can exist e.g. there are only three possible monosubstituted pyrenes.

Symmetry also plays an important part in the determination of the structure of molecules. Here, a great deal of the evidence comes from the measurement of crystal structures, infra-red spectra, ultra-violet spectra, dipole moments, and optical activities. All of these are properties which depend on molecular symmetry. In connection with the spectroscopic evidence, it is interesting to note that in the preface to his famous book on group theory, Wigner writes:

> I like to recall his [M. von Laue's] question as to which results derived in the present volume I considered most important. My answer was that the explanation of Laporte's rule (the concept of parity) and the quantum theory of the vector addition model appeared to me most significant. Since that time, I have come to agree with his answer that the recognition that almost all rules of spectroscopy follow from the symmetry of the problem is the most remarkable result.

Of course, the basis for our *understanding* of molecular structure (rather than simply its determination) lies in quantum mechanics and therefore any consideration of the role of symmetry in chemistry is basically a consideration of its role in quantum mechanics. The link between symmetry and quantum mechanics is provided by that part of mathematics known as group theory.

1-4. Historical sketch

In spite of the title, most of the mathematics which occurs in this book is in fact only a small part of the subject known as group theory. We will, however, now briefly sketch the history of this theory.

No one person was responsible for the group idea but the figure which looms largest is that of the man who gave the concept its name: Évariste Galois (1811–32). Galois had a short if action-packed life, and he was probably the youngest mathematician ever to make such significant discoveries. He was born in 1811 at Bourg-la-Reine just outside Paris, and by the age of sixteen he had read and understood the works of the great mathematicians of his day. However, despite his genius for mathematics, he failed twice in the entrance examinations to the École Polytechnique which was in those days the Mecca for French mathematicians. Finally, in 1830 he was accepted at the École Normal, only to be expelled the same year for a newspaper letter concerning the actions of the school's director during the July Revolution. Galois had always been a convinced republican and had a strong hatred for all forms of tyranny, so it is not surprising to find that in 1831 he was arrested for proposing a toast which was interpreted as a threat on the life of King Louis Philippe. He was at first acquitted but then, shortly afterwards, he was arrested again and sentenced to six months in jail for illegally wearing a uniform and carrying weapons. He died on May 31st 1832 when only 20 years old from wounds received from being shot in the intestines during a duel. The duel was fought, under a code of honour, over a 'coquette' but some historians believe it was instigated by an *agent provocateur* on the monarchist side. The night before the duel, Galois with forebodings of death wrote out for posterity notes concerning his most important discoveries, which at that time had not been published. His total work is less than sixty pages.

The concept of a group had been introduced by Galois in his work on the theory of equations and this was followed up by Baron Augustin Louis Cauchy (1789–1857) who went on to originate the theory of permutation groups. Other early workers in group theory were: Arthur Cayley (1821–95) who defined the general abstract group as we now know it† and who at the same time developed the theory of matrices; Camille Marie Ennemond Jordan (1838–1922); Marius Sophus Lie (1842–99) and Ludwig Sylow (1832–1918).

For the chemist, however, the most important part of group theory is *representation theory*. This theory and the idea of group characters were developed almost single-handedly at the turn of the century by the German algebraist George Ferdinand Frobenius (1849–1917). Through a decade nearly every volume of the Berliner Sitzungberichte contained one or other of his beautiful papers on this subject.

One of the earliest applications of the theory of groups was in the study of crystal structure and with the later development of X-ray analysis this application was revised and elaborated.‡ This is, however, purely a matter of geometrical classification and though useful in cataloguing possible types of crystals, it has no profound physical significance. Of much more importance is the work of Hermann Weyl (1885–1955) and Eugene Paul Wigner (1902–) who in the late twenties of this century developed the relationship between group theory and quantum mechanics.

It is interesting that Weyl had a deep conviction that the harmony of nature could be expressed in mathematically beautiful laws and an outstanding characteristic of his work was his ability to unite previously unrelated subjects. He created a general theory of matrix representation of continuous groups and discovered that many of the regularities of quantum mechanics could be best understood by means of group theory.

Wigner's greatest contribution was the application of group theory to atomic and nuclear problems; in 1963 he shared the Nobel Prize for physics with J. H. D. Jensen and M. G. Mayer.

Finally attention is drawn to a germinal paper on the application of group theory to problems concerning the nature of crystals which was published in 1929 by another Nobel Prize winner, the German physicist Hans Albrecht Bethe (1906–).

We will conclude this chapter by noting that it is one of the most extraordinary things in science that something as simple and abstract as the theory of groups should be so useful in the practical and everyday problems of the chemist and it is perhaps worth quoting here the English mathematician and philosopher, A. N. Whitehead (1861–1947) who said 'It is no paradox that in our most theoretical moods we may be nearest to our most practical applications'.

† A. Cayley, *Phil. mag.*, 4th Series, 7, 40 (1854).

‡ Artur Schoenflies, *Theorie der Kristallstruktur*, Borntraeger, Berlin (1923).

2. Symmetry operations

2-1. Introduction

THE purpose of this book is to show how the consideration of molecular symmetry can cut short a lot of the work involved in the quantum mechanical treatment of molecules. Of course, all the problems we will be concerned with could be solved by brute force but the use of symmetry is both more expeditious and more elegant. For example, when we come to consider Hückel molecular orbital theory for the trivinylmethyl radical, we will find that if we take account of the molecule's symmetry, we can reduce the problem of solving a 7×7 determinantal equation to the much easier one of solving one 3×3 and two 2×2 determinantal equations and this leads to having one cubic and two quadratic equations rather than one seventh-order equation to solve. Symmetry will also allow us immediately to obtain useful qualitative information about the properties of molecules from which their structure can be predicted; for example, we will be able to predict the differences in the infra-red and Raman spectra of methane and monodeuteromethane and thereby distinguish between them.

However, to start with we must get a clear idea what it is we mean by the symmetry of a molecule. In the first place it means consideration of the arrangement of the atoms (or, more precisely, the nuclei) in their equilibrium positions. Now, when we look at different nuclear arrangements, it is obvious that we require a much more precise and scientific definition of symmetry than any of those given previously in Chapter 1, for clearly there are many different kinds of symmetry, for example the symmetry of benzene is patently different from that of methane, yet both are in some sense symmetric. Only when we have put the concept of symmetry on a sound basis, will we be able to classify molecules into various symmetry types (see the next chapter).

The way in which we systematize our notion of symmetry is by introducing the concept of a *symmetry operation*, which is an action which moves the nuclear framework into a position indistinguishable from the original one. At first sight it would appear that there are very many such operations possible. We will see, however, that each falls into one of five clearly delineated types: identity, rotation, reflection, rotation–reflection, and inversion.

Related to the symmetry operation will be the *symmetry element*. These two terms are not the same and the reader is warned not to confuse them. The symmetry operation is an action, the symmetry element is a geometrical entity (a point, a line or a plane) about which an action takes place. It is worth stressing here that one of the problems in the theory of symmetry is the confusion over the meaning of words, some of which have a general meaning in everyday life but a very precise one in the theory of this book. Further confusion arises from the use of symbols which have one meaning in arithmetic and another in group theory. The reader is advised to think carefully what the words and symbols in this text really mean and not to jump to conclusions. With this in mind, we discuss initially in this chapter the algebra of operators. An operator is the symbol for an operation (the words operator and operation are often used interchangeably, and though, semantically, they should not be, no great harm comes from doing so). On the surface this algebra appears to be the same as the algebra of numbers but, in fact, it is not so.

At the end of this chapter we will show that knowledge of symmetry can lead us directly to predict whether a molecule can have a dipole moment and whether it can exhibit optical activity.

2-2. The algebra of operators

An operator is the symbol for an operation which produces one function from another. Just as a function f assigns to each x in some range a number $f(x)$, so an operator $\boldsymbol{O}$ assigns to each function f in a certain class, a new function denoted by $\boldsymbol{O}f$. (To distinguish operators from other algebraic symbols we will characterize them by bold face italic letters.) An operator, therefore, is a rule or a means of getting one function from another and at the outset it is important to realize their very great generality; they can, for example, be as simple as multiplication by 2 where in mathematical terms we would write:

$$\boldsymbol{O} = 2 \text{ times}$$

and if $\boldsymbol{O}$ operates on the function $f(x) = x^2$ then the new function would be

$$\boldsymbol{O}f(x) = 2 \text{ times } f(x) = 2x^2.$$

Other examples are: the differential operator $\boldsymbol{O} = \mathrm{d}/\mathrm{d}x$; if the original function is $f(x) = x^2+x$ then $\boldsymbol{O}$ produces the new function

$$\boldsymbol{O}f(x) = \frac{\mathrm{d}}{\mathrm{d}x}(x^2+x) = 2x+1;$$

the squaring operator $\boldsymbol{O} = (\ )^2$; if the original function is $f(x) = x^2+x$ then $\boldsymbol{O}$ produces the new function

$$\boldsymbol{O}f(x) = (x^2+x)^2 = x^4+2x^3+x^2;$$

an operator O which changes the sign of x and y; if the original function is $f(x, y) = \exp(ax)\sin(by)$ then O produces the new function

$$Of(x, y) = \exp(-ax)\sin(-by);$$

and the logarithm operator $O = ln(\ )$; if the original function is $f(x) = \exp(ax)$, then the new function will be

$$Of(x) = ln[\exp(ax)] = ax.$$

Clearly, there are countless examples of operators.

One special kind of operator is the *linear operator*. An operator O will be linear if

$$O(f_1+f_2) = Of_1+Of_2, \tag{2-2.1}$$

where f_1 and f_2 are functions of one or more variables, and if

$$O(kf) = kOf \tag{2-2.2}$$

whenever k is a constant. From the examples given before, it is clear that d/dx is a linear operator since

$$O(f_1+f_2) = \frac{d}{dx}(f_1+f_2) = \frac{df_1}{dx}+\frac{df_2}{dx} = Of_1+Of_2,$$

and

$$O(kf) = \frac{d}{dx}(kf) = k\frac{df}{dx} = kOf,$$

whereas $O = \log(\ )$ is not since

$$O(f_1+f_2) = \log(f_1+f_2) \neq \log f_1+\log f_2 = Of_1+Of_2$$

and

$$O(kf) = \log(kf) \neq k\log f = kOf.$$

Another special type of operator is the unitary operator, but we will leave discussion of this type until later (§ 5-7).

The algebra of linear operators consists of a definition of (1) a sum law, (2) a product law, (3) an associative law, and (4) a distributive law.

(1) The sum law. The sum of two linear operators O_1 and O_2 is defined by the equation:

$$(O_1+O_2)f = O_1f+O_2f \tag{2-2.3}$$

e.g.

$$(d/dx+3)x^2 = dx^2/dx+3x^2.$$

(2) The product law. The product of two linear operators O_1 and O_2 is defined by the equation

$$O_1O_2f = O_1(O_2f), \tag{2-2.4}$$

that is O_2 first operates on the function f to produce a new function and then O_1 operates on this new function to produce a final function. It should be noticed that the order of the operators in a product is important (in this respect their algebra is different from the algebra of numbers), and that $\boldsymbol{O}_1\boldsymbol{O}_2$ is not necessarily the same operator as $\boldsymbol{O}_2\boldsymbol{O}_1$ e.g. if $\boldsymbol{O}_1 = \mathrm{d}/\mathrm{d}x$ and $\boldsymbol{O}_2 = x^2$ times, then

$$\boldsymbol{O}_1\boldsymbol{O}_2 f = \frac{\mathrm{d}}{\mathrm{d}x}(x^2 f) = 2xf + x^2\frac{\mathrm{d}f}{\mathrm{d}x}$$

and

$$\boldsymbol{O}_2\boldsymbol{O}_1 f = x^2\left(\frac{\mathrm{d}f}{\mathrm{d}x}\right) = x^2\frac{\mathrm{d}f}{\mathrm{d}x}$$

and symbolically we write:

$$\boldsymbol{O}_1\boldsymbol{O}_2 \neq \boldsymbol{O}_2\boldsymbol{O}_1 \tag{2-2.5}$$

and we say that $\boldsymbol{O}_1$ and $\boldsymbol{O}_2$ do not *commute*. Incidentally, the reader must constantly be aware of the fact that when one writes a product of two operators, $\boldsymbol{O}_1\boldsymbol{O}_2$ it does not mean $\boldsymbol{O}_1$ multiplied by $\boldsymbol{O}_2$ although on paper it might appear that way.

(3) The associative law. This can be expressed by the equation:

$$\boldsymbol{O}_1(\boldsymbol{O}_2\boldsymbol{O}_3) = (\boldsymbol{O}_1\boldsymbol{O}_2)\boldsymbol{O}_3 \tag{2-2.6}$$

which means that combining $\boldsymbol{O}_2$ and $\boldsymbol{O}_3$ (see eqn (2-2.4)) first and then this product with $\boldsymbol{O}_1$ is the same as combining $\boldsymbol{O}_1$ and $\boldsymbol{O}_2$ first and then this product with $\boldsymbol{O}_3$, e.g. if $\boldsymbol{O}_1 = x^2$ times, $\boldsymbol{O}_2 = \mathrm{d}/\mathrm{d}x$ and $\boldsymbol{O}_3 = \mathrm{d}^2/\mathrm{d}x^2$ then:

$$\boldsymbol{O}_1(\boldsymbol{O}_2\boldsymbol{O}_3)f = x^2(\mathrm{d}^3/\mathrm{d}x^3)f = x^2\frac{\mathrm{d}^3 f}{\mathrm{d}x^3},$$

and

$$(\boldsymbol{O}_1\boldsymbol{O}_2)\boldsymbol{O}_3 f = \left(x^2\frac{\mathrm{d}}{\mathrm{d}x}\right)\frac{\mathrm{d}^2 f}{\mathrm{d}x^2} = x^2\frac{\mathrm{d}^3 f}{\mathrm{d}x^3}$$

and eqn (2-2.6) is satisfied.

(4) The distributive law. This law is given by the equations:

$$\begin{aligned} \boldsymbol{O}_1(\boldsymbol{O}_2+\boldsymbol{O}_3) &= \boldsymbol{O}_1\boldsymbol{O}_2+\boldsymbol{O}_1\boldsymbol{O}_3 \\ \text{and}\quad (\boldsymbol{O}_2+\boldsymbol{O}_3)\boldsymbol{O}_1 &= \boldsymbol{O}_2\boldsymbol{O}_1+\boldsymbol{O}_3\boldsymbol{O}_1. \end{aligned} \tag{2-2.7}$$

2-3. Symmetry operations

A symmetry operation is an operation which when applied to a molecule (by which we mean the nuclear framework) moves it in such a way that its final position is *physically* indistinguishable from its initial position. It should be pointed out that such an operation can have no effect on any physical property of the molecule. Also, in this text, we will establish the convention that the operation is applied to the molecule itself and not to some set of spatial axes. The symbol for such an operation is called a symmetry operator (for which bold-face italic type will be used). For every symmetry operation there is a corresponding symmetry element (a point, a line, or a plane) with respect to which the operation is carried out. There are five different kinds of symmetry operation.

The identical operation

This is the operation of doing nothing (leaving the molecule unchanged) and, at first sight, may seem somewhat gratuitous; its inclusion, however, is necessary for the group theory that comes later. The corresponding symmetry element is called the identity and it has the symbol E (from the German word *Einheit* meaning unity).

The rotation operation

This is the operation of rotating a molecule clockwise about an axis. If a rotation by $2\pi/n$ brings the nuclear framework into coincidence with itself, the molecule is said to have as a symmetry element an n-fold axis of symmetry (other terms are n-fold proper axis and n-fold rotation axis). Necessarily n is an integer. The symbol for this element is C_n and for the operator $\boldsymbol{C}_n$. If rotation by $2\pi/n$ produces coincidence then clearly so will rotation by k times $2\pi/n$ (where k is an integer), such an axis, which will coincide with C_n, is given the symbol C_n^k. The corresponding operator $\boldsymbol{C}_n^k$ can be interpreted in one of two ways: a rotation by $k2\pi/n$ or the application of $\boldsymbol{C}_n$ k times. It is apparent that $\boldsymbol{C}_n^n = \boldsymbol{E}$ since a rotation by $n2\pi/n = 2\pi$ is equivalent to doing nothing, and hence n must be an integer. It must also be true that for a molecule containing the symmetry element C_n an *anti-clockwise* rotation by $k2\pi/n$ must also be a symmetry operation and this is denoted by $\boldsymbol{C}_n^{-k}$, from which we see that $\boldsymbol{C}_n^k = \boldsymbol{C}_n^{-(n-k)}$. The axis having the largest n value is called the principal axis.

In Fig. 2-3.1 we illustrate these definitions for a square-based pyramid by labelling the four corners of the base. This labelling is merely to enable us to see that an operation has taken place and it has no physical significance: the whole point of the symmetry operation is that the final orientation is indistinguishable from the original one. Of the operations shown in Fig. 2-3.1, only three, excluding $\boldsymbol{E}$, are distinct: C_4, $\boldsymbol{C}_2$, and $\boldsymbol{C}_4^3$. It is conventional when choosing the symbol for a rotational operation to do so in such a way that n is as small as possible, e.g. $\boldsymbol{C}_2$ is used in preference to $\boldsymbol{C}_4^2$. Finally, it is apparent that quite often symmetry elements will coincide and in such cases we will link the symmetry elements e.g. the C_4, C_2, and C_4^3 axes in Fig. 2-3.1 will be written as C_4–C_2–C_4^3.

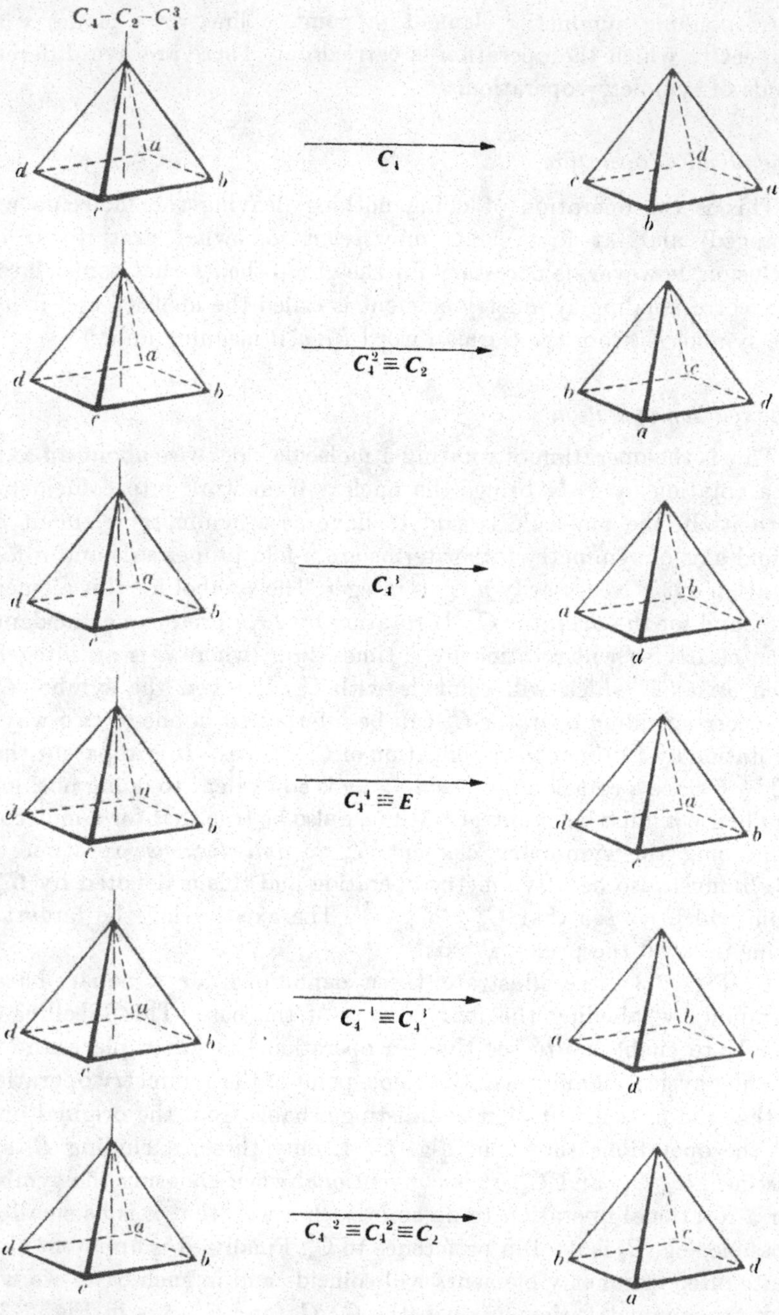

FIG. 2-3.1. Rotation.

The reflection operation

This is the operation of reflection about a plane. If the reflection brings the nuclear framework into coincidence with itself, the molecule is said to have a plane of symmetry as a symmetry element. The symbol given this element is σ (after the German word *Spiegel*, meaning mirror). If such a plane is perpendicular to the principal axis it is labelled σ_h (h = horizontal) and if it contains the principal axis σ_v (v = vertical), if the plane contains the principal axis and bisects the angle between two two-fold axes of symmetry which are perpendicular to the principal axis, it is labelled σ_d (d = diagonal or dihedral), this latter plane is just a special kind of σ_v. We notice that reflecting a molecule twice in the same plane brings it back to its original position and we can write $\boldsymbol{\sigma}^2 = E$. In Fig. 2-3.2 we illustrate these planes for an octahedron and a symmetric tripod.

The rotation–reflection operation

This is the operation of clockwise rotation by $2\pi/n$ about an axis followed by reflection in a plane perpendicular to that axis (or vice versa, the order is not important). If this brings the molecule into coincidence with itself, the molecule is said to have a n-fold alternating axis of symmetry (or improper axis, or rotation–reflection axis) as a symmetry element. It is the 'knight's move' of symmetry. It is symbolized by S_n and illustrated for a tetrahedral molecule in Fig. 2-3.3.†

It is clear that if a molecule has a C_n axis and a plane of symmetry perpendicular to that axis, the C_n axis is also a S_n axis. It is easily seen that the application of S_n twice is the same as the application of C_n twice (the reflection part of S_n is simply annulled); this is written as

$$S_n^2 = C_n^2.$$

In general, k applications of S_n will give

$$S_n^k = \sigma_h C_n^k \quad \text{if } k \text{ is odd}$$

and

$$S_n^k = C_n^k \quad \text{if } k \text{ is even.}$$

Consequently S_n^k can only be interpreted as a rotation C_n^k followed by a reflection in the horizontal plane if k is odd; the opposite is also true e.g. a rotation by $2.2\pi/3$ plus reflection is written as S_3^5 and not as S_3^2 (which would simply be C_3^2). Furthermore, simple arguments lead to

† In this Figure the tetrahedral structure is shown by the cube which circumscribes it and the tetrahedral corners are the alternate corners of the cube. We shall frequently display tetrahedra in this way.

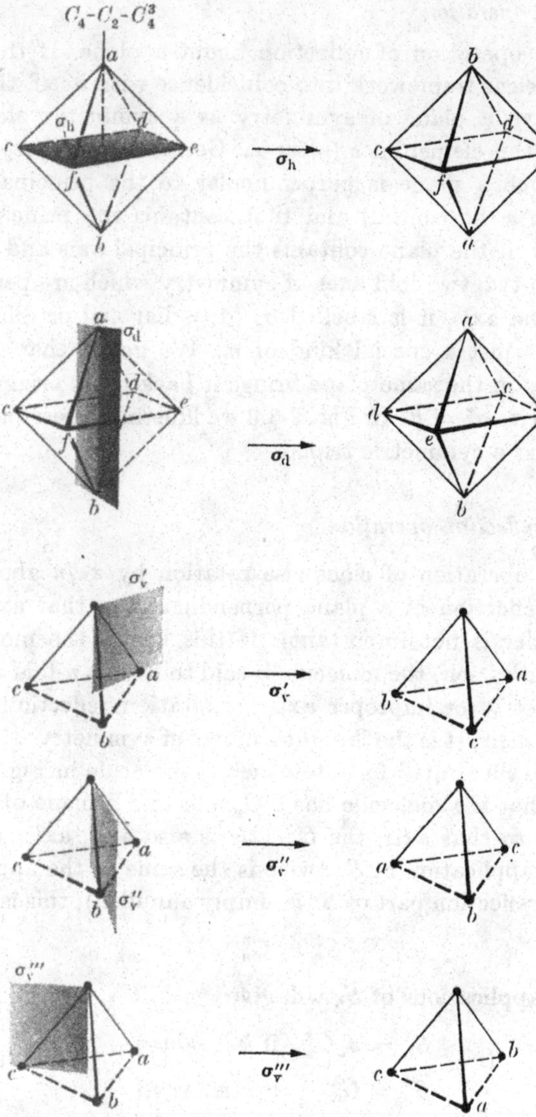

FIG. 2-3.2. Reflection.

the equations:

$$S_1 = \sigma$$

$$S_n^n = \sigma_h \quad \text{if } n \text{ is odd}$$

and

$$S_n^n = E \quad \text{if } n \text{ is even.}$$

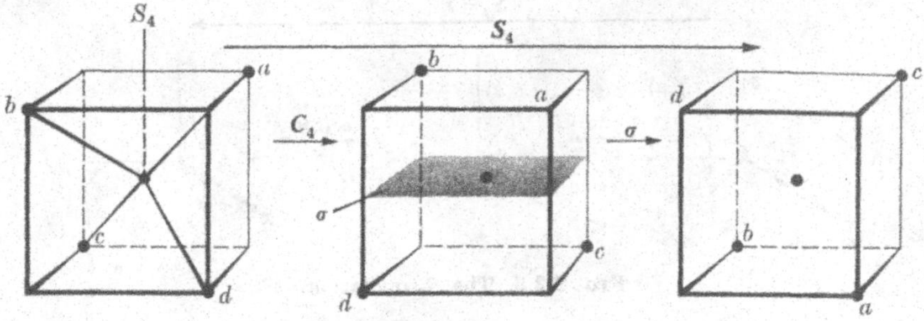

FIG. 2-3.3. Rotation–reflection.

The inverse operation

This is the operation of inverting all points in a body about some centre, i.e. if the centre is O, then any point A is moved to A′ on the line AO such that OA′ = OA or put another way, if a set of Cartesian axes have their origin at O, a point with coordinates (x, y, z) is moved to $(-x, -y, -z)$. If this operation brings the nuclear framework into coincidence with itself, the molecule is said to have a centre of symmetry as a symmetry element and this is symbolized by i (no relation to $\sqrt{-1}$). In Fig. 2-3.4 we show the inversion operation for an octahedral framework.

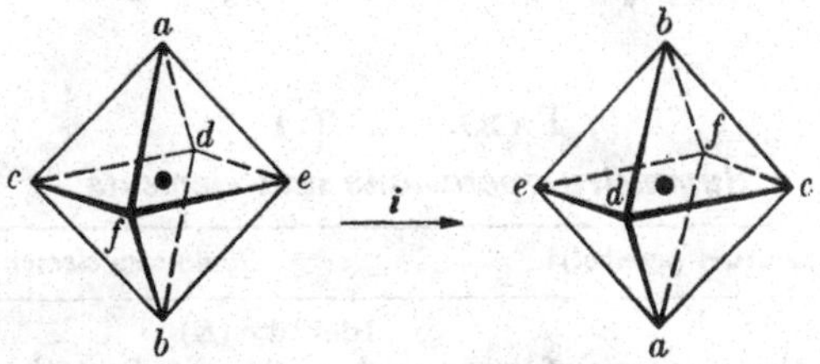

FIG. 2-3.4. Inversion.

It is apparent that S_2 and i are equivalent (see Fig. 2-3.5) and that the application of inversion twice is the same as doing nothing, this is written as $i^2 = E$.

In Table 2-3.1 we summarize the various definitions and symbols which have been discussed.

2-4. The algebra of symmetry operations

Symmetry operations like operators can be combined together and when this is done they produce other symmetry operations, e.g. if P and Q are symbols for any two symmetry operations then PQ is the

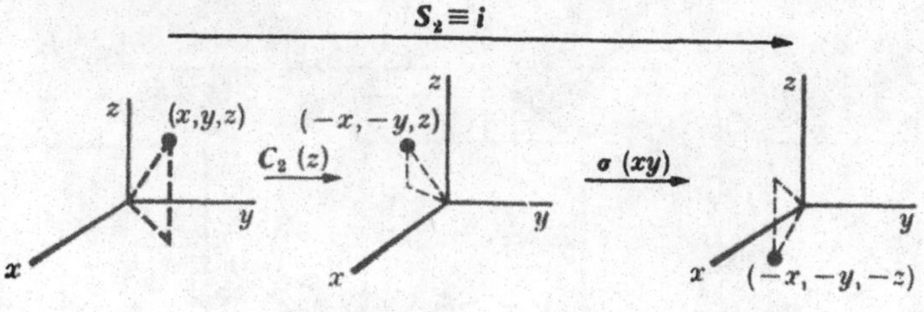

FIG. 2-3.5. The S_2 operation.

operation of first applying Q and then applying P. Notice the convention that the first operation to be carried out is the one on the right. Since both operations, P and Q, leave the molecule coincident with itself, so must their product and therefore the combination of two (or, by extension, more) symmetry operations is itself a symmetry operation. Mathematically, we may write

$$PQ = R \tag{2-4.1}$$

and say that R is also a symmetry operation. The order of the terms in a product is important since symmetry operations do not always commute e.g. for a symmetric tripod (see Fig. 2-4.1)

$$C_3C_3^{-1} = E \quad \text{and} \quad C_3^{-1}C_3 = E \tag{2-4.2}$$

TABLE 2-3.1
Symmetry operations and elements

Symmetry operation (symbol)	*Symmetry element (symbol)*
No change (E)	Identity (E)
Rotation by $2\pi/n$ about an axis of symmetry (C_n)	A n-fold axis of rotation (C_n)
Reflection in a plane of symmetry perpendicular to the principal axis of symmetry (σ_h)	A plane of symmetry perpendicular to the principal axis of symmetry (σ_h)
Reflection in a plane of symmetry containing the principal axis of symmetry (σ_v)	A plane of symmetry containing the principal axis of symmetry (σ_v)
Reflection in a plane of symmetry containing the principal axis of symmetry and bisecting the angle between two 2-fold axes of symmetry which are perpendicular to the principal axis (σ_d)	A plane of symmetry containing the principal axis of symmetry and bisecting the angle between two 2-fold axes of symmetry which are perpendicular to the principal axis (σ_d)
Rotation by $2\pi/n$ about an axis followed by reflection in a plane perpendicular to that axis (S_n)	The n-fold alternating axis (S_n)
Inversion in a centre of symmetry (i)	The centre of symmetry (i)

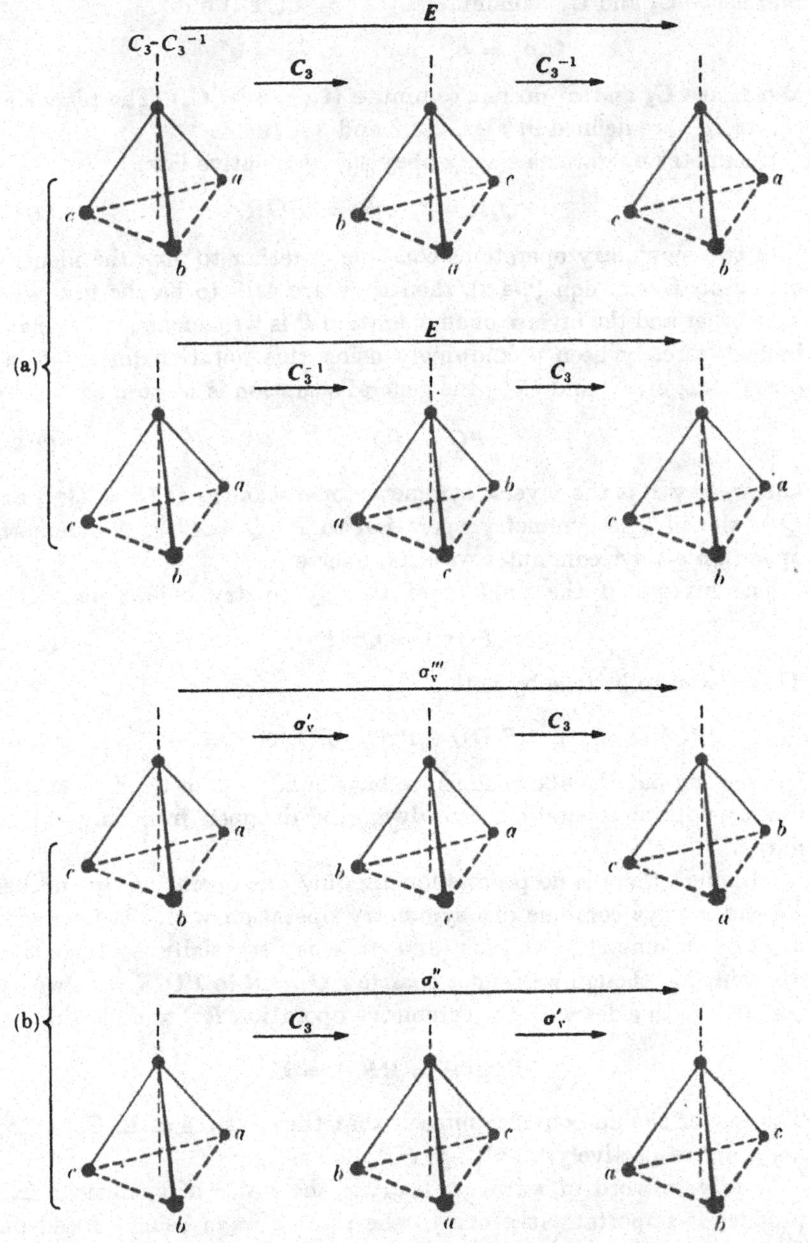

FIG. 2-4.1. (a) Commutation between C_3 and C_3^{-1}; (b) non-commutation between C_3 and σ_v'.

and hence C_3 and C_3^{-1} commute ($C_3C_3^{-1} = C_3^{-1}C_3$) but

$$C_3\sigma'_v = \sigma'''_v \quad \text{and} \quad \sigma'_vC_3 = \sigma''_v$$

and hence C_3 and σ'_v do not commute ($C_3\sigma'_v \neq \sigma'_vC_3$). The planes σ'_v, σ''_v, and σ'''_v are defined in Figs. 2-3.2 and 3-4.1.

Symmetry operations clearly obey the associative law:

$$(PQ)R = P(QR) = PQR. \tag{2-4.3}$$

If two symmetry operations combine together to give the identical operation E, e.g. eqn (2-4.2), then they are said to be the inverse of each other and the inverse of an operation P is written as P^{-1} (we have, in fact, already been unknowingly using this notation for rotational operations, cf: C_3 and C_3^{-1}), the general situation is written as:

$$PQ = QP = E \tag{2-4.4}$$

and we say P is the inverse symmetry operation to Q ($P = Q^{-1}$) and Q is the inverse symmetry operation to P ($Q = P^{-1}$). A symmetry operation always commutes with its inverse.

The inverse of the product of two symmetry operations PQ is $Q^{-1}P^{-1}$:

$$(PQ)^{-1} = Q^{-1}P^{-1}. \tag{2-4.5}$$

This is seen to be true by noting that

$$(PQ)(Q^{-1}P^{-1}) = P(QQ^{-1})P^{-1} = P(E)P^{-1} = PP^{-1} = E.$$

In carrying out the above steps we have made use of the fact that E, the 'do nothing' operation, can always be dropped from any combination.

Although there is no process for dividing one operation by another, we can always combine one symmetry operation with the inverse of another symmetry operation and this is essentially equivalent to division, e.g. though we cannot change $PQ = R$ to $PQ/R = 1$, we can combine both sides with the symmetry operation R^{-1} and obtain

$$PQR^{-1} = RR^{-1} = E.$$

The reader should convince himself that the inverses of E, C_n, σ, S_n, and i are respectively, E, C_n^{-1}, σ, S_n^{-1}, and i.

Finally, a word of warning; because the order of operations in a product is important (the one to the right always being carried out first), one must be careful when manipulating equations involving symmetry operations, for example, if $PQ = R$ then we can write $TPQ = TR$ (combining T on the left of *both* sides) or $PQT = RT$

(combining T on the right of *both* sides) but we cannot write

$$PQT = TR \quad \text{or} \quad TPQ = RT.$$

2-5. Dipole moments

One application of symmetry operations that we can make right away concerns dipole moments. The use of symmetry arguments can tell us whether a molecule has a dipole moment and in many cases along which line it lies. Since a symmetry operation leaves a molecule in a configuration *physically* indistinguishable from the one before the operation, the direction of the dipole moment vector must also remain unchanged after a symmetry operation. Therefore if a molecule has a n-fold axis of rotation C_n the dipole moment must lie along this axis but if we have two or more non-coincident symmetry axes, the molecule cannot have a dipole moment because it cannot lie on two axes at the same time. Methane CH_4 has four non-coincident C_3 axes and therefore has no dipole moment (see Fig. 2-5.1). If there is a plane of symmetry σ, the dipole moment must lie in this plane and if there are several symmetry planes, the dipole moment must lie along their intersection. In ammonia NH_3 the dipole moment lies along the C_3

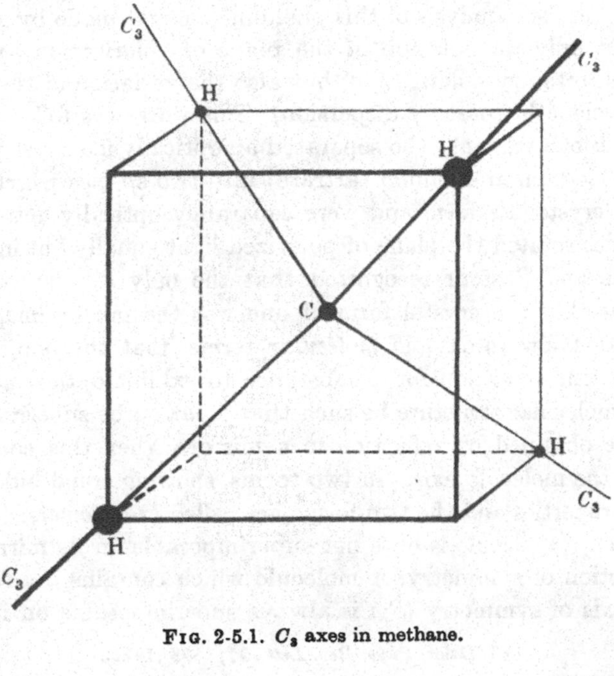

FIG. 2-5.1. C_3 axes in methane.

axis which is also the intersection of three symmetry planes (see Fig. 2-4.1). A molecule containing a centre of symmetry i cannot have a dipole moment, since inversion reverses the direction of any vector.

Our conclusions about dipole moments are all within the context of the Born–Oppenheimer approximation and methane, for example, has in reality a small permanent moment whose magnitude is the order of 10^{-5} to 10^{-6} D. This moment is caused by centrifugal distortion effects.†

2-6. Optical activity

The origins of the subject of optical activity go back to 1690 when Huygens discovered that light could be polarized by a doubly-refracting crystal of Iceland spar (calcite). Detailed descriptions of light polarization are available in many books and for our purposes it is only necessary to recall that in circular polarization, the electric vector associated with the light describes a right- or a left-handed helix and that if a right and a left circularly polarized ray of the same frequency are superimposed, the resultant electric field vector is a sine wave function along a single direction (plane) in space. Such light is said to be plane polarized.

Many substances can rotate the plane of polarization of a ray of plane polarized light. These substances are said to be optically active. The first detailed analysis of this phenomenon was made by Biot, who found not only the rotation of the plane of polarization by various materials (rotatory polarization) but also the variation of the rotation with wavelength (rotatory dispersion). This work was followed up by Pasteur, Biot's student, who separated an optically inactive crystalline material (sodium ammonium tartrate) into two species which were of different crystalline form and were separately optically active. These two species rotated the plane of polarized light equally but in opposite directions and Pasteur recognized that the only difference between them was that the crystal form of one was the mirror image of the other. We know to-day, in *molecular* terms, that the one necessary and sufficient condition for a substance to exhibit optical activity is that its molecular structure be such that it *cannot* be superimposed on its image obtained by reflection in a mirror. When this condition is satisfied the molecule exists in two forms, showing equal but opposite optical properties and the two forms are called *enantiomers*.

Whether a molecule is or is not superimposable on its mirror image is a question of symmetry. A molecule which contains a n-fold alternating axis of symmetry (S_n) is always superimposable on its mirror

† I. Ozier, *Phys. Rev. Lett.*, **27**, 1329 (1971).

image. This is true because the operation S_n consists of two parts: a rotation C_n and a reflection σ. Since a reflection creates the mirror image, the operation S_n is equivalent to rotating in space the mirror image. By definition, a molecule containing a S_n axis is brought into coincidence with itself by the operation S_n and hence its mirror image, after rotation, is superimposable. The reader is reminded that $S_1 = \sigma$ and $S_2 = i$, so that a molecule with either a plane or a centre of symmetry is also optically inactive. However, the most general rule is: a molecule with a S_n axis is optically inactive. Conversely, it can be shown that a molecule without a S_n axis is, in principle, optically active.

Planes and centres of symmetry are easily identified and molecules having these symmetry elements are readily classified as inactive; alternating axes of symmetry (with $n > 2$) can be harder to spot. The first example of a molecule with a S_n axis ($n > 2$) to be experimentally studied was 3,4,3',4'-tetramethyl-spiro-(1,1')-bipyrrolidinium ion in 1955 (see Fig. 2-6.1). This ion has a S_4 axis and was, as expected, found to be inactive.

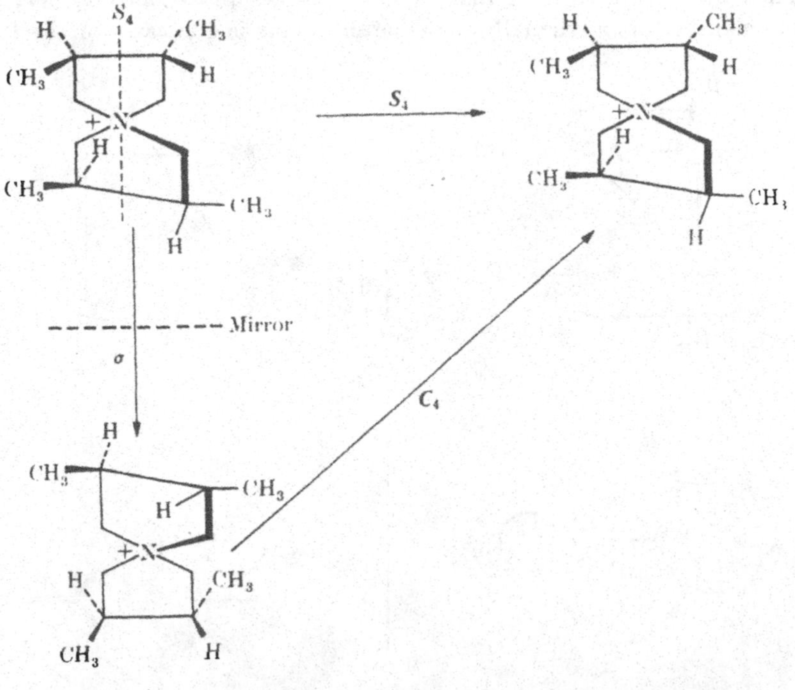

FIG. 2-6.1. 3,4,3',4'-tetramethyl-spiro-(1,1')-bipyrrolidinium ion (inactive).

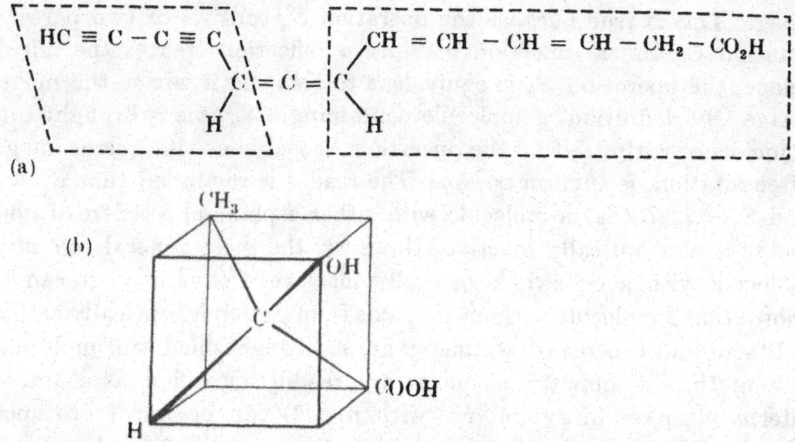

FIG. 2-6.2. (a) Mycomycin, a naturally-occurring optically-active allene; (b) one form of lactic acid (optically active).

In principle, the lack of a S_n axis dictates the existence of optical activity and two examples are given in Fig. 2-6.2, however, in practice, it may not always be possible to actually demonstrate the activity. There are two reasons for this. One is that the optical activity may be so small that it is virtually undetectable; this is the case for the two

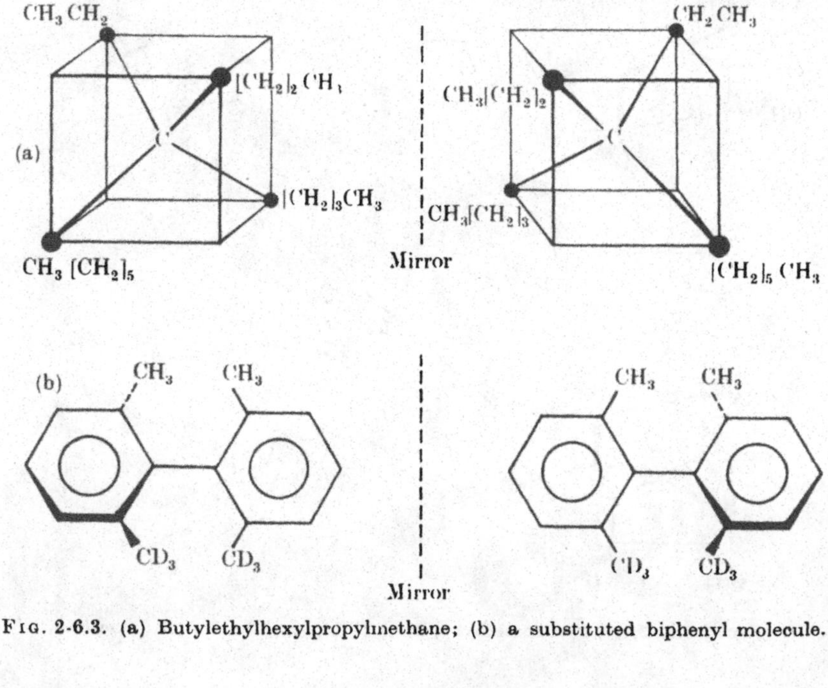

FIG. 2-6.3. (a) Butylethylhexylpropylmethane; (b) a substituted biphenyl molecule.

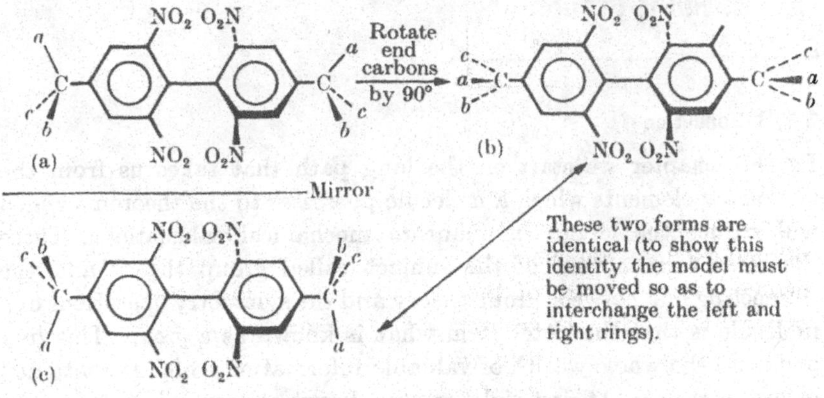

FIG. 2-6.4. Lack of activity due to free rotation.

forms of butylethylhexylpropylmethane or for the two forms of the substituted biphenyl in Fig. 2-6.3. The lack of activity in these molecules can be ascribed to the fact that there are only small differences between the substituent groups. The second reason is that free rotation can prevent a distinction being made. For example, the substituted biphenyl molecule in Fig. 2-6.4 has rotation about the central bond restricted by the nitro-groups but nevertheless rotation of both end groups as shown in conformation (a) gives rise to conformation (b) which is completely superimposable on (c) the mirror image of (a). In other words, though (a) and (c) are not superimposable, (a) can be converted to (b) = (c) simply by free rotation.

The reader is advised that optical activity is alluded to again at the end of the next chapter in problem 3.7.

PROBLEMS

2.1. Give all the symmetry elements of H_2O, NH_3 and CH_4. For each molecule list the symmetry operations which commute.

2.2. On the basis of symmetry, which of the following molecules cannot have a dipole moment: CH_4, CH_3Cl, CH_2D_2, H_2S, SF_6?

2.3. Which of the following molecules cannot be optically active: CHFClBr, H_2O_2, $Co(en)_3^{3+}$, *cis*-$Co(en)_2(NH_3)_2^{3+}$, *trans*-$Co(en)_2(NH_3)_2^{3+}$?

3. Point groups

3-1. Introduction

In this chapter we start on the long path that takes us from the symmetry elements which a molecule possesses to the theorems which will reduce the labour in quantum mechanical calculations. These theorems form a part of the subject called group theory and the connecting link between group theory and the symmetry operations of a molecule is that the latter form what is known as a *group*. This link opens up the whole wealth of valuable information which is contained in group theory. At first sight, group theory appears so abstract and so unrelated to physical reality that it seems amazing that it should be the powerful *practical* tool which it is.

Having defined a group and given some examples, we will consider the specific type of group which is of interest to us: the *point group*. We will then introduce the notation which must be mastered and which allows us to classify molecules according to the symmetry elements they possess and we will give a simple scheme for determining the point group to which a given molecule belongs. Once this classification of molecules has been made, we will no longer need to consider specific molecules but only bodies having certain symmetry properties, i.e. belonging to a particular point group.

3-2. Definition of a group

A group is any set or collection of *elements* which together with some well-defined combining operation obey a certain set of rules. The meaning of the word 'elements' is very general. The elements could, for example, be numbers, matrices, vectors, roots of an equation or symmetry operations. It is important to remember that the definition of a group requires specifying a combining operation. This too is quite general. It could be, depending on the particular group, ordinary addition, ordinary multiplication, matrix multiplication, vector addition, or one operation followed by another.

The rules which the elements of a group must obey are as follows.

The 'product' or combination of any two elements of the group must produce an element which is also a member of the group, i.e. if P and Q are symbols for two members of the group and $PQ = R$, then R must be a member of the group. Notice that PQ does *not* mean P multiplied by Q but rather P combined with Q according to the defined combining rule.

The group must contain the identity element, which is given the symbol E, and is such that when combined with *any* element in the group R it leaves that element unchanged i.e. $RE = ER = R$. Notice that E commutes with all elements of the group.

The associative law must hold for all elements of the group, i.e. $P(QR) = (PQ)R$, or, in words, the combining of P with the combination QR must be the same as the combining of the combination PQ with R.

Every element R must have an inverse element R^{-1} which must also be a member of the group. The inverse is defined by the equation:

$$RR^{-1} = R^{-1}R = E.$$

These four rules are put into compact form in Table 3-2.1.

TABLE 3-2.1

Definition of a group

1. $PQ = R$	R in the group
2. $RE = ER = R$	E in the group
3. $P(QR) = (PQ)R$	For all elements
4. $RR^{-1} = R^{-1}R = E$	R^{-1} in the group

3-3. Some examples of groups

The preceding definition of a group seems rather abstract and exceedingly general and in order to bring things down to earth, we give in this section some concrete examples.

All positive and negative whole numbers together with zero form a group whose combining rule is algebraic addition. The first rule of a group is obeyed because the addition of any two whole numbers produces another whole number which is, by definition, also an element of the group. If we add zero to any whole number, the number is unchanged and so for this group the identity element is zero, and this too is a member of the group. The associative rule clearly holds and the inverse of any number is simply the negative of that number and this is also an element of the group i.e. $a+(-a) = (-a)+a = 0$. There are an infinite number of elements in this group and for this reason it is called an infinite group.

Another infinite group consists of the vectors $\mathbf{r} = a\mathbf{i}+b\mathbf{j}+c\mathbf{k}$, where $\mathbf{i}$, $\mathbf{j}$, and $\mathbf{k}$ are non-coplanar vectors and a, b, and c are positive or negative whole numbers or zero. The combining operation is vector addition and the identity element is the null vector: $a = 0$, $b = 0$, $c = 0$.

Another group consists of the elements 1, −1, i and −i (where $i = \sqrt{-1}$) and the combining operation of algebraic multiplication. Combination, or multiplication, of any two elements produces one of the four elements; 1 is the identity element; the associative rule holds; and the inverse of 1 is 1, of −1 is −1, of i is −i, of −i is i. This is an example of a finite group.

The roots of the equation $x^3 = 1$ are 1, $(i\sqrt{3}-1)/2$, and $-(i\sqrt{3}+1)/2$ and if the combining rule is algebraic multiplication, these three numbers form a group for which $E = 1$. The inverse of 1 is 1 and the other two elements are inverses of each other.

All powers of two, $\ldots 2^{-2}, 2^{-1}, 2^0, 2^1, 2^2 \ldots$, form an infinite group if the combining rule is algebraic multiplication.

The four matrices

$$\begin{Vmatrix} 1 & 0 & 0 & 0 \\ 0 & 1 & 0 & 0 \\ 0 & 0 & 1 & 0 \\ 0 & 0 & 0 & 1 \end{Vmatrix}, \quad \begin{Vmatrix} 0 & 1 & 0 & 0 \\ 1 & 0 & 0 & 0 \\ 0 & 0 & 0 & 1 \\ 0 & 0 & 1 & 0 \end{Vmatrix}, \quad \begin{Vmatrix} 0 & 0 & 0 & 1 \\ 0 & 0 & 1 & 0 \\ 0 & 1 & 0 & 0 \\ 1 & 0 & 0 & 0 \end{Vmatrix}, \quad \begin{Vmatrix} 0 & 0 & 1 & 0 \\ 0 & 0 & 0 & 1 \\ 1 & 0 & 0 & 0 \\ 0 & 1 & 0 & 0 \end{Vmatrix}$$

form a finite group if the combining rule is matrix multiplication. In this example, the first matrix is the identity element E. The multiplication of any two matrices produces one of the four. The algebra of matrices is discussed in the next chapter.

3-4. Point groups

From our point of view, the most important type of group is the one which consists of all the symmetry operations (*not* the symmetry elements) pertaining to a molecular structure. For such a group the combining rule is one operation followed by another. Since the application of any symmetry operation leaves a molecule physically unchanged and with the same orientation in space, its centre of mass must also remain fixed in space under all symmetry operations. From this it follows that all the axes and planes of symmetry of a molecule must intersect at at least one common point. Such groups are called *point groups*. (For a crystal of infinite size we can have symmetry operations, e.g. translations, that leave *no* point fixed in space; these give rise to *space groups*.)

That the symmetry operations of a molecule obey the four rules for a group is easily verified. The combination of any two symmetry operations must produce another symmetry operation (see eqn (2-4.1)), the identity element is the 'do nothing' operation E, the associative law holds (see eqn (2-4.3)) and corresponding to each symmetry operation there is an inverse operation which annuls its effect and which is also a symmetry operation and therefore a member of the group.

To illustrate these concepts let us consider the symmetric tripod framework (this has the same symmetry as NH_3). In Fig. 3-4.1 we show the following six symmetry operations:

(1) do nothing ($\boldsymbol{E}$),
(2) reflection in the σ_v' plane ($\boldsymbol{\sigma}_v'$),
(3) reflection in the σ_v'' plane ($\boldsymbol{\sigma}_v''$),
(4) reflection in the σ_v''' plane ($\boldsymbol{\sigma}_v'''$),
(5) clockwise rotation by $2\pi/3$ about the C_3 axis ($\boldsymbol{C}_3$),
(6) clockwise rotation by $4\pi/3$ about the C_3 axis or anti-clockwise rotation by $2\pi/3$ ($\boldsymbol{C}_3^2 \equiv \boldsymbol{C}_3^{-1}$).

Notice that the symmetry elements with respect to which the operations are carried out, remain fixed in space; that is, if we introduce a fixed set of laboratory axes (x, y, z), then the operations can be defined with respect to these axes e.g. σ_v' is the yz plane, σ_v'' is the plane containing the z axis and 30° clockwise from the xz plane, etc. It is also important to understand that the labels (a, b, c) on the feet of the tripod have no physical significance; they are only a convenient way of identifying which symmetry operation has been carried out.

In the light of what comes later, we will always consider rotations in their clockwise sense e.g. we will interpret $\boldsymbol{C}_3^{-1}$ as a clockwise rotation of $4\pi/3$. Furthermore, in all operations it will be the *body* which is moved *not* the set of laboratory axes. The reader is cautioned that some text books take the opposite convention and keep the body fixed while moving the laboratory axes; it all comes down to the same thing in the long run but intermediate steps will look different, especially with regard to the signs in the corresponding equations. For this reason, when consulting another book, always check the convention being used.

The six operations for the symmetric tripod form a particular point group and the way in which they combine together is conveniently summarized by what is known as a *group table* (Table 3-4.1). In this table the operation to be first carried out is given in the first row and the second operation to be carried out in the first column, the combination falls in the body of the table at the intersection of the appropriate row and column, e.g. $\boldsymbol{\sigma}_v'\boldsymbol{\sigma}_v'' = \boldsymbol{C}_3$, that is the operation $\boldsymbol{\sigma}_v''$ *followed* by the operation $\boldsymbol{\sigma}_v'$ is identical to the single operation $\boldsymbol{C}_3$ (see Fig. 3-4.2).

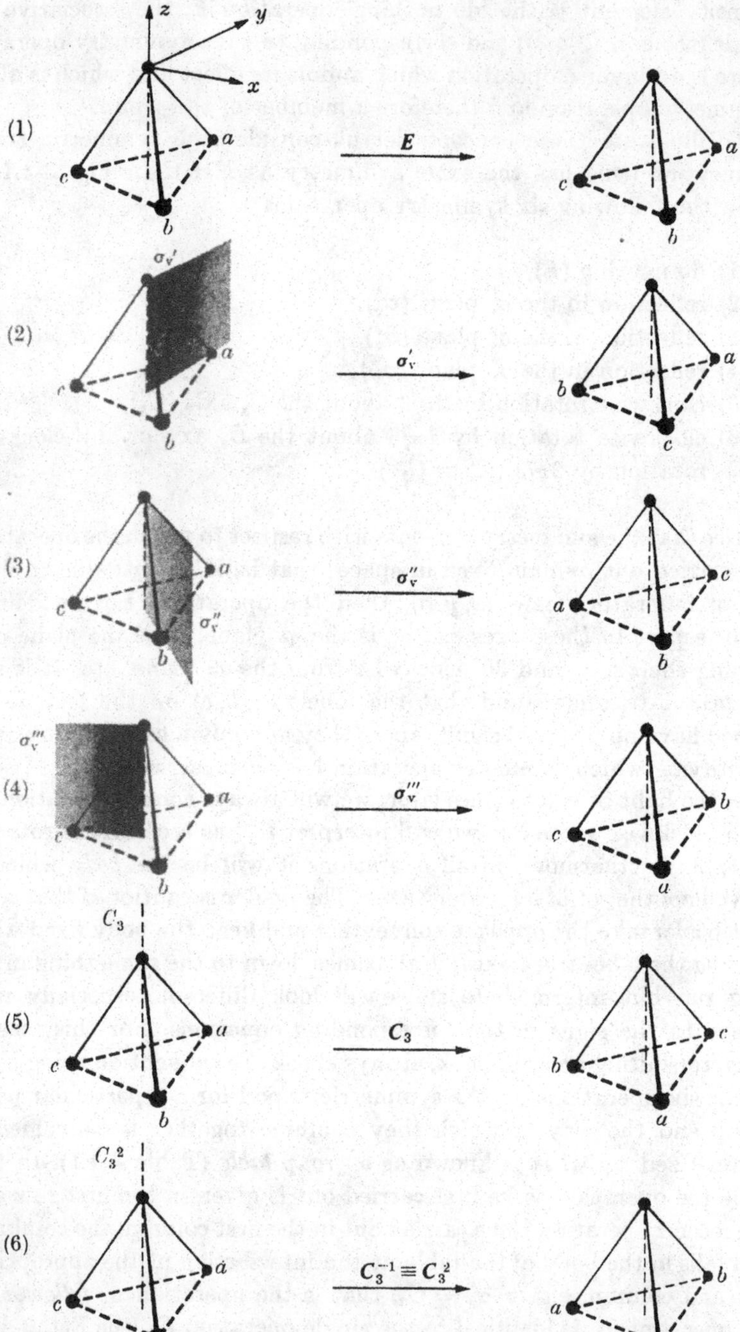

FIG. 3-4.1. Symmetry operations for a symmetric tripod.

TABLE 3-4.1

Group table for a symmetric tripod†

	E	σ_v'	σ_v''	σ_v'''	C_3	C_3^{-1}
E	E	σ_v'	σ_v''	σ_v'''	C_3	C_3^{-1}
σ_v'	σ_v'	E	C_3	C_3^{-1}	σ_v''	σ_v'''
σ_v''	σ_v''	C_3^{-1}	E	C_3	σ_v'''	σ_v'
σ_v'''	σ_v'''	C_3	C_3^{-1}	E	σ_v'	σ_v''
C_3	C_3	σ_v'''	σ_v'	σ_v''	C_3^{-1}	E
C_3^{-1}	C_3^{-1}	σ_v''	σ_v'''	σ_v'	E	C_3

† $C_3^{-1} \equiv C_3^2$

Essentially a group table shows how the first rule for a group is obeyed.

Inspection of Table 3-4.1 shows that each row or column contains each element once and once only. This is true for *all* group tables and the proof, the Rearrangement Theorem, is given in Appendix A.3-1.

An important concept concerning groups is *isomorphism*. Two groups, $\mathscr{G}$ with the elements A, B, ... and $\mathscr{G}'$ with the elements A', B', ..., are said to exhibit *isomorphism* if a one to one correspondence

$$A \leftrightarrow A', \; B \leftrightarrow B', \ldots$$

can be established between their elements such that

$$AB = C \quad \text{implies} \quad A'B' = C'$$

and vice versa, or more briefly if

$$(AB)' = A'B'.$$

Paraphrasing this definition, we may say that *isomorphic* groups have group tables of the same structure or form although they may differ in respect of the notation and nature of their elements and differ in

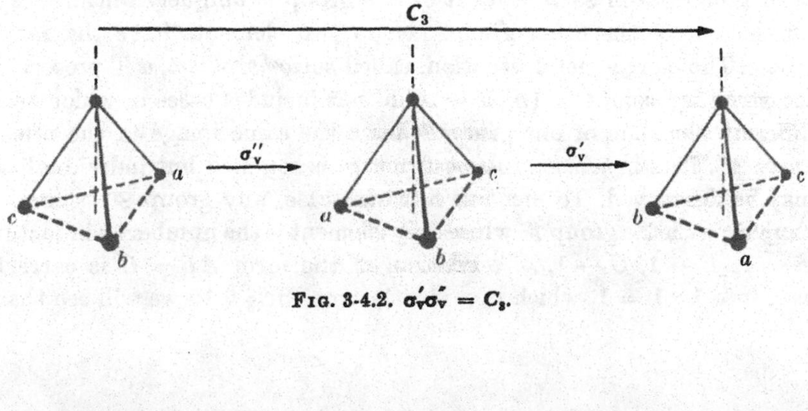

FIG. 3-4.2. $\sigma_v'\sigma_v'' = C_3$.

their combining rules. The following two groups of order 4 are isomorphic, the combining rule for each being stated in brackets:

$\mathscr{G}$: 1, i, −1, −i (ordinary multiplication)

$$\mathscr{G}': \quad \begin{Vmatrix} 1 & 0 \\ 0 & 1 \end{Vmatrix}, \begin{Vmatrix} 0 & 1 \\ -1 & 0 \end{Vmatrix}, \begin{Vmatrix} -1 & 0 \\ 0 & -1 \end{Vmatrix}, \begin{Vmatrix} 0 & -1 \\ 1 & 0 \end{Vmatrix}$$

(matrix multiplication).

Indeed if the elements of each set are renamed E, A, B, C (in this order), their common group table is seen to be

	E	*A*	*B*	*C*
E	*E*	*A*	*B*	*C*
A	*A*	*B*	*C*	*E*
B	*B*	*C*	*E*	*A*
C	*C*	*E*	*A*	*B*

Likewise, if the elements E, A, B, C, D, F of some group combine according to Table 3-4.2, which has the same structure as Table 3-4.1,

TABLE 3-4.2

A group table with the same structure as Table 3-4.1

	E	*A*	*B*	*C*	*D*	*F*
E	*E*	*A*	*B*	*C*	*D*	*F*
A	*A*	*E*	*D*	*F*	*B*	*C*
B	*B*	*F*	*E*	*D*	*C*	*A*
C	*C*	*D*	*F*	*E*	*A*	*B*
D	*D*	*C*	*A*	*B*	*F*	*E*
F	*F*	*B*	*C*	*A*	*E*	*D*

then this group is said to be isomorphic with the group of symmetry operations for the symmetric tripod.

In isomorphism each element of one group is uniquely mirrored by an element of the other group, i.e. no two elements have the same image. The more general situation, called *homomorphism*, still preserves the structure condition $(AB)' = A'B'$ but includes cases in which two different elements of one group $\mathscr{G}$ have the same image in the other group $\mathscr{G}'$. Thus in *homomorphism* structure is retained but individuality may be destroyed. To mention a trivial case, any group $\mathscr{G}$ is homomorphic with the group $\mathscr{G}'$ whose only element is the number 1; in fact if $A \to 1$, $B \to 1$, $C \to 1$, ... a relation of the form $AB = C$ is carried over into $1 \times 1 = 1$, which is evidently true. In § 5-10 we will see that for a group of square matrices there is always a homomorphic correspondence between each matrix and its determinant.

3-5. Some properties of groups

The number of elements in a group is called the *order* of the group and is usually given the symbol g e.g. for the point group for the symmetric tripod g is 6 and for an infinite group g is ∞.

It is clear that the elements of point groups do not *necessarily* commute, that is the order in which one combines two symmetry operations can be important (see, for example, Fig. 2-4.1 and Table 3-4.1 where for the symmetric tripod $C_3\sigma_v' \neq \sigma_v' C_3$). A group for which all the elements *do* commute is called an *Abelian* group.

One can, of course, combine more than two elements of a group; one simply uses the given combining rule more than once, e.g. PQR is obtained by combining Q with R to obtain some other element of the group and then combining P with this element.

If P and Q are elements of a group, then so, by the definition of a group, is the inverse of Q, Q^{-1}, and consequently $Q^{-1}PQ$. If this latter combination is identical with the element R we can write

$$R = Q^{-1}PQ \tag{3-5.1}$$

and we say that R is the *transform* of P by Q or that P and R are *conjugate* to each other. The criterion, therefore, for two elements P and R to be conjugate to each other is that eqn (3-5.1) holds for at least one element Q of the group. We can combine Q and Q^{-1} on the left and right hand sides respectively of both sides of eqn (3-5.1) and produce

$$QRQ^{-1} = QQ^{-1}PQQ^{-1} = EPE = P. \tag{3-5.2}$$

Either of eqns (3-5.1) and (3-5.2) acts as a definition of conjugation between two elements of a group. Furthermore, if P is conjugate to Q and Q is conjugate to R, then P is conjugate to R. This is demonstrated by the following equations:

$$P = X^{-1}QX,$$
$$Q = Y^{-1}RY,$$

therefore

$$P = X^{-1}Y^{-1}RYX$$
$$= (YX)^{-1}R(YX), \quad \text{(see eqn (2-4.5))}$$

and as YX must be some element of the group say Z we have

$$P = Z^{-1}RZ. \tag{3-5.3}$$

The elements of a group which are conjugate to each other are said to form a *class* and the number of elements in the ith class is given the symbol g_i. For the symmetric tripod point group the elements fall into three classes: E; σ_v', σ_v'', σ_v'''; C_3, C_3^{-1} with $g_1 = 1$, $g_2 = 3$ and $g_3 = 2$. For this and all groups E is conjugate with itself: $X^{-1}EX = E$ (all X) and hence E is always in a class of its own; we will always consider E to be the first class. The reader, with the aid of Table 3-4.1, can confirm for himself that if $Q = \sigma_v'$ or σ_v'' or σ_v''' then with any of the six elements of the group X the transform $X^{-1}QX$ is either σ_v' or σ_v'' or σ_v''' and that if $Q = C_3$ or C_3^{-1} then with any element of the group X the transform $X^{-1}QX$ is either C_3 or C_3^{-1}.

Any element will always be conjugate with itself as E is always in the group and $P = E^{-1}PE$.

If P and Q are conjugate, then so are their inverses; this follows using eqn (2-4.5) from

$$\begin{aligned} P^{-1} &= (X^{-1}QX)^{-1} = (QX)^{-1}(X^{-1})^{-1} \\ &= (QX)^{-1}X = X^{-1}Q^{-1}X. \end{aligned} \tag{3-5.4}$$

Hence the inverses of the elements of a class belong to a class (it is sometimes the same one) and the order of a class of elements and the order of the class of their inverses is the same.

In an Abelian group,

$$X^{-1}PX = X^{-1}XP = P,$$

and each element is in a class of its own and the number of classes is the same as the order of the group g.

The technique that has been given for dividing any group into classes can be replaced by a simpler one in the case of point groups. This is valuable since the testing of eqn (3-5.1) for all possible combinations of Q and P of a given point group can be quite time consuming, e.g. an equilateral triangle belongs to a point group which has 12 symmetry operations and therefore for this point group we would have to work out $Q^{-1}PQ$ 144 times. That there should be a simpler way is indicated by the fact that in the example of the symmetric tripod point group given above, the classes appear to be a 'natural' sub-division of the point group and to contain operations which are 'similar'.

The following simple rules can be established for point groups:

(1) The symmetry operations E, i, and σ_h are each in a class by themselves.

(2) The rotation operation C_n^k and its inverse C_n^{-k} will belong to the same class (a different class for each value of k) provided there is

either a plane of symmetry containing the C_n^k axis or a C_2 axis at right angles to the C_n^k axis; if not, C_n^k and C_n^{-k} are in classes by themselves. The same is true for the rotation–reflection operations S_n^k and S_n^{-k}.

(3) Two reflection operations σ and σ' will belong to the same class provided there is a symmetry operation in the point group which moves all the points on the σ' symmetry plane into corresponding positions on the σ symmetry plane. A similar rule holds true for two rotational operations C_n^k and $C_n^{k'}$ (or S_n^k and $S_n^{k'}$) about different rotational axes, i.e. the two operations belong to the same class provided there is a symmetry operation in the point group which moves all the points on the $C_n^{k'}$ (or $S_n^{k'}$) axis to corresponding positions on the C_n^k (or S_n^k) axis.

The proof of these three rules is based on their relationship with eqn (3-5.1). The symmetry operations E, i, and σ_h commute with all the other symmetry operations of the point group of which they are a part and hence, if $P = E, i,$ or σ_h and Q is any symmetry operation of the given point group, $Q^{-1}PQ = Q^{-1}QP = P$ and P is in a class by itself (Rule (1)).

The proof of Rules (2) and (3) depends on the fact that the symmetry operation $Q^{-1}PQ$ must be of the same general type, rotation or reflection, as P (this is easily demonstrated). Now consider the rotations C_n^k and C_n^{-k} for a point group which contains a reflection operation σ where the plane of symmetry σ contains the C_n^k axis. The symmetry operation $\sigma^{-1}C_n^k\sigma$, or $\sigma C_n^k\sigma$ since $\sigma^{-1} = \sigma$, must be a rotation about the C_n^k axis since points on this axis are unmoved under the three component operations (σ, C_n^k, σ^{-1}) and the only question remaining is the magnitude of this rotation. In Fig. 3-5.1 it is shown what happens to the rectangle OABC under the symmetry operation $\sigma^{-1}C_n^k\sigma$ if C_n^k is a clockwise rotation by θ degrees and σ is the xz plane; it is clear that OABC is rotated anti-clockwise by θ degrees. Hence the rotation $\sigma^{-1}C_n^k\sigma$ must be C_n^{-k} and C_n^k and C_n^{-k} must belong to the same class. In Fig. 3-5.2 the operation $C_2^{-1}C_n^kC_2 (= C_2C_n^kC_2)$ is carried out. This operation must also be a rotation about C_n^k since points on this axis end up by being unmoved. What happens to the rectangle OABC demonstrates that $C_2^{-1}C_n^kC_2 = C_n^{-k}$. These results, together with their extension to rotation-reflection operations form the basis of Rule (2).

To prove Rule (3) first consider the reflection operation $Q^{-1}\sigma Q$ where Q is a symmetry operation which moves all the points on some symmetry plane σ' to corresponding positions on the symmetry plane σ. Taking the components of $Q^{-1}\sigma Q$ one by one: Q will move the

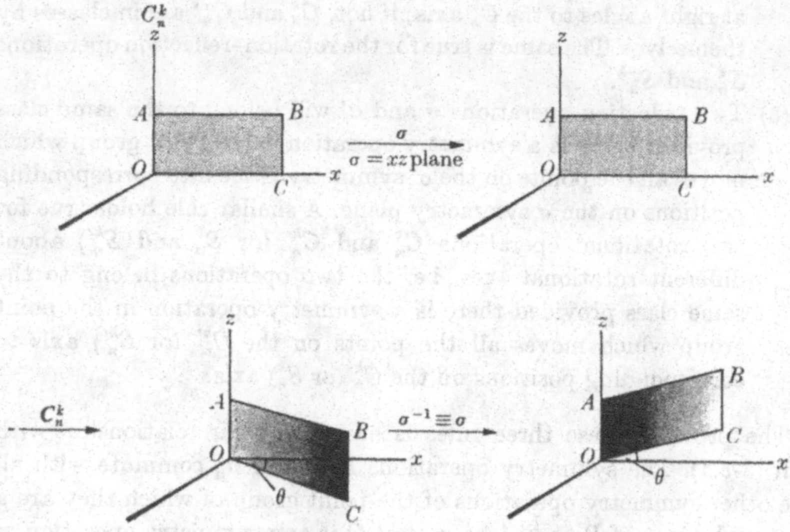

FIG. 3-5.1. The effect of $\sigma^{-1}C_n^k\sigma$ on a rectangle which contains the C_n^k axis, under the conditions of Rule (2).

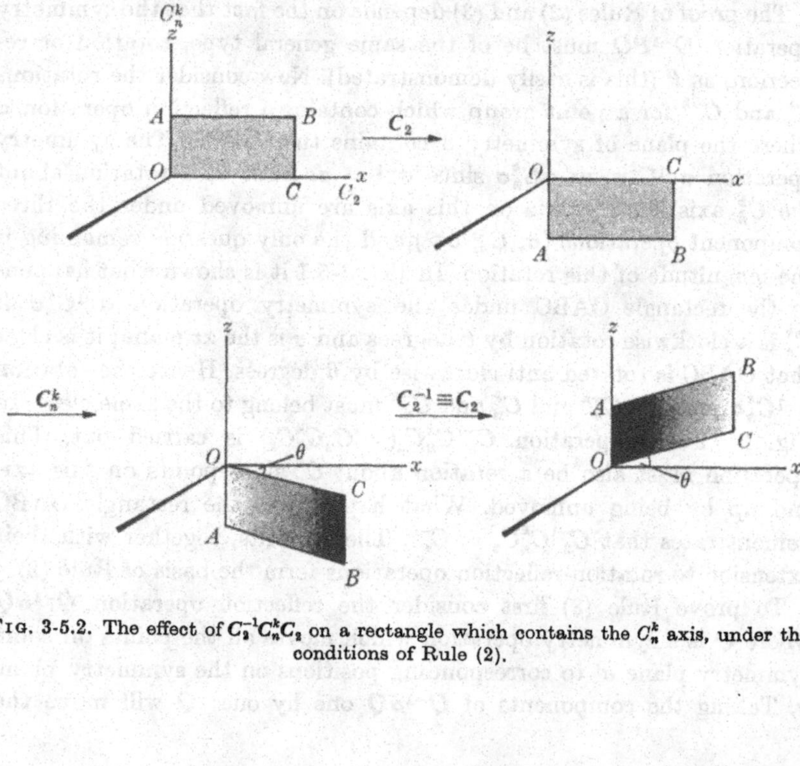

FIG. 3-5.2. The effect of $C_2^{-1}C_n^kC_2$ on a rectangle which contains the C_n^k axis, under the conditions of Rule (2).

points initially in the σ' plane to the σ plane, $\boldsymbol{\sigma}$ will leave these points unchanged and Q^{-1} will move the points back to the σ' plane. Hence, $Q^{-1}\boldsymbol{\sigma}Q$ leaves the points in the σ' plane unmoved and must therefore be a reflection in that plane, i.e. $Q^{-1}\boldsymbol{\sigma}Q = \boldsymbol{\sigma}'$ and $\boldsymbol{\sigma}$ and $\boldsymbol{\sigma}'$ belong to the same class. Now consider the similar rule for rotations: if Q moves the points on the $C_n^{k'}$ axis to corresponding positions on the C_n^k axis, then C_n^k will leave them unchanged and Q^{-1} will bring them back to their initial positions on the $C_n^{k'}$ axis, and hence $Q^{-1}C_n^kQ$ is a rotation

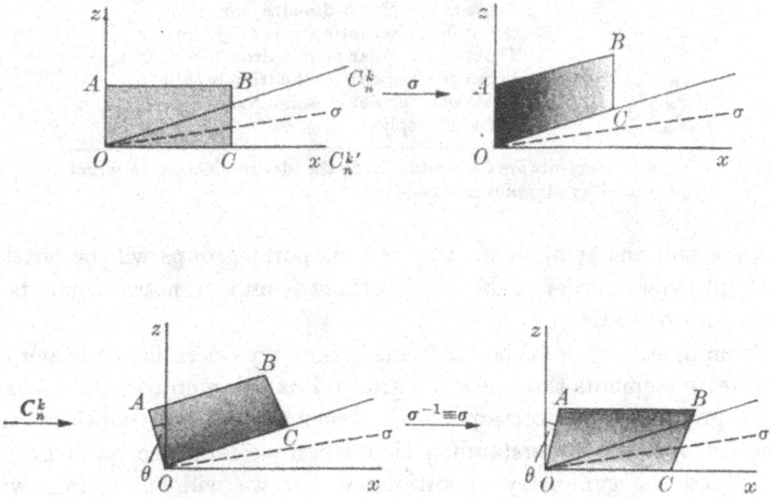

FIG. 3-5.3. The effect of $\boldsymbol{\sigma}^{-1}C_n^k\boldsymbol{\sigma}$ on a rectangle which contains the $C_n^{k'}$ axis, under the conditions of Rule (3).

about the $C_n^{k'}$ axis; the magnitude of this rotation can be seen by considering what happens to points in any plane which contains the $C_n^{k'}$ axis. If Q is a rotation, then $Q^{-1}C_n^kQ$ will be a clockwise rotation by $k2\pi/n$ about the $C_n^{k'}$ axis, that is $Q^{-1}C_n^kQ = C_n^{k'}$ and C_n^k and $C_n^{k'}$ belong to the same class. If Q is a reflection, then $Q^{-1}C_n^kQ$ will be an anti-clockwise rotation by $k2\pi/n$ about the $C_n^{k'}$ axis and $Q^{-1}C_n^kQ = C_n^{-k'}$ (see Fig. 3-5.3), but since for all the point groups which have this property, the symmetry operations $C_n^{-k'}$ and $C_n^{k'}$ fall into the same class, C_n^k, $C_n^{-k'}$ and $C_n^{k'}$ will *all* be in the same class and Rule (3) will still hold true. A similar proof exists for rotation–reflection operations.

3-6. Classification of point groups

We are now in a position to describe and classify the various point groups that exist. The notation we will use is known as the *Schoenflies*

TABLE 3-6.1

The point groups and their essential symmetry elements

Point group	*Essential symmetry elements*†
$\mathscr{C}_s$	One symmetry plane
$\mathscr{C}_i$	A centre of symmetry
$\mathscr{C}_n$	One n-fold axis of symmetry
$\mathscr{D}_n$	One C_n axis plus n C_2 axes perpendicular to it
$\mathscr{C}_{nv}$	One C_n axis plus n vertical planes σ_v
$\mathscr{C}_{nh}$	One C_n axis plus a horizontal plane σ_h
$\mathscr{D}_{nh}$	Those of $\mathscr{D}_n$ plus a horizontal plane σ_h
$\mathscr{D}_{nd}$	Those of $\mathscr{D}_n$ plus n dihedral planes σ_d
$\mathscr{S}_n$ (n even)	One n-fold alternating axis of symmetry
$\mathscr{T}_d$	Those of a regular tetrahedron
$\mathscr{O}_h$	Those of a regular octahedron or cube
$\mathscr{I}_h$	Those of a regular icosahedron
$\mathscr{K}_h$	Those of a sphere

† These elements are all in addition to the identity element E which is possessed by all point groups.

notation and the symbols for the various point groups will be written in script type in order to distinguish them from symmetry elements or symmetry operations.

Though, strictly speaking, it is the symmetry operations and *not* the symmetry elements that form the group, it is common to describe each point group by the corresponding elements. We will continue this practice with the understanding that when we later use point groups it will be the symmetry operations which we will be dealing with rather than the symmetry elements.

In Table 3-6.1 we give the *essential* symmetry elements for the various point groups. We use the word essential since some of the symmetry elements listed in this Table for a given point group will necessarily imply the existence of others which are not listed. In Table 3-6.2 some alternative symbols are shown. An *exhaustive* list of

TABLE 3-6.2

Alternative symbols

$$\mathscr{C}_s \equiv \mathscr{C}_{1v} \equiv \mathscr{C}_{1h} \equiv \mathscr{S}_1$$
$$\mathscr{C}_i \equiv \mathscr{S}_2$$
$$\mathscr{C}_{nh} \equiv \mathscr{S}_n \quad (n \text{ odd})$$
$$\mathscr{D}_2 \equiv \mathscr{V}\dagger$$
$$\mathscr{D}_{2h} \equiv \mathscr{V}_h$$
$$\mathscr{D}_{2d} \equiv \mathscr{V}_d$$
$$\mathscr{D}_{3d} \equiv \mathscr{S}_{6v}$$
$$\mathscr{D}_{4d} \equiv \mathscr{S}_{8v}$$

† From the German word *Vierergruppe*.

TABLE 3-6.3

The point groups and all their symmetry elements†

Point group	*Symmetry elements*
$\mathscr{C}_s$	E, σ_h
$\mathscr{C}_i$	E, i
$\mathscr{C}_1$	E
$\mathscr{C}_2$	E, C_2
$\mathscr{C}_3$	E, C_3–C_3^2
$\mathscr{C}_4$	E, C_4–C_2–C_4^3
$\mathscr{C}_5$	E, C_5–C_5^2–C_5^3–C_5^4
$\mathscr{C}_6$	E, C_6–C_3–C_2–C_3^2–C_6^5
$\mathscr{C}_7$	E, C_7–C_7^2–C_7^3–C_7^4–C_7^5–C_7^6
$\mathscr{C}_8$	E, C_8–C_4–C_8^3–C_2–C_8^5–C_4^3–C_8^7
$\mathscr{D}_2$	E, three C_2 (mutually perpendicular)
$\mathscr{D}_3$	E, C_3–C_3^2, three C_2 (perpendicular to C_3)
$\mathscr{D}_4$	E, C_4–C_2–C_4^3, four C_2 (perpendicular to C_4)‡
$\mathscr{D}_5$	E, C_5–C_5^2–C_5^3–C_5^4, five C_2 (perpendicular to C_5)
$\mathscr{D}_6$	E, C_6–C_3–C_2–C_3^2–C_6^5, six C_2 (perpendicular to C_6)‡
$\mathscr{C}_{1v}$	same as $\mathscr{C}_s$
$\mathscr{C}_{2v}$	E, C_2, two σ_v
$\mathscr{C}_{3v}$	E, C_3–C_3^2, three σ_v
$\mathscr{C}_{4v}$	E, C_4–C_2–C_4^3, four σ‡
$\mathscr{C}_{5v}$	E, C_5–C_5^2–C_5^3–C_5^4, five σ_v
$\mathscr{C}_{6v}$	E, C_6–C_3–C_2–C_3^2–C_6^5, six σ‡
$\mathscr{C}_{\infty v}$	E, infinite number of coincidental rotational axes, infinite number of σ_v
$\mathscr{C}_{1h}$	same as $\mathscr{C}_s$
$\mathscr{C}_{2h}$	E, C_2, i, σ_h
$\mathscr{C}_{3h}$	E, C_3–C_3^2–S_3–S_3^5,§ σ_h
$\mathscr{C}_{4h}$	E, C_4–C_2–C_4^3–S_4–S_4^3, σ_h, i
$\mathscr{C}_{5h}$	E, C_5–C_5^2–C_5^3–C_5^4–S_5–S_5^7–S_5^3–S_5^9,§ σ_h
$\mathscr{C}_{6h}$	E, C_6–C_3–C_2–C_3^2–C_6^5–S_6–S_3–S_3^5–S_6^5,§ σ_h, i
$\mathscr{D}_{2h}$	E, three C_2 (mutually perpendicular), i, three σ (mutually perpendicular)
$\mathscr{D}_{3h}$	E, C_3–C_3^2–S_3–S_3^5,§ three C_2 (perpendicular to C_3), σ_h, three σ_v
$\mathscr{D}_{4h}$	E, C_4–C_2–C_4^3–S_4–S_4^3, four C_2 (perpendicular to C_4),‡ i, σ_h, four σ‡
$\mathscr{D}_{5h}$	E, C_5–C_5^2–C_5^3–C_5^4–S_5–S_5^7–S_5^3–S_5^9,§ five C_2 (perpendicular to C_5), σ_h, five σ_v
$\mathscr{D}_{6h}$	E, C_6–C_3–C_2–C_3^2–C_6^5–S_6–S_3–S_3^5–S_6^5,§ six C_2 (perpendicular to C_6),‡ i, σ_h, six σ‡
$\mathscr{D}_{\infty h}$	E, infinite number of coincidental rotational C and alternating S axes, ($S_1 = \sigma_h$), infinite number of σ_v, i, infinite number of C_2 axes
$\mathscr{D}_{2d}$	E, C_2–S_4–S_4^3, two C_2 (perpendicular to each other and to the other C_2), two σ_d (through S_4)
$\mathscr{D}_{3d}$	E, C_3–C_3^2–S_6–S_6^5, three C_2 (perpendicular to C_3), i, three σ_d
$\mathscr{D}_{4d}$	E, C_4–C_2–C_4^3–S_8–S_8^3–S_8^5–S_8^7, four C_2 (perpendicular to C_4), four σ_d
$\mathscr{D}_{5d}$	E, C_5–C_5^2–C_5^3–C_5^4–S_{10}–S_{10}^3–S_{10}^7–S_{10}^9, five C_2 (perpendicular to C_5), i, five σ_d
$\mathscr{D}_{6d}$	E, C_6–C_3–C_2–C_3^2–C_6^5–S_{12}–S_4–S_{12}^5–S_{12}^7–S_4^3–S_{12}^{11}, six C_2 (perpendicular to C_6), six σ_d
$\mathscr{S}_1$	same as $\mathscr{C}_s$
$\mathscr{S}_2$	same as $\mathscr{C}_i$
$\mathscr{S}_3$	same as $\mathscr{C}_{3h}$
$\mathscr{S}_4$	E, C_2–S_4–S_4^3
$\mathscr{S}_5$	same as $\mathscr{C}_{5h}$
$\mathscr{S}_6$	E, C_3–C_3^2–S_6–S_6^5, i
$\mathscr{S}_7$	same as $\mathscr{C}_{7h}$

TABLE 3-6.3 (*Cont.*)

Point group	*Symmetry elements*
$\mathscr{S}_8$	E, C_4–C_2–C_4^3–S_8–S_8^3–S_8^5–S_8^7
$\mathscr{T}_d$	E, four C_3–C_3^2, three C_2–S_4–S_4^3 (mutually perpendicular), six σ_d (see Fig. 3-6.2)
$\mathscr{O}_h$	E, four C_3–C_3^2–S_6–S_6^5, three C_4–C_2–C_4^3–S_4–S_4^3 (mutually perpendicular), six C_2, i, three σ_h, six σ_d (See Fig. 3-6.3)

† Axes which coincide are linked, e.g. C_3–C_3^2–S_6–S_6^5.

‡ For reasons which will become clear later on, the σ planes in $\mathscr{C}_{4v}$, $\mathscr{C}_{6v}$, $\mathscr{D}_{4h}$, and $\mathscr{D}_{6h}$ and the C_2 axes in $\mathscr{D}_4$, $\mathscr{D}_6$, $\mathscr{D}_{4h}$, and $\mathscr{D}_{6h}$ are conventionally separated into two types: the planes into σ_v and σ_d planes and the axes into C_2' and C_2'' axes. These distinctions are shown in Fig. 3-6.1, where it is apparent that in $\mathscr{C}_{4v}$ and $\mathscr{C}_{6v}$, the planes labelled σ_d do not fulfill the requirement of Table 2-3.1. Nonetheless, the notation given is that recommended by R. S. Mulliken, (*Journal of Chemical Physics* **23**, 1997 (1955)). We will not bother with the differences between the C_2' and C_2'' axes in $\mathscr{D}_4$ and $\mathscr{D}_6$ as these point groups have no chemical significance.

§ S_3^5 is the element corresponding to rotation about an axis by 5·2π/3 (or 2·2π/3) followed by a reflection in the plane perpendicular to that axis. We cannot use the symbol S_3^2 as this is identical with C_3^2 (see § 2–3 page 13). A similar argument holds for S_5^7 and S_5^9.

the symmetry elements for each point group is given in Table 3-6.3 and in Fig. 3-6.4 we show some molecular examples.

3-7. Determination of molecular point groups

With experience one comes to recognize the point group to which a molecule belongs simply by analogy with some other known molecule. However, until one builds up a memory file of the point groups of representative molecules, it is best to use some systematic method. A scheme which will enable the reader to do this is shown in Table 3-7.1 and to illustrate how it works, we will consider three typical cases.

Take, for example, the bent triatomic molecule B—A—B (say, H_2O). Following Table 3-7.1, it is not linear, it does not have two or more C_n with $n \geq 3$, it does have a C_2 axis but there are not n C_2 axes perpendicular to this axis, it does possess two σ_v (vertical) planes, it therefore must belong to the $\mathscr{C}_{2v}$ point group.

Next, take the square planar molecule AB_4 (say, $PtCl_4^{2-}$), it is not linear, it does not have two or more C_n with $n \geq 3$ (though it does have one), its principal axis is C_4 and there are four C_2 axes perpendicular to this axis, the plane of the molecule is a σ_h plane and therefore it belongs to the $\mathscr{D}_{4h}$ point group. Notice that this molecule also possesses σ_d planes, but the σ_h plane is enough to associate it with the $\mathscr{D}_{4h}$ point group.

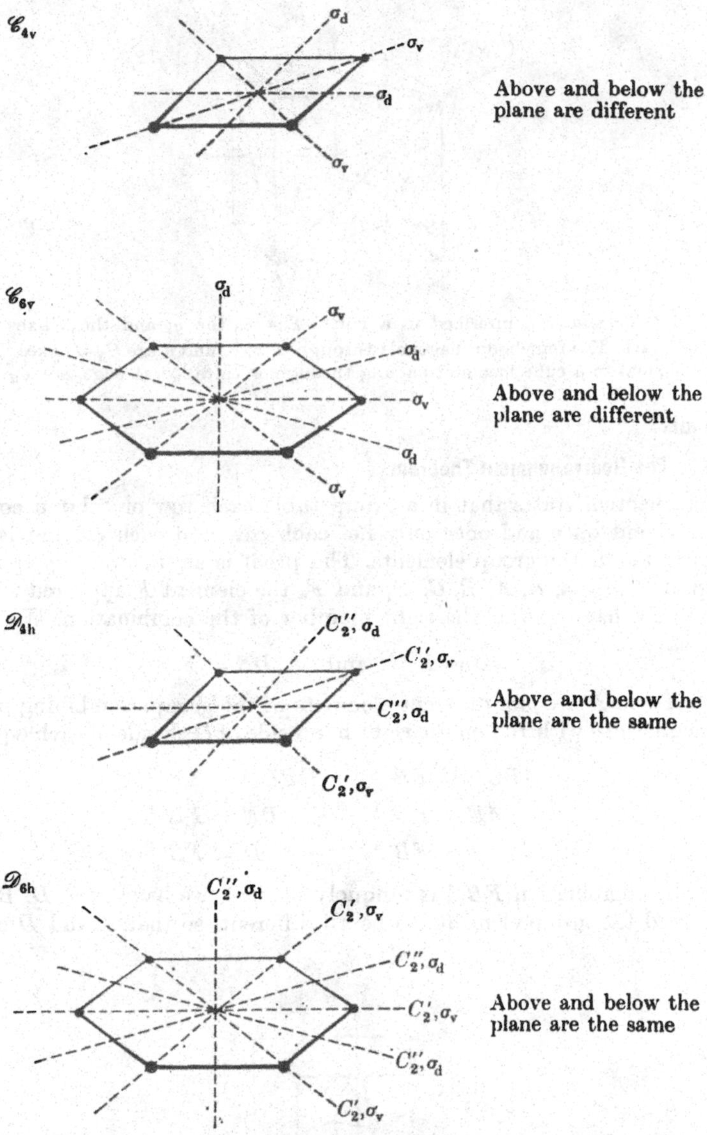

FIG. 3-6.1. Some special notations for the point groups $\mathscr{C}_{4v}$, $\mathscr{C}_{6v}$, $\mathscr{D}_{4h}$, and $\mathscr{D}_{6h}$.

Finally, consider the puckered octagon (say, S_8), it is not linear, does not have two or more C_n with $n \geq 3$, its principal axis is C_4 and there are four C_2 axes perpendicular to this axis, there is no σ_h plane, there are, however, four σ_d planes, the molecule consequently belongs to the $\mathscr{D}_{4d}$ point group.

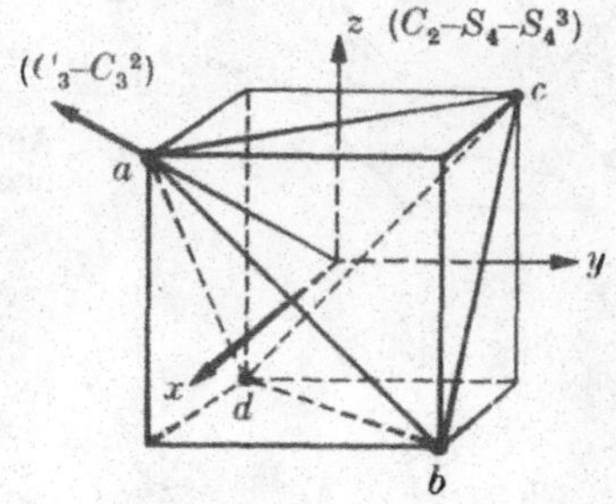

FIG. 3-6.2. Tetrahedron inscribed in a cube. The x, the y, and the z axes are all C_2–S_4–S_4^3 axes. The four body diagonals through a, b, c, and d are C_3–C_3^2 axes. The six planes normal to a cube face and passing through a tetrahedral edge are σ_d planes.

Appendix

A. 3-1. The Rearrangement Theorem

This theorem states that in a group table each row or column contains each element once and once only i.e. each row and each column is some permutation of the group elements. The proof is as follows: suppose for a group of elements, E, A, B, C, D, and F, the element F appeared twice in the column having B as the right member of the combination. We would have, say,

$$AB = F \qquad \text{and} \qquad DB = F$$

where A and D are two different elements of the group. Combining each of these equations with B^{-1} on the right hand side of each side of each equation gives:

$$ABB^{-1} = FB^{-1} \qquad DBB^{-1} = FB^{-1}$$
$$AE = FB^{-1} \qquad DE = FB^{-1}$$
$$A = FB^{-1} \qquad D = FB^{-1}$$

Since the combination FB^{-1} is uniquely defined, we have $A = D$. But we postulated the group elements to be all different, so that A and D cannot

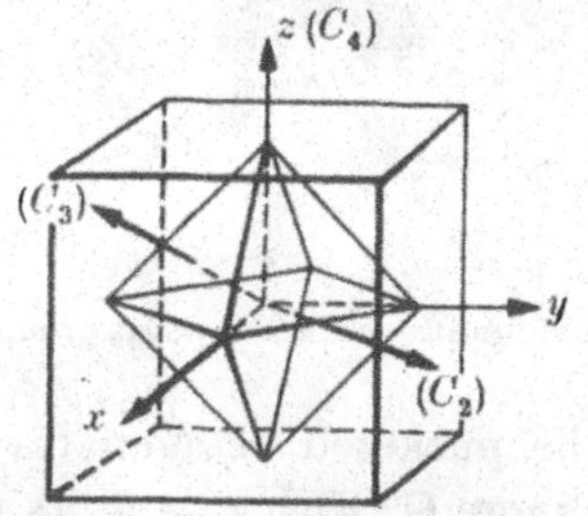

FIG. 3-6.3. Octahedron inscribed in a cube. The x, the y, and the z axes are all C_4–C_2–C_4^3–S_4–S_4^3 axes. The four body diagonals are C_3–C_3^2–S_6–S_6^5 axes. The six axes through the origin parallel to face diagonals are C_2 axes. The xy, xz, and yz planes are σ_h planes. The six planes normal to a cube face and passing through a diagonal are σ_d planes

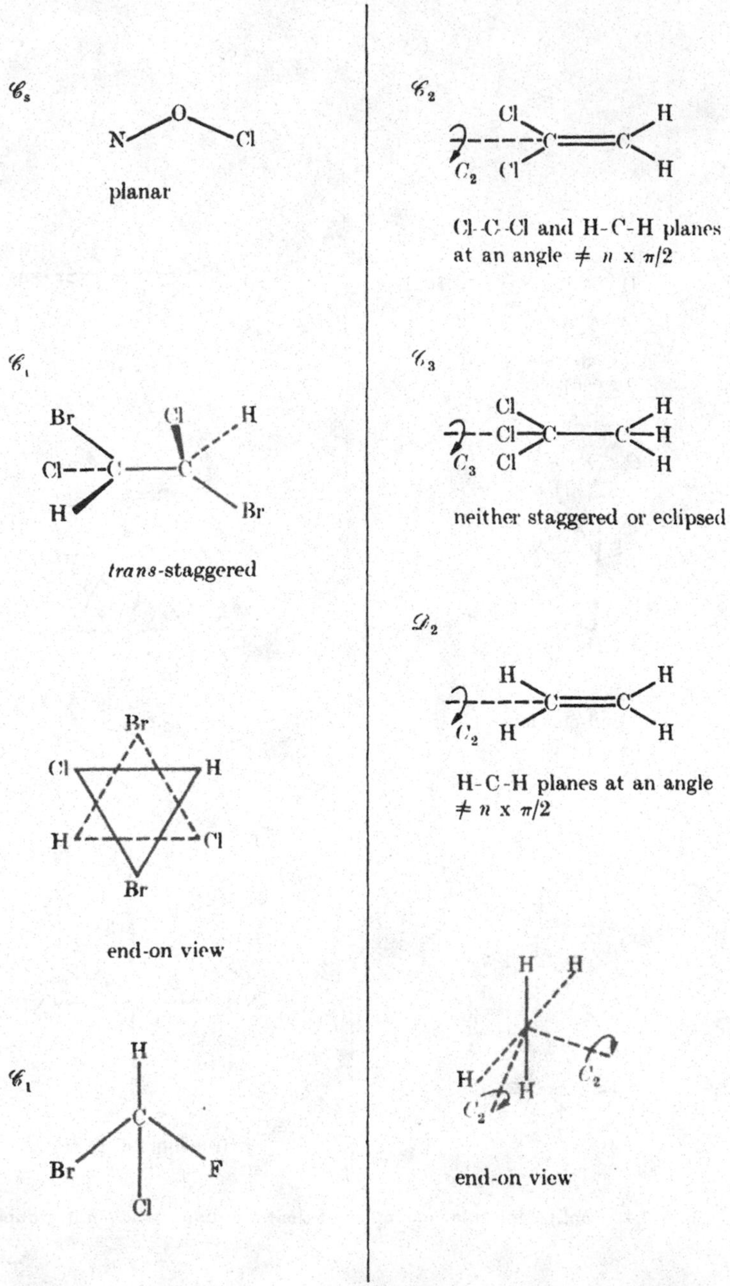

FIG. 3-6.4. Molecular examples of the more important point groups.

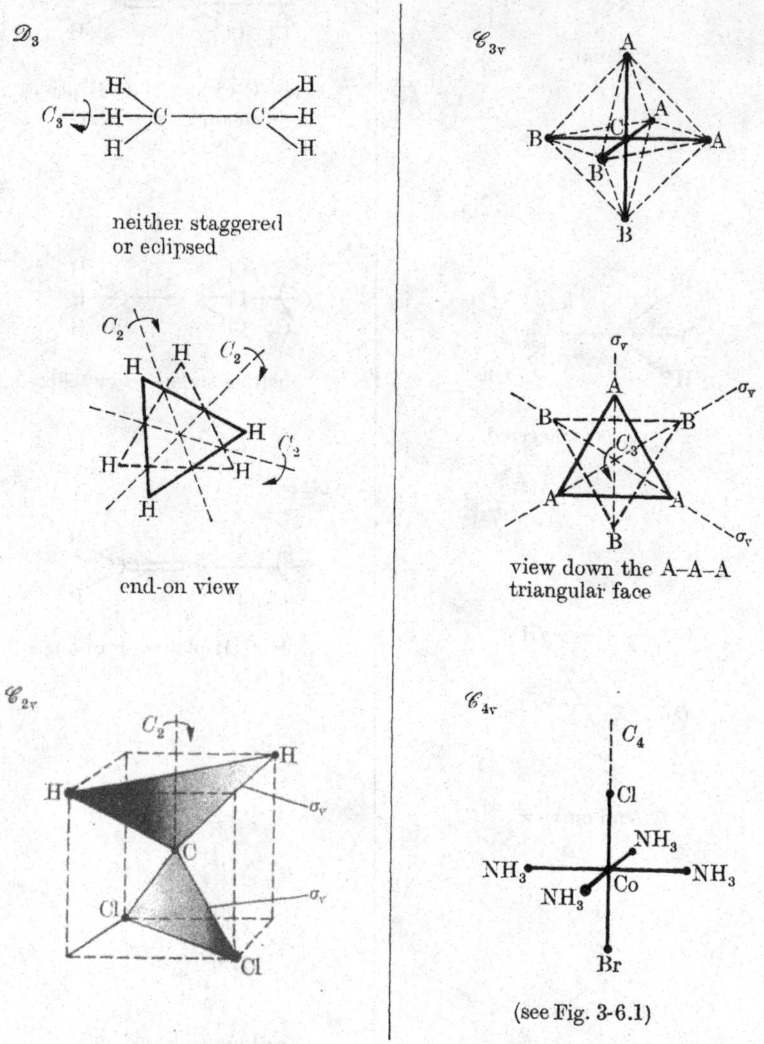

FIG. 3-6.4 (*Cont.*). Molecular examples of the more important point groups.

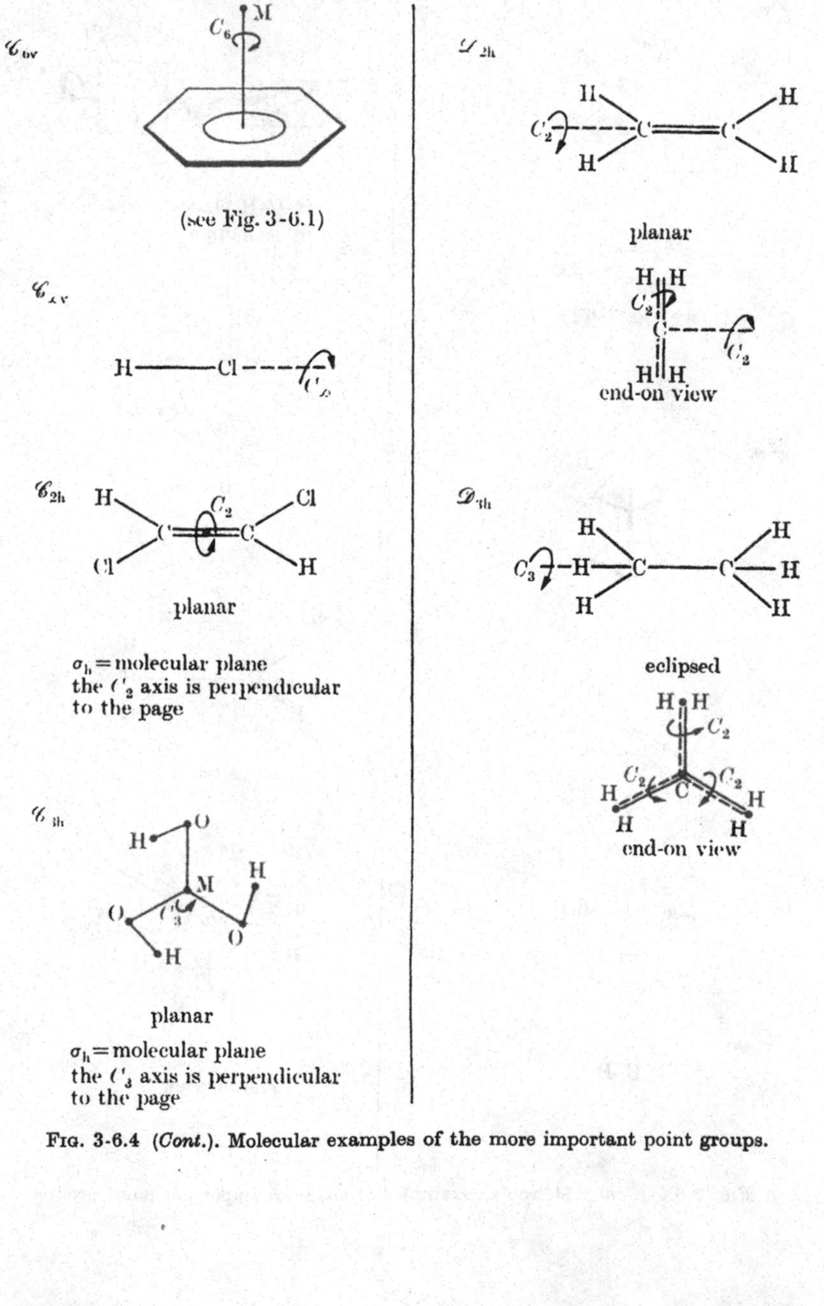

Fig. 3-6.4 (*Cont.*). Molecular examples of the more important point groups.

$\mathscr{D}_{4h}$

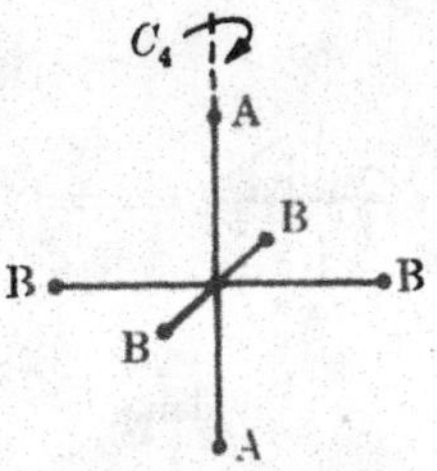

(see Fig. 3-6·1)

$\mathscr{D}_{5h}$

$\mathscr{D}_{6h}$

(see Fig. 3-6·1)

$\mathscr{D}_{\infty h}$

H–H

$\mathscr{D}_{2d}$

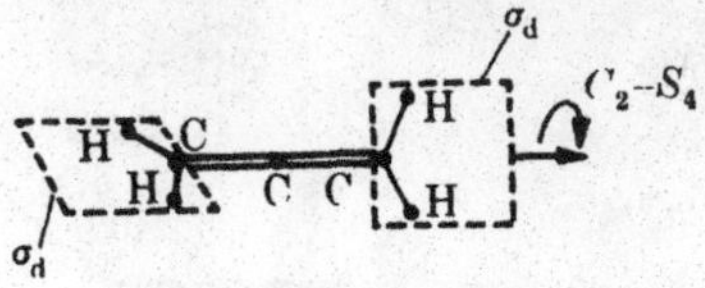

H-C-H planes at $\pi/2$ to each other

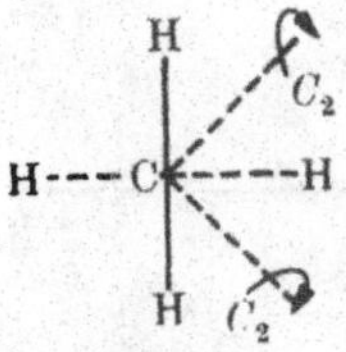

end-on view

$\mathscr{D}_{3d}$

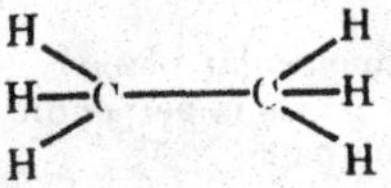

staggered

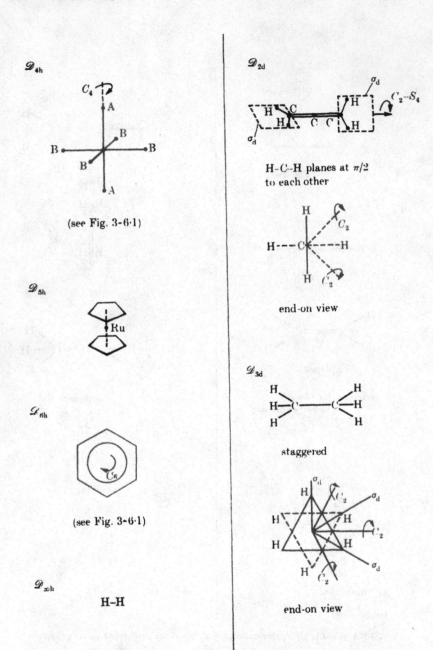

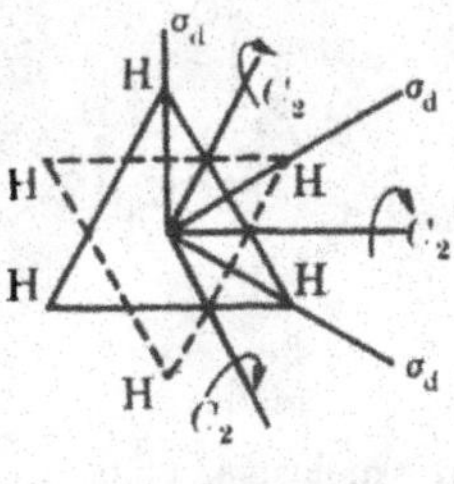

end-on view

FIG. 3-6.4 (*Cont.*). Molecular examples of the more important point groups.

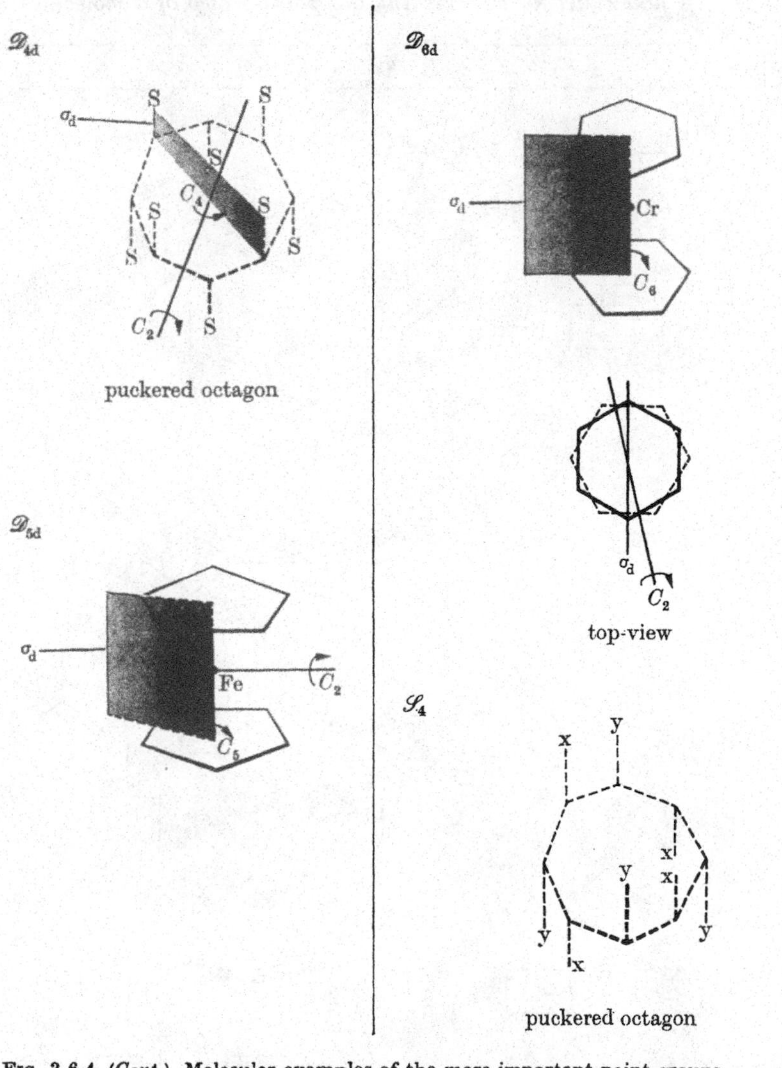

Fig. 3-6.4 (*Cont.*). **Molecular examples of the more important point groups.**

TABLE 3-7.1

A flow chart for determining the point group of a molecule

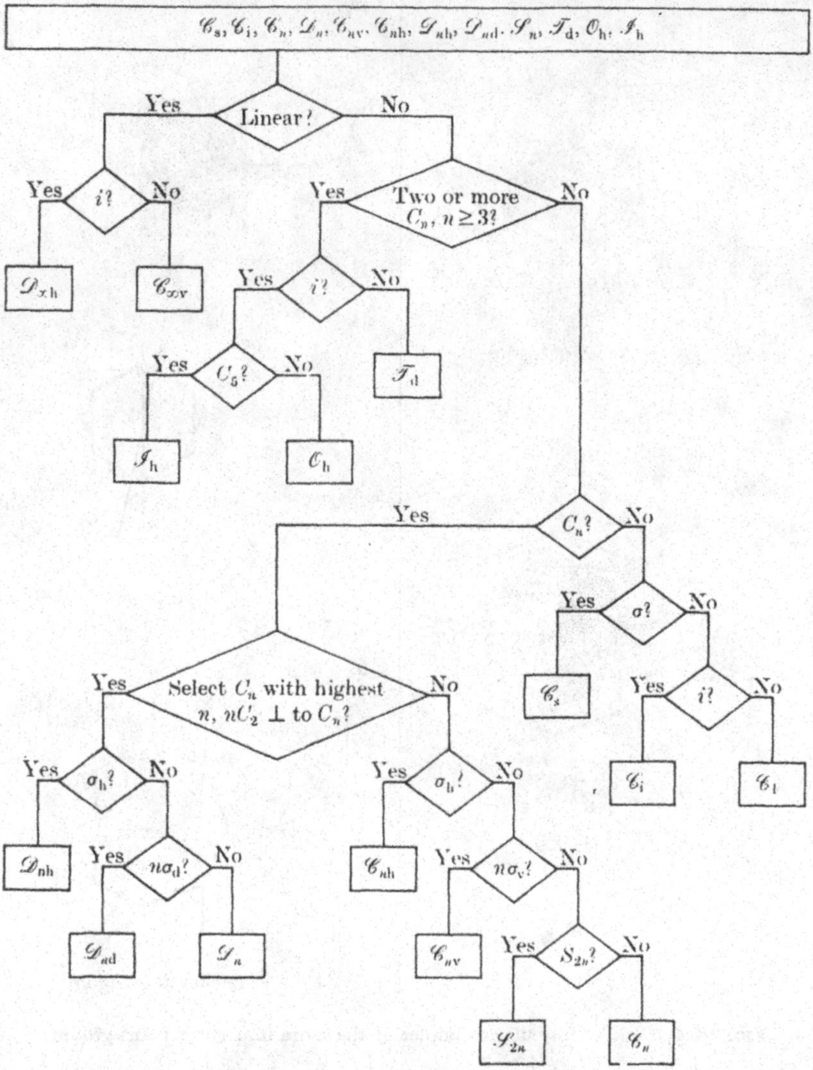

be identical. Hence F or any other element cannot appear twice in the same column and because each row or column contains the same number of elements as there are in the group, each element must appear once.

PROBLEMS

3.1. Determine the point groups of the following: (*a*) CH_2ClF; (*b*) NH_3; (*c*) BCl_3; (*d*) allene; (*e*) 1,3,5-trichlorobenzene, (*f*) *trans*-$Pt(NH_3)_2Cl_2$ (considered as square planar); (*g*) BFClBr.

3.2. Determine the point groups of the following octahedral compounds: (*a*) CoN_6; (*b*) CoN_5A; (*c*) *cis*-CoN_4A_2; (*d*) *trans*-CoN_4A_2; (*e*) *cis-cis*-CoN_3A_3; (*f*) *trans-cis*-CoN_3A_3.

3.3. Determine the point groups of the following: (*a*) chair form of cyclohexane (ignoring the H's); (*b*) boat form of cyclohexane (ignoring the H's); (*c*) staggered C_2H_6; (*d*) eclipsed C_2H_6; (*e*) between staggered and eclipsed C_2H_6.

3.4. Determine the point groups of the following: (*a*) ivy leaf; (*b*) iris; (*c*) starfish; (*d*) ice crystal; (*e*) twin-bladed propellor; (*f*) rectangular bar; (*g*) hexagonal bathroom tile; (*h*) swastika; (*i*) tennis ball (with seam); (*j*) Chinese abacus (counters all in their lowest positions); (*k*) ying–yang.

3.5. Determine the point groups of the following: (*a*) a square-based pyramid; (*b*) a right circular cone; (*c*) a square lamina; (*d*) a square lamina with the top and bottom sides painted differently; (*e*) a right circular cylinder; (*f*) a right circular cylinder with the two ends painted differently; (*g*) a right circular cylinder with a stripe painted parallel to the axis.

3.6. What is the point group for the tris(ethylenediamine)cobalt(III) ion?

3.7. For which point groups can a molecule (*a*) have a dipole moment, (*b*) be optically active?

4. Matrices

4-1. Introduction

We have shown how the symmetry operations for a molecule form a point group and we have introduced the notation which allows us to classify point groups. The next step which leads to further progress is to find sets of *matrices* which behave in a fashion similar to the symmetry operations; that is, matrices which are *homomorphic* with the symmetry operations (see § 3-4). These matrices will be said to *represent* the symmetry operations. Matrix representations of molecular point groups paraphrase in a mathematical way the symmetry of a molecule and are central to all our applications of group theory to chemistry. There are many theorems in mathematics concerning matrix representations and once the link between representations and point groups has been made, these theorems can immediately be invoked for solving chemical problems as different as the molecular orbital theory of benzene and the classification of the vibrational levels of methane. But before we can discuss these matrix representations it is necessary to understand something about the properties of matrices themselves. Therefore in this chapter we define a matrix, describe the algebra of matrices, and introduce the matrix eigenvalue equation and the subject of similarity transformations. The reader who is in awe of the large number of definitions and theorems in this chapter may, if he wishes, go straight to Chapter 5 and then pick up the material on matrices in slow stages as it is required in the succeeding chapters.

The various theorems for matrices are proven in the appendices to this chapter but these proofs are not essential for the understanding of future chapters and may be ignored by the reader who finds the mathematics too heavy going. In Appendix A.4-1 certain special matrices and terms are defined and these will crop up in the future; the reader is therefore advised to make himself familiar with them. They are summarized in Table 4-1.1.

4-2. Definitions (matrices and determinants)

A *matrix* is a rectangular array of terms (numbers or symbols) called *elements* which are written between parentheses or double lines, e.g.

$$\begin{pmatrix} a_1 & b_1 & c_1 \\ a_2 & b_2 & c_2 \end{pmatrix} \quad \text{or} \quad \left\| \begin{matrix} a_1 & b_1 & c_1 \\ a_2 & b_2 & c_2 \end{matrix} \right\|.$$

TABLE 4-1.1

Special matrices and terms

Symbols (see also List of symbols, p. xv)

$A^* =$ conjugate complex of matrix A
$\tilde{A} =$ transpose of matrix A
$A^\dagger =$ adjoint of matrix A
$A^{-1} =$ inverse of matrix A
$\det(A) =$ determinant of matrix A
$\mathrm{Trace}(A) =$ sum of diagonal elements of matrix A
$\delta_{ij} =$ Kronecker delta (equals 0 if $i \neq j$, equals 1 if $i = j$)

Definitions‡

$$(A^*)_{ij} = (A_{ij})^*$$
$$(\tilde{A})_{ij} = A_{ji}$$
$$(A^\dagger)_{ij} = (A_{ji})^*$$
$$A^{-1}A = AA^{-1} = E$$

Special matrices

Identity matrix	$A_{ij} = \delta_{ij}$
Null matrix	$A_{ij} = 0$
Diagonal matrix	$A_{ij} = d_i\delta_{ij},\ d_i \neq 0$
A real matrix	$A = A^*$
A symmetric matrix	$A = \tilde{A}$
An Hermitian matrix	$A = A^\dagger$
A unitary matrix	$A^\dagger = A^{-1}$
An orthogonal matrix	$\tilde{A} = A^{-1}$

‡ A_{ij} or $(A)_{ij}$ is the element in the ith row and jth column of matrix A.

There is a definite algebra associated with matrices (see § 4-3). In this book we will only be concerned with square matrices (where the number of rows equals the number of columns) or with single row or single column matrices. We will use the double line notation, e.g.

$$\left\| \begin{matrix} A_{11} & A_{12} & A_{13} \\ A_{21} & A_{22} & A_{23} \\ A_{31} & A_{32} & A_{33} \end{matrix} \right\|, \quad \left\| \begin{matrix} -1 & \mathrm{i} & 0 \\ 2 & 3 & \mathrm{i} \\ -1 & 0 & 1 \end{matrix} \right\|.$$

Row and column matrices, e.g.

$$\| A_1 \quad A_2 \quad A_3 \|, \quad \left\| \begin{matrix} a \\ b \\ c \end{matrix} \right\|,$$

$$\| x_{11} \quad x_{12} \quad x_{13} \quad x_{14} \|, \quad \left\| \begin{matrix} x_{11} \\ x_{21} \end{matrix} \right\|,$$

can be used to define the components of a vector. In general we will use a capital italic letter to symbolize a matrix and an element in the ith row and jth column of matrix A will be written as A_{ij}, e.g. if

$$A = \begin{Vmatrix} 1 & 2 & 3 \\ 4 & 7 & 8 \\ 9 & 5 & 6 \end{Vmatrix}$$

then $A_{23} = 8$.

The number of rows (or columns) in a square matrix is called the order of the matrix.

The reader must not confuse square matrices with *determinants*. A determinant is a *square* array of elements symbolizing the sum of certain products of the elements. Unlike a matrix, a determinant has a definite quantitative value. A single straight vertical line on either side of an array indicates a determinant, e.g. $\begin{vmatrix} a_1 & b_1 \\ a_2 & b_2 \end{vmatrix}$ is a determinant of second order whose value is $a_1b_2-a_2b_1$:

$$\begin{vmatrix} a_1 & b_1 \\ a_2 & b_2 \end{vmatrix} = a_1b_2-a_2b_1 \tag{4-2.1}$$

and $\begin{vmatrix} a_1 & b_1 & c_1 \\ a_2 & b_2 & c_2 \\ a_3 & b_3 & c_3 \end{vmatrix}$ is a determinant of third order whose value is

$$a_1b_2c_3+a_2b_3c_1+a_3b_1c_2-a_3b_2c_1-a_2b_1c_3-a_1b_3c_2.$$

In general, a determinant is equal to the sum of the products of the elements in any given column (or row) with their corresponding cofactors; the cofactor of an element, $\mathscr{A}_{ij}$, is the determinant, of next lower order, obtained by striking out the row i and the column j in which the element lies, multiplied by $(-1)^{i+j}$, e.g.

$$\begin{vmatrix} a_1 & b_1 & c_1 \\ a_2 & b_2 & c_2 \\ a_3 & b_3 & c_3 \end{vmatrix} = a_1\mathscr{A}_{11}+a_2\mathscr{A}_{21}+a_3\mathscr{A}_{31} = a_1\times(-1)^2\times\begin{vmatrix} b_2 & c_2 \\ b_3 & c_3 \end{vmatrix}$$

$$+a_2\times(-1)^3\times\begin{vmatrix} b_1 & c_1 \\ b_3 & c_3 \end{vmatrix}+a_3\times(-1)^4\times\begin{vmatrix} b_1 & c_1 \\ b_2 & c_2 \end{vmatrix}$$

$$= a_1(b_2c_3-b_3c_2)-a_2(b_1c_3-b_3c_1)+a_3(b_1c_2-b_2c_1) \tag{4-2.2}$$

or $$\begin{vmatrix} a_1 & b_1 & c_1 \\ a_2 & b_2 & c_2 \\ a_3 & b_3 & c_3 \end{vmatrix} = a_1\mathscr{A}_{11}+b_1\mathscr{A}_{12}+c_1\mathscr{A}_{13} = a_1\times(-1)^2\times\begin{vmatrix} b_2 & c_2 \\ b_3 & c_3 \end{vmatrix}$$

$$+b_1\times(-1)^3\times\begin{vmatrix} a_2 & c_2 \\ a_3 & c_3 \end{vmatrix}+c_1\times(-1)^4\times\begin{vmatrix} a_2 & b_2 \\ a_3 & b_3 \end{vmatrix} = \text{etc.}$$

Hence, we may successively break down any determinant into products involving lower order determinants, until second order determinants are reached, the values of which are given by eqn (4-2.1).

The determinant of a square matrix is the determinant obtained by considering the array of elements in the matrix as a determinant; if the matrix is A we will write the determinant as det(A) i.e. if

$$A = \left\| \begin{matrix} A_{11} & A_{12} & \dots & A_{1n} \\ A_{21} & A_{22} & \dots & A_{2n} \\ \cdot & \cdot & & \cdot \\ \cdot & \cdot & & \cdot \\ \cdot & \cdot & & \cdot \\ A_{n1} & A_{n2} & \dots & A_{nn} \end{matrix} \right\|,$$

then

$$\det(A) = \left| \begin{matrix} A_{11} & A_{12} & \dots & A_{1n} \\ A_{21} & A_{22} & \dots & A_{2n} \\ \cdot & \cdot & & \cdot \\ \cdot & \cdot & & \cdot \\ \cdot & \cdot & & \cdot \\ A_{n1} & A_{n2} & \dots & A_{nn} \end{matrix} \right|.$$

4-3. Matrix algebra

The algebra of matrices gives rules for (1) equality, (2) addition and subtraction, (3) multiplication, and (4) 'division' as well as (5) an associative and a distributive law. It also includes definitions of (6) a transpose, adjoint and inverse of a matrix.

(1) Equality. Two matrices A and B are equal if, and only if, $A_{ij} = B_{ij}$ for all i and j,

$$\text{e.g. if } A = \left\| \begin{matrix} 1 & 2 \\ 3 & 4 \end{matrix} \right\| \text{ and } A = B, \text{ then } B = \left\| \begin{matrix} 1 & 2 \\ 3 & 4 \end{matrix} \right\|.$$

(2) Addition and subtraction. Matrices may be added and subtracted only if they are of the same dimensions. Under these circumstances, the sum of A and B is given by the matrix C,

$$A + B = C,$$

where $C_{ij} = A_{ij} + B_{ij}$ for all i and j, e.g.

$$\left\| \begin{matrix} 1 & 2 \\ 3 & 4 \end{matrix} \right\| + \left\| \begin{matrix} 5 & 6 \\ 7 & 8 \end{matrix} \right\| = \left\| \begin{matrix} 6 & 8 \\ 10 & 12 \end{matrix} \right\|.$$

Under the same circumstances, the subtraction of B from A produces the matrix C,

$$A - B = C,$$

where $C_{ij} = A_{ij} - B_{ij}$ for all i and j, e.g.

$$\begin{Vmatrix} 1 & 2 \\ 3 & 4 \end{Vmatrix} - \begin{Vmatrix} 5 & 6 \\ 7 & 8 \end{Vmatrix} = \begin{Vmatrix} -4 & -4 \\ -4 & -4 \end{Vmatrix}.$$

From this, it follows that multiplying a matrix A by a number c produces a matrix B,

$$B = cA,$$

whose elements are given by $B_{ij} = cA_{ij}$, for all i and j,

$$\text{e.g. } 3\begin{Vmatrix} 1 & 2 \\ 3 & 4 \end{Vmatrix} = \begin{Vmatrix} 3 & 6 \\ 9 & 12 \end{Vmatrix}.$$

(3) Multiplication. Two matrices A and B may only be multiplied together (called matrix multiplication) if the number of columns in A, say n, equals the number of rows in B; the product is then defined as the matrix C,

$$C = AB,$$

whose elements are given by the equation:

$$C_{ij} = \sum_{k=1}^{n} A_{ik} B_{kj} \tag{4-3.1}$$

for all i and j. If the matrix A has m rows and n columns (an $m \times n$ matrix) and the matrix B has n rows and p columns (an $n \times p$ matrix,) the matrix C will have m rows and p columns (an $m \times p$ matrix). The following equations are examples of matrix multiplication:

$$\begin{Vmatrix} 1 & 2 \\ 3 & 4 \end{Vmatrix} \begin{Vmatrix} 5 & 6 \\ 7 & 8 \end{Vmatrix} = \begin{Vmatrix} 19 & 22 \\ 43 & 50 \end{Vmatrix}, \tag{4-3.2}$$

$$\begin{Vmatrix} 5 & 6 \\ 7 & 8 \end{Vmatrix} \begin{Vmatrix} 1 & 2 \\ 3 & 4 \end{Vmatrix} = \begin{Vmatrix} 23 & 34 \\ 31 & 46 \end{Vmatrix}, \tag{4-3.3}$$

$$\begin{Vmatrix} 1 & 2 & 3 \\ 4 & 5 & 6 \\ 7 & 8 & 9 \end{Vmatrix} \begin{Vmatrix} 1 \\ 2 \\ 3 \end{Vmatrix} = \begin{Vmatrix} 14 \\ 32 \\ 50 \end{Vmatrix}, \tag{4-3.4}$$

$$\begin{Vmatrix} 1 & 2 & 3 \end{Vmatrix} \begin{Vmatrix} 1 & 2 & 3 \\ 4 & 5 & 6 \\ 7 & 8 & 9 \end{Vmatrix} = \begin{Vmatrix} 30 & 36 & 42 \end{Vmatrix}. \tag{4-3.5}$$

An easy way of remembering how to carry out matrix multiplication is to realize that in doing so, one takes succeeding *rows* in the *first* matrix multiplied vectorially by succeeding *columns* in the *second* matrix, the ith row and jth column producing the i, j element in the product; or schematically:

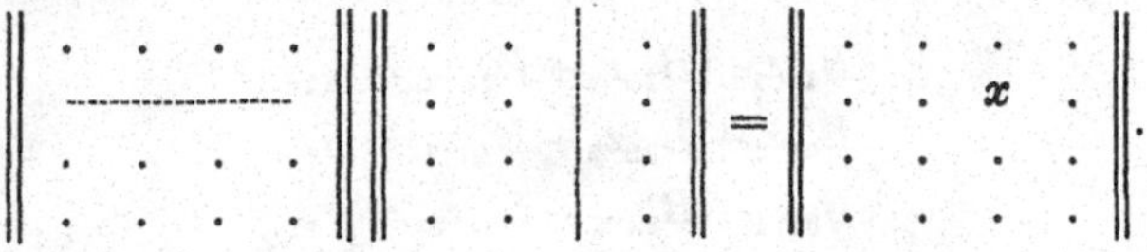

More than two matrices can be multiplied together; one simply uses the multiplication rule more than once, multiplying pairs of matrices at a time (see eqn (4-3.12)). For the product of three matrices,

$$D = ABC,$$

the general element of the product is given by:

$$D_{ij} = \sum_{k}^{r} \sum_{m}^{s} A_{ik} B_{km} C_{mj} \tag{4-3.6}$$

for all i and j, where r is the number of columns in A which must be the same as the number of rows in B and s is the number of columns in B which must be the same as the number of rows in C (see problem 4.3). Notice the restrictions which apply to the number of rows and columns in matrices which are to be multiplied together.

It is apparent from the examples in eqn (4-3.2) and (4-3.3), that, in general, matrices do not *commute*, i.e. $AB \neq BA$. For this reason care must be taken when matrix equations are being manipulated and one must remember that, like operators, the *order* in which matrices are multiplied together can be significant.

One of the uses of matrices is to express sets of linear equations in a compact form, for example with the above definition of matrix multiplication, it is possible to write the equations:

$$\begin{aligned} A_{11}y_1 + A_{12}y_2 + A_{13}y_3 &= x_1 \\ A_{21}y_1 + A_{22}y_2 + A_{23}y_3 &= x_2 \\ A_{31}y_1 + A_{32}y_2 + A_{33}y_3 &= x_3 \end{aligned} \tag{4-3.7}$$

in the form:

$$\begin{Vmatrix} A_{11} & A_{12} & A_{13} \\ A_{21} & A_{22} & A_{23} \\ A_{31} & A_{32} & A_{33} \end{Vmatrix} \begin{Vmatrix} y_1 \\ y_2 \\ y_3 \end{Vmatrix} = \begin{Vmatrix} x_1 \\ x_2 \\ x_3 \end{Vmatrix}$$

or

$$AY = X \tag{4-3.8}$$

where

$$A = \begin{Vmatrix} A_{11} & A_{12} & A_{13} \\ A_{21} & A_{22} & A_{23} \\ A_{31} & A_{32} & A_{33} \end{Vmatrix}, \quad Y = \begin{Vmatrix} y_1 \\ y_2 \\ y_3 \end{Vmatrix}, \quad X = \begin{Vmatrix} x_1 \\ x_2 \\ x_3 \end{Vmatrix}.$$

Furthermore, if, coupled with these equations, there is also the set of equations

$$\begin{aligned} B_{11}z_1+B_{12}z_2+B_{13}z_3 &= y_1 \\ B_{21}z_1+B_{22}z_2+B_{23}z_3 &= y_2 \\ B_{31}z_1+B_{32}z_2+B_{33}z_3 &= y_3 \end{aligned}$$

then

$$Y = BZ,$$

where

$$B = \begin{Vmatrix} B_{11} & B_{12} & B_{13} \\ B_{21} & B_{22} & B_{23} \\ B_{31} & B_{32} & B_{33} \end{Vmatrix} \quad \text{and} \quad Z = \begin{Vmatrix} z_1 \\ z_2 \\ z_3 \end{Vmatrix}.$$

Therefore

$$(AB)Z = X.$$

This result can be expressed in the following way: if the transformation of z's to y's is defined by a matrix B and that of y's to x's by a matrix A, then the transformation of z's to x's is defined by the matrix AB.

(4) 'Division'. As with operators, 'division' can only be accomplished through an inverse process. Every matrix A which has a non-zero determinant,

$$\det(A) \neq 0,$$

is said to be non-singular. For such, and only such, matrices, an *inverse* A^{-1} can be defined by the following equation:

$$AA^{-1} = A^{-1}A = E, \tag{4-3.9}$$

where E is the identity or unit matrix (see Appendix A.4-1(a)). The matrix operation which is equivalent to division is matrix multiplication by an inverse, e.g. if

$$AB = C \tag{4-3.10}$$

we may write

$$ABB^{-1} = CB^{-1} \tag{4-3.11}$$

or

$$AE = CB^{-1}$$

or

$$A = CB^{-1}.$$

Notice that when we multiply eqn (4-3.10) by B^{-1} to produce eqn (4-3.11), as matrices do not necessarily commute, we must do so on the right-hand side of *both* sides of the equation. A method for finding the inverse of a non-singular matrix is given in Appendix A.4-2. From this method it is apparent why A^{-1} is only defined when $\det(A) \neq 0$.

This also implies that only square matrices can have an inverse, since only square matrices have a determinant.

(5) Associative and distributive laws. These laws are respectively:

$$A(BC) = (AB)C, \tag{4-3.12}$$

and

$$A(B+C) = AB+AC. \tag{4-3.13}$$

(6) Transpose, adjoint, and inverse of a matrix. The inverse has been defined in (4) above and the transpose and adjoint are defined in Appendix A.4-1 and Table 4-1.1. The reader is left to prove (see problem 4.1) that the transpose, adjoint, and inverse of the product of two matrices are given by:

$$\widetilde{AB} = \tilde{B}\tilde{A}, \tag{4-3.14}$$

$$(AB)^{\dagger} = B^{\dagger}A^{\dagger}, \tag{4-3.15}$$

$$(AB)^{-1} = B^{-1}A^{-1}. \tag{4-3.16}$$

4-4. The matrix eigenvalue equation

For every square matrix A of order n there is an *eigenvalue* equation of the form:

$$Ax = \lambda x, \tag{4-4.1}$$

where x is a column matrix (with dimensions $n \times 1$) and λ is a number or *scalar*. The solutions of this equation, and in general there will be n which are distinct, are the values of λ (called the *eigenvalues*) and the corresponding column matrices x (called the *eigenvectors*). The equation can be expressed in words by saying that matrix A multiplied on the right by the column matrix x produces the same column matrix simply multiplied by a number.

We can distinguish between the several solutions of eqn (4-4.1) with subscripts and write the different eigenvectors which correspond to the different eigenvalues $\lambda_1, \lambda_2, \ldots, \lambda_n$ as $x_1, x_2, \ldots, x_n$ or

$$\left\|\begin{matrix} x_{11} \\ x_{21} \\ \cdot \\ \cdot \\ \cdot \\ x_{n1} \end{matrix}\right\|, \left\|\begin{matrix} x_{12} \\ x_{22} \\ \cdot \\ \cdot \\ \cdot \\ x_{n2} \end{matrix}\right\|, \ldots, \left\|\begin{matrix} x_{1n} \\ x_{2n} \\ \cdot \\ \cdot \\ \cdot \\ x_{nn} \end{matrix}\right\|$$

and we can write eqn (4-4.1) as

$$Ax_i = \lambda_i x_i \quad i = 1, 2, \ldots, n. \tag{4-4.2}$$

It is customary to require the eigenvectors to be normalized (see Appendix A.4-1(g)), that is:

$$x_i^\dagger x_i = \|1\| \qquad i = 1, 2 \ldots n$$

or

$$\| x_{1i}^* \, x_{2i}^* \ldots x_{ni}^* \| \left\| \begin{matrix} x_{1i} \\ x_{2i} \\ \cdot \\ \cdot \\ \cdot \\ x_{ni} \end{matrix} \right\| = \|1\| \quad i = 1, 2 \ldots n$$

or

$$\sum_{k=1}^{n} x_{ki}^* x_{ki} = 1 \qquad i = 1, 2 \ldots n. \tag{4-4.3}$$

This restriction cuts out those superfluous eigenvectors which differ merely by a constant factor.

An alternative form for eqn (4-4.2) is

$$(A - \lambda_i E)x_i = 0 \qquad i = 1, 2 \ldots n \tag{4-4.4}$$

where E is the identity matrix and 0 the null matrix (see Appendix A.4-1(f)). For this equation to have non-trivial solutions (that is, excluding solutions of the form $x_i = 0$), it is necessary that the eigenvalues λ_i obey the determinantal equation:

$$\det(A - \lambda E) = 0. \tag{4-4.5}$$

The proof of this is given in Appendix A.4-3. Eqn (4-4.5) is often called the characteristic equation of matrix A and it is essentially a polynomial equation in λ with n roots: $\lambda_1, \lambda_2, \ldots \lambda_n$. These roots or eigenvalues, which guarantee non-trivial eigenvectors, once found can be used, one by one, in the solution of eqn (4-4.4) coupled with eqn (4-4.3); each eigenvalue λ_i leading to a corresponding normalized eigenvector x_i (see Appendix A.4-5).

There are two important theorems for the eigenvalues and eigenvectors of eqns (4-4.4). If a matrix A is Hermitian ($A = A^\dagger$, see Appendix A.4-1(e)), its eigenvalues are real and its eigenvectors are orthogonal to each other if they correspond to different eigenvalues i.e. if the eigenvalues are *non-degenerate* ($\lambda_k \neq \lambda_l$). This result can be expressed as

$$\lambda_i = \lambda_i^* \qquad i = 1, 2, \ldots, n$$

$$x_k^\dagger x_l = 0 \qquad \text{or} \qquad \sum_{j=1}^{n} x_{jk}^* x_{jl} = 0 \qquad (\lambda_k \neq \lambda_l) \tag{4-4.6}$$

and, if we combine eqn (4-4.6) with eqn (4-4.3), we have

$$x_k x_l^\dagger = \|\delta_{kl}\| \qquad \begin{matrix} k = 1, 2, \ldots n \\ l = 1, 2, \ldots n \end{matrix} \tag{4-4.7}$$

provided that A is Hermitian and the roots of eqn (4-4.5) are distinct. Eqns (4-4.7) define a set of normalized orthogonal eigenvectors.

The other theorem states that the matrix X formed by using the eigenvectors of a Hermitian matrix as its columns is unitary (for the definition of a unitary matrix, see Appendix A.4-1(g)). The proof of these two theorems is given in Appendix A.4-3.

The matrix X formed from the eigenvectors of a matrix A may be used to combine the solutions of $Ax = \lambda x$ into a single equation:

$$AX = X\Lambda \tag{4-4.8}$$

where

$$\Lambda = \begin{Vmatrix} \lambda_1 & 0 & \ldots & 0 \\ 0 & \lambda_2 & \ldots & 0 \\ \cdot & \cdot & & \cdot \\ \cdot & \cdot & & \cdot \\ \cdot & \cdot & & \cdot \\ 0 & 0 & \ldots & \lambda_n \end{Vmatrix}.$$

Notice the order is $X\Lambda$ *not* ΛX; the reader is left to confirm this fact for himself. In eqn (4-4.8), if A is Hermitian then of course X will be unitary and if A is symmetric then X will be orthogonal.

4-5. Similarity transformations

If a matrix Q exists such that

$$Q^{-1}AQ = B, \tag{4-5.1}$$

then the matrices A and B are said to be related by a *similarity transformation*. Such transformations will be very important in what follows later. It is immediately apparent that eqn (4-5.1) parallels the relation that exists between two symmetry operations that belong to the same class (see § 3-5). If Q is a unitary matrix (see Appendix A.4-1(g)), then A and B are said to be related by *a unitary transformation*.

There are a number of useful theorems concerning matrices which are related by a similarity transformation. If the matrices A and B are related by a similarity transformation, then their determinants, eigenvalues, and traces (sum of the diagonal elements) will be identical:

$$\det(A) = \det(B), \tag{4-5.2}$$

$$\lambda\text{'s of } A = \lambda\text{'s of } B, \tag{4-5.3}$$

$$\text{Trace}(A) = \text{Trace}(B). \tag{4-5.4}$$

If $A' = Q^{-1}AQ$, $B' = Q^{-1}BQ$, $C' = Q^{-1}CQ$ etc., then any relationship between A, B, C etc. is also satisfied by A', B', C' etc.

If the result of a similarity transformation is to produce a diagonal matrix (see Appendix A.4-1(b)), then the process is called *diagonalization*. If the matrices A and B can be diagonalized by the *same* matrix, then A and B commute.

If X is the matrix formed from the eigenvectors of a matrix A, then the similarity transformation $X^{-1}AX$ will produce a *diagonal* matrix whose elements are the eigenvalues of A. Furthermore, if A is Hermitian, then X will be unitary and therefore we can see that a Hermitian matrix can always be diagonalized by a unitary transformation, and a symmetric matrix by an orthogonal transformation.

The final theorem is that a unitary transformation leaves a unitary matrix unitary.

These theorems are proven in Appendix A.4-4 and, in Appendix A.4-5, we show, with an example, how a matrix can be diagonalized.

Appendices

A.4-1. Special matrices

(*a*) *Identity matrix*. This is a square matrix whose diagonal elements are all unity and whose off-diagonal elements are all zero:

$$E = \begin{Vmatrix} 1 & 0 & 0 & . & . & . \\ 0 & 1 & 0. & & . & . \\ 0 & 0 & 1 & . & . & . \\ . & . & . & . & . & . \\ . & . & . & . & . & . \\ . & . & . & . & . & . \end{Vmatrix} \tag{A.4-1.1}$$

i.e.

$$E_{ij} = \delta_{ij} = \begin{cases} 0 & i \neq j \\ 1 & i = j. \end{cases}$$

(The symbol δ_{ij} is called Kronecker delta.) Other symbols for the identity matrix are I and **1** and it is sometimes called the unit matrix. It can be of any order and can be inserted anywhere in any matrix equation.

(*b*) *Diagonal matrix*. Any square matrix for which all the off-diagonal elements are zero and at least one of the diagonal elements is non-zero is said to be diagonal:

$$D = \begin{Vmatrix} d_1 & 0 & 0 & . & . & . \\ 0 & d_2 & 0 & . & . & . \\ 0 & 0 & d_3 & . & . & . \\ . & . & . & . & . & . \end{Vmatrix} \tag{A.4-1.2}$$

i.e.

$$D_{ij} = 0 \text{ if } i \neq j,\ D_{ij} \neq 0 \text{ if } i = j.$$

(c) *Real matrix.* The conjugate complex f^* of a number or function f is obtained by replacing i everywhere in f by $-$i. The *conjugate complex* of matrix A is written as A^* and the elements of A^* are the conjugate complexes of the elements of A, i.e. $(A^*)_{ij} = (A_{ij})^*$. For a real matrix all the elements are real and therefore

$$A = A^* \tag{A.4-1.3}$$

i.e.

$$A_{ij} = A_{ij}^* \text{ for all } i \text{ and } j.$$

(d) *Symmetric matrix.* The *transpose* of a matrix A is obtained by changing rows into columns (or vice versa) and is given the symbol $\tilde{A}$ (or, sometimes, A'), e.g. the transpose of

$$A = \begin{Vmatrix} 1 & 2 & 3 \\ 4 & 5 & 6 \\ 7 & 8 & 9 \end{Vmatrix} \quad \text{is} \quad \tilde{A} = \begin{Vmatrix} 1 & 4 & 7 \\ 2 & 5 & 8 \\ 3 & 6 & 9 \end{Vmatrix}.$$

For a symmetric matrix:

$$A = \tilde{A}, \tag{A.4-1.4}$$

i.e. $A_{ij} = A_{ji}$ for all i and j. The matrix

$$A = \begin{Vmatrix} 1 & 2 & 3 \\ 2 & 4 & 5 \\ 3 & 5 & 6 \end{Vmatrix}$$

is symmetric. All symmetric matrices must be square.

(e) *Hermitian matrix.* The *adjoint* of a matrix A is obtained by taking the complex conjugate of the transpose and is given the symbol $A^\dagger$ i.e. $A^\dagger = \tilde{A}^*$. (Mathematicians frequently call $A^\dagger$ the Hermitian conjugate and reserve the term adjoint for the transposed matrix of the cofactors of A.) The adjoint of

$$A = \begin{Vmatrix} 1 & 4 & \mathrm{i} \\ \mathrm{e}^{\mathrm{i}} & 2 & -\mathrm{i} \\ 3 & \mathrm{e}^{2\mathrm{i}} & 1 \end{Vmatrix} \quad \text{is} \quad A^\dagger = \begin{Vmatrix} 1 & \mathrm{e}^{-\mathrm{i}} & 3 \\ 4 & 2 & \mathrm{e}^{-2\mathrm{i}} \\ -\mathrm{i} & \mathrm{i} & 1 \end{Vmatrix}.$$

A Hermitian matrix is one which obeys the equation

$$A = A^\dagger \tag{A.4-1.5}$$

i.e. $A_{ij} = A_{ji}^*$ for all i and j. For example,

$$A = \begin{Vmatrix} 1 & \mathrm{i} & \mathrm{e}^{-\mathrm{i}} \\ -\mathrm{i} & 2 & 4 \\ \mathrm{e}^{\mathrm{i}} & 4 & 3 \end{Vmatrix}$$

is Hermitian. All Hermitian matrices must be square. For real matrices the criterion for being Hermitian or symmetric is the same.

(*f*) *Null* (*or zero*) *matrix.* A matrix of *any* dimension whose elements are all zero:

$$0 = \begin{Vmatrix} 0 & 0 & . & . & . \\ 0 & 0 & . & . & . \\ . & . & . & . & . \\ . & . & . & . & . \\ . & . & . & . & . \end{Vmatrix}, \tag{A.4-1.6}$$

i.e. $0_{ij} = 0$ for all i and j.

(*g*) *Unitary matrix.* A matrix is unitary if its adjoint (see (*e*) above) is equal to its inverse:

$$A^{\dagger} = A^{-1},$$

or

$$A^{\dagger} A = E, \tag{A.4-1.7}$$

or

$$AA^{\dagger} = E.$$

The columns (or rows) of a unitary matrix are related to a set of orthogonal normalized vectors in a general vector space. If, for example,

$$A = \begin{Vmatrix} A_{11} & A_{12} & \dots & A_{1n} \\ A_{21} & A_{22} & \dots & A_{2n} \\ . & . & & . \\ . & . & & . \\ . & . & & . \\ A_{n1} & A_{n2} & \dots & A_{nn} \end{Vmatrix}$$

is unitary then $A^{\dagger}A = E$ and

$$\sum_{k=1}^{n} A^{*}_{ki} A_{kj} = \delta_{ij} \qquad \begin{matrix} i = 1, 2, \dots n \\ j = 1, 2 \dots n \end{matrix} \tag{A.4-1.8}$$

and this is the requirement, by definition, for the general vectors:

$$\mathbf{r}_j = \sum_{k=1}^{n} A_{kj} \mathbf{e}_k \qquad j = 1, 2 \dots n \tag{A.4-1.9}$$

to be orthogonal and normalized i.e. the columns of A form the orthogonal, normalized vectors $\mathbf{r}_j$. In eqn (A.4-1.9) the $\mathbf{e}_k$ ($k = 1, 2 \dots n$) are unit orthogonal base vectors.

Alternatively, using $AA^{\dagger} = E$, we have

$$\sum_{k=1}^{n} A_{ik} A^{*}_{jk} = \delta_{ij} \qquad \begin{matrix} i = 1, 2 \dots n \\ j = 1, 2 \dots n \end{matrix} \tag{A.4-1.10}$$

and this is the requirement, by definition, for the general vectors:

$$\mathbf{s}_i = \sum_{k=1}^{n} A_{ik} \mathbf{e}_k \qquad i = 1, 2 \dots n \tag{A.4-1.11}$$

to be orthogonal and normalized, i.e. the rows of A form the orthogonal, normalized vectors $\mathbf{s}_i$.

All unitary matrices are square.

$$A = \begin{Vmatrix} 1 & 0 & 0 \\ 0 & 0 & \omega \\ 0 & \omega^2 & 0 \end{Vmatrix}$$

where $\omega = \exp(2\pi i/3)$ is an example of a unitary matrix. For this matrix the adjoint is

$$A^\dagger = \begin{Vmatrix} 1 & 0 & 0 \\ 0 & 0 & \omega \\ 0 & \omega^2 & 0 \end{Vmatrix}$$

and it is left to the reader to show that $AA^\dagger = A^\dagger A = E$.

(*h*) *Orthogonal matrix*. A matrix is orthogonal if its transpose (see (*d*) above) is equal to its inverse:

$$\tilde{A} = A^{-1}. \tag{A.4-1.12}$$

For real matrices (see (*c*) above) the criterion for being orthogonal or unitary is the same. All orthogonal matrices are square and

$$A = \begin{Vmatrix} \cos\theta & \sin\theta & 0 \\ -\sin\theta & \cos\theta & 0 \\ 0 & 0 & 1 \end{Vmatrix}$$

is an example of an orthogonal matrix.

A.4-2. Method for determining the inverse of a matrix

Consider the n equations:

$$\begin{aligned} y_1 &= A_{11}x_1 + A_{12}x_2 \dots + A_{1n}x_n \\ y_2 &= A_{21}x_1 + A_{22}x_2 \dots + A_{2n}x_n \\ &\cdot \quad \cdot \quad \cdot \quad \cdot \\ &\cdot \quad \cdot \quad \cdot \quad \cdot \\ &\cdot \quad \cdot \quad \cdot \quad \cdot \\ y_n &= A_{n1}x_1 + A_{n2}x_2 \dots + A_{nn}x_n \end{aligned} \tag{A.4-2.1}$$

they may be written in matrix notation as

$$\begin{Vmatrix} y_1 \\ y_2 \\ \cdot \\ \cdot \\ \cdot \\ y_n \end{Vmatrix} = \begin{Vmatrix} A_{11} & A_{12} & \dots & A_{1n} \\ A_{21} & A_{22} & \dots & A_{2n} \\ \cdot & \cdot & & \cdot \\ \cdot & \cdot & & \cdot \\ \cdot & \cdot & & \cdot \\ A_{n1} & A_{n2} & \dots & A_{nn} \end{Vmatrix} \begin{Vmatrix} x_1 \\ x_2 \\ \cdot \\ \cdot \\ \cdot \\ x_n \end{Vmatrix}$$

or $Y = AX$. Multiplying both sides of this equation by A^{-1}, the inverse of A, we get $X = A^{-1}Y$ or, if we let

$$A^{-1} = \begin{Vmatrix} A'_{11} & A'_{12} & \cdots & A'_{1n} \\ A'_{21} & A'_{22} & \cdots & A'_{2n} \\ \cdot & \cdot & & \cdot \\ \cdot & \cdot & & \cdot \\ \cdot & \cdot & & \cdot \\ A'_{n1} & A'_{n2} & \cdots & A'_{nn} \end{Vmatrix},$$

$$\begin{Vmatrix} x_1 \\ x_2 \\ \cdot \\ \cdot \\ \cdot \\ x_n \end{Vmatrix} = \begin{Vmatrix} A'_{11} & A'_{12} & \cdots & A'_{1n} \\ A'_{21} & A'_{22} & \cdots & A'_{2n} \\ \cdot & \cdot & & \cdot \\ \cdot & \cdot & & \cdot \\ \cdot & \cdot & & \cdot \\ A'_{n1} & A'_{n2} & \cdots & A'_{nn} \end{Vmatrix} \begin{Vmatrix} y_1 \\ y_2 \\ \cdot \\ \cdot \\ \cdot \\ y_n \end{Vmatrix}. \tag{A.4-2.2}$$

Now the determinant of A can be written as

$$\det(A) = \begin{vmatrix} A_{11} & A_{12} & \cdots & A_{1n} \\ A_{21} & A_{22} & \cdots & A_{2n} \\ \cdot & \cdot & & \cdot \\ \cdot & \cdot & & \cdot \\ \cdot & \cdot & & \cdot \\ A_{n1} & A_{n2} & \cdots & A_{nn} \end{vmatrix}$$

$$= A_{11}\mathscr{A}_{11} + A_{21}\mathscr{A}_{21} + \cdots + A_{n1}\mathscr{A}_{n1}$$
$$= A_{12}\mathscr{A}_{12} + A_{22}\mathscr{A}_{22} + \cdots + A_{n2}\mathscr{A}_{n2}$$
$$= \quad \cdots \qquad \cdots \qquad \cdots$$
$$= A_{1n}\mathscr{A}_{1n} + A_{2n}\mathscr{A}_{2n} + \cdots + A_{nn}\mathscr{A}_{nn}$$

where $\mathscr{A}_{ij}$ is the cofactor of A_{ij} (see § 4-2). If we multiply the first of eqns (A.4-2.1) by $\mathscr{A}_{11}$, the second by $\mathscr{A}_{21}, \ldots$, the nth by $\mathscr{A}_{n1}$ and add, we get:

$$\mathscr{A}_{11}y_1 + \mathscr{A}_{21}y_2 \cdots + \mathscr{A}_{n1}y_n$$
$$= (A_{11}\mathscr{A}_{11} + A_{21}\mathscr{A}_{21} + \cdots + A_{n1}\mathscr{A}_{n1})x_1$$
$$+ (A_{12}\mathscr{A}_{11} + A_{22}\mathscr{A}_{21} + \cdots + A_{n2}\mathscr{A}_{n1})x_2$$
$$+ \cdots$$
$$+ (A_{1n}\mathscr{A}_{11} + A_{2n}\mathscr{A}_{21} + \cdots + A_{nn}\mathscr{A}_{n1})x_n$$
$$= \det(A)x_1$$

$$+ \begin{vmatrix} A_{12} & A_{12} & \cdots & A_{1n} \\ A_{22} & A_{22} & \cdots & A_{2n} \\ \cdot & \cdot & & \cdot \\ \cdot & \cdot & & \cdot \\ \cdot & \cdot & & \cdot \\ A_{n2} & A_{n2} & \cdots & A_{nn} \end{vmatrix} x_2 \cdots + \begin{vmatrix} A_{1n} & A_{12} & \cdots & A_{1n} \\ A_{2n} & A_{22} & \cdots & A_{2n} \\ \cdot & \cdot & & \cdot \\ \cdot & \cdot & & \cdot \\ \cdot & \cdot & & \cdot \\ A_{nn} & A_{n2} & \cdots & A_{nn} \end{vmatrix} x_n.$$

Since determinants with two or more identical columns or rows are zero, this equation becomes

$$\mathscr{A}_{11}y_1+\mathscr{A}_{21}y_2 \ldots +\mathscr{A}_{n1}y_n = \det(A)x_1 \tag{A.4-2.3}$$

or

$$x_1 = \frac{\mathscr{A}_{11}}{\det(A)}y_1+\frac{\mathscr{A}_{21}}{\det(A)}y_2 \ldots +\frac{\mathscr{A}_{n1}}{\det(A)}y_n. \tag{A.4-2.4}$$

In the same fashion, with the other cofactors, we can obtain

$$x_2 = \frac{\mathscr{A}_{12}}{\det(A)}y_1+\frac{\mathscr{A}_{22}}{\det(A)}y_2 \ldots +\frac{\mathscr{A}_{n2}}{\det(A)}y_n \tag{A.4-2.4}$$

$$\cdot$$
$$\cdot$$
$$\cdot$$

$$x_n = \frac{\mathscr{A}_{1n}}{\det(A)}y_1+\frac{\mathscr{A}_{2n}}{\det(A)}y_2 \ldots +\frac{\mathscr{A}_{nn}}{\det(A)}y_n. \tag{A.4-2.4}$$

Comparison of these equations with eqn (A.4-2.2) shows that

$$A^{-1} = \left\| \begin{matrix} \dfrac{\mathscr{A}_{11}}{\det(A)} & \dfrac{\mathscr{A}_{21}}{\det(A)} & \cdots & \dfrac{\mathscr{A}_{n1}}{\det(A)} \\ \dfrac{\mathscr{A}_{12}}{\det(A)} & \dfrac{\mathscr{A}_{22}}{\det(A)} & \cdots & \dfrac{\mathscr{A}_{n2}}{\det(A)} \\ \cdot & \cdot & & \cdot \\ \cdot & \cdot & & \cdot \\ \cdot & \cdot & & \cdot \\ \dfrac{\mathscr{A}_{1n}}{\det(A)} & \dfrac{\mathscr{A}_{2n}}{\det(A)} & \cdots & \dfrac{\mathscr{A}_{nn}}{\det(A)} \end{matrix} \right\|. \tag{A.4-2.5}$$

So that for any square matrix A

$$(A^{-1})_{ij} = \mathscr{A}_{ji}/\det(A), \tag{A.4-2.6}$$

where $(A^{-1})_{ij}$ is the element in the ith row and jth column of the inverse of matrix A and $\mathscr{A}_{ji}$, the cofactor of A_{ji}, is $(-1)^{i+j}$ times the determinant of the lower order matrix obtained from A by striking out the jth row and ith column.

Clearly, eqn (A.4-2.6), and therefore an inverse, cannot be defined if $\det(A)$ is zero (i.e. if A is singular); also, for $\det(A)$ to exist, A must be square.

Eqns. (A.4-2.4) give the solution of any set of n equations in n variables; the method is known as Cramer's rule.

A.4-3. Theorems for eigenvectors

(*a*) *Theorem.* If λ_i are chosen as the roots of the equation: $\det(A-\lambda E) = 0$, then the equations:

$$(A-\lambda_i E)x_i = 0 \qquad i = 1, 2 \ldots n \tag{A.4-3.1}$$

will give non-trivial eigenvectors (column matrices) x_i.

Proof. $(A-\lambda_i E)x_i = 0$ is simply the matrix notation for a set of simultaneous equations which are equivalent to eqns (A.4-2.1) with $y_i = 0$. Consequently, if the elements of x_i are $x_{1i}, x_{2i}, \dots x_{ni}$, we have, by comparison with eqn (A.4-2.3),

$$\det(A-\lambda_i E)x_{1i} = 0,$$
$$\det(A-\lambda_i E)x_{2i} = 0,$$
$$\cdot \quad \cdot \quad \cdot$$
$$\cdot \quad \cdot \quad \cdot$$
$$\cdot \quad \cdot \quad \cdot$$
$$\det(A-\lambda_i E)x_{ni} = 0.$$

If the column matrix x_i is to be non-zero, then at least one of its elements, say x_{ki}, must be non-zero and since $\det(A-\lambda_i E)x_{ki} = 0$, this implies that $\det(A-\lambda_i E) = 0$. Hence, for non-trivial solutions ($x_i \neq 0$) the λ_i must be the roots of the equation $\det(A-\lambda E) = 0$.

(*b*) *Theorem.* If A is Hermitian, its eigenvalues are real and its eigenvectors are orthogonal to each other provided they correspond to non-degenerate eigenvalues.

Proof. Consider eqn (4-4.2):

$$Ax_i = \lambda_i x_i, \tag{A.4-3.2}$$

taking the adjoint of this equation (see eqn (4-3.15)), we have

$$x_i^\dagger A^\dagger = \lambda_i^* x_i^\dagger \qquad (\text{since } \lambda_i \text{ is a scalar, } \lambda_i^\dagger = \lambda_i^*)$$

or

$$x_i^\dagger A = \lambda_i^* x_i^\dagger \qquad (\text{since } A \text{ is Hermitian, } A^\dagger = A)$$

and multiplying by x_i produces

$$x_i^\dagger A x_i = \lambda_i^* x_i^\dagger x_i. \tag{A.4-3.3}$$

But from eqn (A.4-3.2)

$$x_i^\dagger A x_i = \lambda_i x_i^\dagger x_i \quad (\text{being a scalar, } \lambda_i \text{ commutes}) \tag{A.4-3.4}$$

and subtraction of eqn (A.4-3.4) from eqn (A.4-3.3) gives $(\lambda_i^* - \lambda_i)x_i^\dagger x_i = 0$. Since $x_i^\dagger x_i \neq 0$ (see eqn (4-4.3)),

$$\lambda_i^* = \lambda_i \tag{A.4-3.5}$$

or λ_i is real.

Now consider the pair of equations

$$Ax_k = \lambda_k x_k,$$
$$Ax_l = \lambda_l x_l.$$

Taking the adjoint of the first and multiplying on the right hand side by x_l gives

$$x_k^\dagger A x_l = \lambda_k^* x_k^\dagger x_l$$
$$= \lambda_k x_k^\dagger x_l \quad (\text{since } \lambda_k \text{ is real}). \tag{A.4-3.6}$$

Multiplying the second equation on the left hand side by $x_k^\dagger$ and subtracting it from eqn (A.4-3.6) gives

$$(\lambda_k - \lambda_l)x_k^\dagger x_l = 0 \quad \text{and therefore, if} \quad \lambda_k \neq \lambda_l, \qquad x_k^\dagger x_l = 0, \tag{A.4-3.7}$$

i.e. the eigenvectors are orthogonal.

(c) *Theorem.* The matrix X formed by using the eigenvectors of a Hermitian matrix as its columns is unitary.

Proof. The eigenvectors of a Hermitian matrix, if normalized and non-degenerate, obey the relation $x_k^\dagger x_l = \|\delta_{kl}\|$ and consequently we can write:

$$\begin{Vmatrix} x_{11}^* & x_{21}^* & \dots & x_{n1}^* \\ x_{12}^* & x_{22}^* & \dots & x_{n2}^* \\ \cdot & \cdot & & \cdot \\ \cdot & \cdot & & \cdot \\ \cdot & \cdot & & \cdot \\ x_{1n}^* & x_{2n}^* & \dots & x_{nn}^* \end{Vmatrix} \begin{Vmatrix} x_{11} & x_{12} & \dots & x_{1n} \\ x_{21} & x_{22} & \dots & x_{2n} \\ \cdot & \cdot & & \cdot \\ \cdot & \cdot & & \cdot \\ \cdot & \cdot & & \cdot \\ x_{n1} & x_{n2} & \dots & x_{nn} \end{Vmatrix} = \begin{Vmatrix} 1 & 0 & \dots & 0 \\ 0 & 1 & \dots & 0 \\ \cdot & \cdot & & \cdot \\ \cdot & \cdot & & \cdot \\ \cdot & \cdot & & \cdot \\ 0 & 0 & \dots & 1 \end{Vmatrix}$$

or $X^\dagger X = E$ and therefore $X^\dagger = X^{-1}$ and X is unitary.

Corollary. The matrix X, formed by using the eigenvectors of a symmetric matrix as its columns is orthogonal. This follows from the fact that the eigenvectors of a symmetric matrix obey the relationship $\tilde{x}_k x_l = \|\delta_{kl}\|$.

A.4-4. Theorems for similarity transformations

(a) *Theorem.* If $Q^{-1}AQ = B$, then $\det(A) = \det(B)$.

Proof. Since $\det(XY) = \det(X)\det(Y)$ (see Appendix A.4-6) we have:

$$\begin{aligned} \det(B) &= \det(Q^{-1})\det(AQ) \\ &= \det(Q^{-1})\det(A)\det(Q) \\ &= \det(Q^{-1})\det(Q)\det(A) \\ &= \det(Q^{-1}Q)\det(A) \\ &= \det(E)\det(A) \\ &= \det(A). \end{aligned}$$

(b) *Theorem.* If $Q^{-1}AQ = B$, then the eigenvalues of A and B are identical.

Proof. Since,

$$\begin{aligned} (B-\lambda E) &= (Q^{-1}AQ-\lambda E) \\ &= Q^{-1}(A-\lambda E)Q \end{aligned}$$

we have

$$\begin{aligned} \det(B-\lambda E) &= \det(Q^{-1})\det(A-\lambda E)\det(Q) \\ &= \det(Q^{-1}Q)\det(A-\lambda E) \\ &= \det(A-\lambda E). \end{aligned}$$

The roots of $\det(A-\lambda E) = 0$ and $\det(B-\lambda E) = 0$ must be identical, since the equations are identical.

(c) *Theorem.* If $Q^{-1}AQ = B$, then

$$\text{Trace}(A) = \text{Trace}(B).$$

Proof.

$$\begin{aligned}\text{Trace}(B) &= \sum_i B_{ii} \\ &= \sum_i \sum_j \sum_k (Q^{-1})_{ik} A_{kj} Q_{ji} \quad \text{(see eqn (4-3.6))} \\ &= \sum_j \sum_k A_{kj} \sum_i Q_{ji}(Q^{-1})_{ik} \\ &= \sum_j \sum_k A_{kj}(QQ^{-1})_{jk} \\ &= \sum_j \sum_k A_{kj}\, \delta_{jk} \\ &= \sum_k A_{kk} \\ &= \text{Trace}(A).\end{aligned}$$

(d) *Theorem.* If $A' = Q^{-1}AQ$, $B' = Q^{-1}BQ$, $C' = Q^{-1}CQ$ etc., then any relationship between A, B, C, etc. is also satisfied by A', B', C' etc.

Proof. Consider, as an example,

$$D = ABC$$

then,

$$\begin{aligned}D' &= Q^{-1}DQ \\ &= Q^{-1}ABCQ \\ &= Q^{-1}AQQ^{-1}BQQ^{-1}CQ \\ &= A'B'C'.\end{aligned}$$

(e) *Theorem.* If A and B are two matrices which can be diagonalized by the same matrix Q then A and B commute.

Proof. Since,

$$Q^{-1}AQ = D_a$$
$$Q^{-1}BQ = D_b$$

where D_a and D_b are diagonal matrices, then

$$\begin{aligned}Q^{-1}(AB)Q &= (Q^{-1}AQ)(Q^{-1}BQ) \\ &= D_aD_b \qquad \text{(diagonal matrices commute)} \\ &= D_bD_a \\ &= (Q^{-1}BQ)(Q^{-1}AQ) \\ &= Q^{-1}(BA)Q\end{aligned}$$

and therefore $AB = BA$ and A and B commute.

(f) *Theorem.* If X is the matrix formed from the eigenvectors of A, then $X^{-1}AX$ is a diagonal matrix Λ composed of the eigenvalues of A.

Proof. From eqn (4-4.8) we have

$$AX = X\Lambda$$

where Λ is the diagonal matrix composed of the eigenvalues of A and multiplying on the left of both sides of this equation by X^{-1} we obtain directly

$$X^{-1}AX = \Lambda. \tag{A.4-4.1}$$

If A is Hermitian, then X will be unitary (Appendix A.4-3(c)) and the transformation eqn (A.4-4.1) will be unitary. If A is symmetric, X and the transformation will be orthogonal.

(*g*) *Theorem.* A unitary transformation leaves a unitary matrix unitary.

Proof. Since $X = U^{-1}AU$, where A and U are unitary, we have:

$$\begin{aligned} X^{-1} &= (AU)^{-1}U \qquad \text{(see eqn (4-3.16))} \\ &= U^{-1}A^{-1}U \end{aligned}$$

and

$$\begin{aligned} X^{\dagger} &= (AU)^{\dagger}(U^{-1})^{\dagger} \qquad \text{(see eqn (4-3.15))} \\ &= U^{\dagger}A^{\dagger}(U^{-1})^{\dagger}. \end{aligned}$$

But, by definition, $U^{\dagger} = U^{-1}$, $A^{\dagger} = A^{-1}$, and $(U^{-1})^{\dagger} = (U^{\dagger})^{\dagger} = U$. Thus $X^{-1} = X^{\dagger}$ and X is unitary.

Likewise, if A and U are orthogonal (real as well as unitary) then so is X.

A.4-5. The diagonalization of a matrix or how to find the eigenvalues and eigenvectors of a matrix

Consider the diagonalization of the matrix:

$$A = \begin{Vmatrix} -1 & 0 & -4 \\ 0 & 2 & 0 \\ 2 & 0 & 5 \end{Vmatrix}.$$

This may be accomplished with a matrix X which is composed of the eigenvectors of A. The diagonal matrix created will consist of the eigenvalues of A. The steps involved are: (1) the determination of the eigenvalues λ_1, λ_2, and λ_3 from $\det(A - \lambda E) = 0$, (2) the determination (using these eigenvalues) of the eigenvectors

$$\begin{Vmatrix} x_{11} \\ x_{21} \\ x_{31} \end{Vmatrix}, \quad \begin{Vmatrix} x_{12} \\ x_{22} \\ x_{32} \end{Vmatrix}, \quad \text{and} \quad \begin{Vmatrix} x_{13} \\ x_{23} \\ x_{33} \end{Vmatrix} \quad \text{from the equations}$$

$$(A - \lambda_i E)\begin{Vmatrix} x_{1i} \\ x_{2i} \\ x_{3i} \end{Vmatrix} = \begin{Vmatrix} 0 \\ 0 \\ 0 \end{Vmatrix}, \quad \text{and} \quad \sum_{k=1}^{3} x_{ki}^{*}x_{ki} = 1 \quad (i = 1, 2, 3)$$

and (3) the determination of X^{-1} from X.

(1) The determinantal equation is:

$$\det(A-\lambda E) = \begin{vmatrix} -1-\lambda & 0 & -4 \\ 0 & 2-\lambda & 0 \\ 2 & 0 & 5-\lambda \end{vmatrix} = 0$$

or $\lambda^3-6\lambda^2+11\lambda-6 = 0$. The roots of this equation are: $\lambda_1 = 1$, $\lambda_2 = 2$, and $\lambda_3 = 3$. These are the eigenvalues.

(2) With these roots, we can now be assured of non-trivial eigenvectors.

$$\underline{\lambda_1 = 1}$$

$$\begin{Vmatrix} -2 & 0 & -4 \\ 0 & 1 & 0 \\ 2 & 0 & 4 \end{Vmatrix} \begin{Vmatrix} x_{11} \\ x_{21} \\ x_{31} \end{Vmatrix} = \begin{Vmatrix} 0 \\ 0 \\ 0 \end{Vmatrix}$$

$$\begin{aligned} -2x_{11}-4x_{31} &= 0 \\ x_{21} &= 0 \\ 2x_{11}+4x_{31} &= 0 \quad \text{(redundant)} \\ x_{11}^2+x_{21}^2+x_{31}^2 &= 1 \quad \text{(normalization).} \end{aligned}$$

Therefore,

$$x_{11} = -2/\sqrt{5}, \quad x_{21} = 0, \quad \text{and} \quad x_{31} = 1/\sqrt{5}$$

$$\underline{\lambda_2 = 2}$$

$$\begin{Vmatrix} -3 & 0 & -4 \\ 0 & 0 & 0 \\ 2 & 0 & 3 \end{Vmatrix} \begin{Vmatrix} x_{12} \\ x_{22} \\ x_{32} \end{Vmatrix} = \begin{Vmatrix} 0 \\ 0 \\ 0 \end{Vmatrix}$$

$$\begin{aligned} -3x_{12}-4x_{32} &= 0 \\ 0 &= 0 \quad \text{(redundant)} \\ 2x_{12}+3x_{32} &= 0 \\ x_{12}^2+x_{22}^2+x_{32}^2 &= 1 \quad \text{(normalization).} \end{aligned}$$

Therefore,

$$x_{12} = 0, \quad x_{22} = 1, \quad \text{and} \quad x_{32} = 0.$$

$$\underline{\lambda_3 = 3}$$

$$\begin{Vmatrix} -4 & 0 & -4 \\ 0 & -1 & 0 \\ 2 & 0 & 2 \end{Vmatrix} \begin{Vmatrix} x_{13} \\ x_{23} \\ x_{33} \end{Vmatrix} = \begin{Vmatrix} 0 \\ 0 \\ 0 \end{Vmatrix}$$

$$\begin{aligned} -4x_{13}-4x_{33} &= 0 \\ -x_{23} &= 0 \\ 2x_{13}+2x_{33} &= 0 \quad \text{(redundant)} \\ x_{13}^2+x_{23}^2+x_{33}^2 &= 1 \quad \text{(normalization).} \end{aligned}$$

Therefore,

$$x_{13} = 1/\sqrt{2}, \quad x_{23} = 0, \quad \text{and} \quad x_{33} = -1/\sqrt{2}.$$

Hence

$$X = \begin{Vmatrix} -2/\sqrt{5} & 0 & 1/\sqrt{2} \\ 0 & 1 & 0 \\ 1/\sqrt{5} & 0 & -1/\sqrt{2} \end{Vmatrix}.$$

(3) $(X^{-1})_{ij} = \mathscr{X}_{ji}/\det(X)$ (see Appendix A.4-2)

$\mathscr{X}_{ji}$ = cofactor of X_{ji}

$$\det(X) = 1/\sqrt{10}$$

$$X^{-1} = \begin{Vmatrix} -\sqrt{5} & 0 & -\sqrt{5} \\ 0 & 1 & 0 \\ -\sqrt{2} & 0 & -2\sqrt{2} \end{Vmatrix}.$$

The reader can confirm that

$$X^{-1}AX = \Lambda$$

i.e.

$$\begin{Vmatrix} -\sqrt{5} & 0 & -\sqrt{5} \\ 0 & 1 & 0 \\ -\sqrt{2} & 0 & -2\sqrt{2} \end{Vmatrix} \begin{Vmatrix} -1 & 0 & -4 \\ 0 & 2 & 0 \\ 2 & 0 & 5 \end{Vmatrix} \begin{Vmatrix} -2/\sqrt{5} & 0 & 1/\sqrt{2} \\ 0 & 1 & 0 \\ 1/\sqrt{5} & 0 & -1/\sqrt{2} \end{Vmatrix}$$

$$= \begin{Vmatrix} 1 & 0 & 0 \\ 0 & 2 & 0 \\ 0 & 0 & 3 \end{Vmatrix}.$$

Note, that had A been Hermitian, step (3) would have been simply $X^{-1} = X^{\dagger}$, since in such a case X would have been unitary. The rearrangement of the columns in X with the parallel rearrangement of the λ_i in Λ and the concomitant changes in X^{-1}, or the changing of the sign of every element in a given column of X with the concomitant changes in X^{-1}, all constitute other valid diagonalizations.

A.4-6. Proof that det *(AB)* = det *(A)* det *(B)*

Necessarily the matrices A and B must be square and of the same order. Consider matrices of the order 2

$$AB = \begin{Vmatrix} A_{11}B_{11}+A_{12}B_{21} & A_{11}B_{12}+A_{12}B_{22} \\ A_{21}B_{11}+A_{22}B_{21} & A_{21}B_{12}+A_{22}B_{22} \end{Vmatrix}$$

$$\det(AB) = \begin{vmatrix} A_{11}B_{11} & A_{11}B_{12} \\ A_{21}B_{11} & A_{21}B_{12} \end{vmatrix} + \begin{vmatrix} A_{11}B_{11} & A_{11}B_{12} \\ A_{22}B_{21} & A_{22}B_{22} \end{vmatrix}$$

$$+ \begin{vmatrix} A_{12}B_{21} & A_{12}B_{22} \\ A_{21}B_{11} & A_{21}B_{12} \end{vmatrix} + \begin{vmatrix} A_{12}B_{21} & A_{12}B_{22} \\ A_{22}B_{21} & A_{22}B_{22} \end{vmatrix}.$$

The first and last determinants vanish since, if the constant factor $A_{11}A_{21}$ is removed from the first determinant, its two rows are identical and removal of the constant factor $A_{12}A_{22}$ from the last determinant leaves it also with two identical rows. Therefore:

$$\begin{aligned}\det(AB) &= A_{11}A_{22}\begin{vmatrix} B_{11} & B_{12} \\ B_{21} & B_{22} \end{vmatrix} + A_{12}A_{21}\begin{vmatrix} B_{21} & B_{22} \\ B_{11} & B_{12} \end{vmatrix} \\ &= (A_{11}A_{22} - A_{12}A_{21})\begin{vmatrix} B_{11} & B_{12} \\ B_{21} & B_{22} \end{vmatrix} \\ &= \det(A)\det(B).\end{aligned}$$

The proof can be extended to matrices of higher order.

PROBLEMS

4.1. Show that for two matrices A and B: (a) $\widetilde{AB} = \tilde{B}\tilde{A}$, (b) $(AB)^\dagger = B^\dagger A^\dagger$, and (c) $(AB)^{-1} = B^{-1}A^{-1}$.

4.2. Prove that (a) the product of two unitary matrices is also unitary and (b) the inverse of a unitary matrix is unitary.

4.3. Show that if four matrices obey the equation $D = ABC$, then

$$D_{ij} = \sum_k \sum_l A_{ik}B_{kl}C_{lj}.$$

4.4. (a) Show that Trace $(AB) = \sum_i \sum_j A_{ij}B_{ji}$. (b) Given two matrices A and B of dimensions $n \times m$ and $m \times n$ respectively, prove Trace (AB) = Trace (BA).

4.5. Prove that for any matrix A: (a) $AA^\dagger$ and $A^\dagger A$ are Hermitian, (b) $(A + A^\dagger)$ and $i(A - A^\dagger)$ are Hermitian.

4.6. If $AB = BA$, show that $QA\tilde{Q}$ and $QB\tilde{Q}$ commute if Q is orthogonal.

4.7. Find the inverse of:

(a) $\begin{Vmatrix} a+ib & c+id \\ -c+id & a-ib \end{Vmatrix}$, (b) $\begin{Vmatrix} 1 & 0 & 0 & 0 \\ 0 & 1 & 0 & 0 \\ 0 & 0 & 1 & 0 \\ 0 & 0 & 0 & 1 \end{Vmatrix}$,

(c) $\begin{Vmatrix} 0 & 0 & a \\ 0 & b & 0 \\ c & 0 & 0 \end{Vmatrix}$, (d) $\begin{Vmatrix} a & -b \\ b & a \end{Vmatrix}$, (e) $\begin{Vmatrix} 0 & -i \\ i & 0 \end{Vmatrix}$, and

(f) $\begin{Vmatrix} 2 & 3 & 1 \\ 3 & 5 & 2 \\ 0 & 0 & 2 \end{Vmatrix}$.

4.8. Show that the matrices:

$$(a)\ \begin{Vmatrix} 1/\sqrt{2} & -1/\sqrt{2} \\ 1/\sqrt{2} & 1/\sqrt{2} \end{Vmatrix}, \qquad (b)\ \begin{Vmatrix} 1/\sqrt{2} & 1/\sqrt{2} \\ 1/\sqrt{2} & -1/\sqrt{2} \end{Vmatrix}, \text{ and}$$

$$(c)\ \begin{Vmatrix} \cos\theta & \sin\theta \\ \sin\theta & -\cos\theta \end{Vmatrix}$$

are orthogonal.

4.9. Show that if $\lambda_1, \lambda_2, \ldots \lambda_n$ are the eigenvalues of A, then $\lambda_1 - k, \lambda_2 - k, \ldots \lambda_n - k$ are the eigenvalues of $A - kE$.

4.10. Obtain the eigenvalues and normalized eigenvectors of:

$$(a)\ \begin{Vmatrix} 1 & -8 \\ 2 & 11 \end{Vmatrix}, \quad (b)\ \begin{Vmatrix} 5 & 10 & 8 \\ 10 & 2 & -2 \\ 8 & -2 & 11 \end{Vmatrix}, \quad (c)\ \begin{Vmatrix} \cos\theta & \sin\theta & 0 \\ -\sin\theta & \cos\theta & 0 \\ 0 & 0 & 1 \end{Vmatrix}.$$

4.11. If

$$A = \begin{Vmatrix} -1/2 & -\sqrt{3}/2 & 0 \\ -\sqrt{3}/2 & 1/2 & 0 \\ 0 & 0 & 2 \end{Vmatrix} \text{ and } Q = \begin{Vmatrix} 1/2 & \sqrt{3}/2 & 0 \\ -\sqrt{3}/2 & 1/2 & 0 \\ 0 & 0 & 1 \end{Vmatrix}$$

show that $Q^{-1}AQ$ is diagonal.

4.12. Diagonalize the following matrices:

$$(a)\ \begin{Vmatrix} 5 & 10 & 8 \\ 10 & 2 & -2 \\ 8 & -2 & 11 \end{Vmatrix}, \quad (b)\ \begin{Vmatrix} \cos\theta & \sin\theta & 0 \\ -\sin\theta & \cos\theta & 0 \\ 0 & 0 & 1 \end{Vmatrix}, \text{ and}$$

$$(c)\ \begin{Vmatrix} 2 & 4-i \\ 4+i & -14 \end{Vmatrix}.$$

5. Matrix representations

5-1. Introduction

THE best way to understand how the symmetry operations of a molecule influence its properties is to study the sets of matrices which mirror, by their group table (see § 3-4), those same operations. Such sets of matrices, homomorphic with the point group, are said to be, or to form, a *representation* of the point group. Essentially, when we introduce a matrix representation, we are replacing the *geometry* of symmetry operations with the *algebra* of matrices. Matrix representations are the crucial link between the symmetry of a molecule and the theorems which determine such practical things as to whether a given infra-red band should be present or not. Mastery of the ideas in this chapter is essential to the proper understanding of subsequent chapters.

There are several different methods of obtaining sets of matrices which are homomorphic with a given point group and in this chapter we discuss these methods in some detail. One way is to consider the effect that a symmetry operation has on the Cartesian coordinates of some point (or, equivalently, on some position vector) in the molecule. Another way is to consider the effect that a symmetry operation has on one or more sets of base vectors (coordinate axes) within the molecule.

A third and more complex way is to first find another set of operators $\boldsymbol{O}_R$, which have certain fundamental properties and are homomorphic with the symmetry operations, and then find a set of matrices which are homomorphic with these new operators. This last step is achieved by consideration of the effect that the $\boldsymbol{O}_R$ have on some 'family' of mathematical functions (a so-called *function space*) e.g. a set of five d-orbitals; it will be seen that the choice of the function space and the choice of the $\boldsymbol{O}_R$ are bound up with each other. It is important to realize that this method involves two steps as opposed to the first two methods which involve only one.

Since it is easy in this subject to 'lose sight of the forest for the trees', the scheme which we are following in these introductory chapters is summarized in Fig. 5-1.1. The reader will probably find it helpful to keep this plan in mind while he pursues the material of this chapter.

For the sake of completeness, various proofs are given in the appendices but, as in Chapter 4, knowledge of them is not critical for the reader who wants to have only a general understanding of the subject.

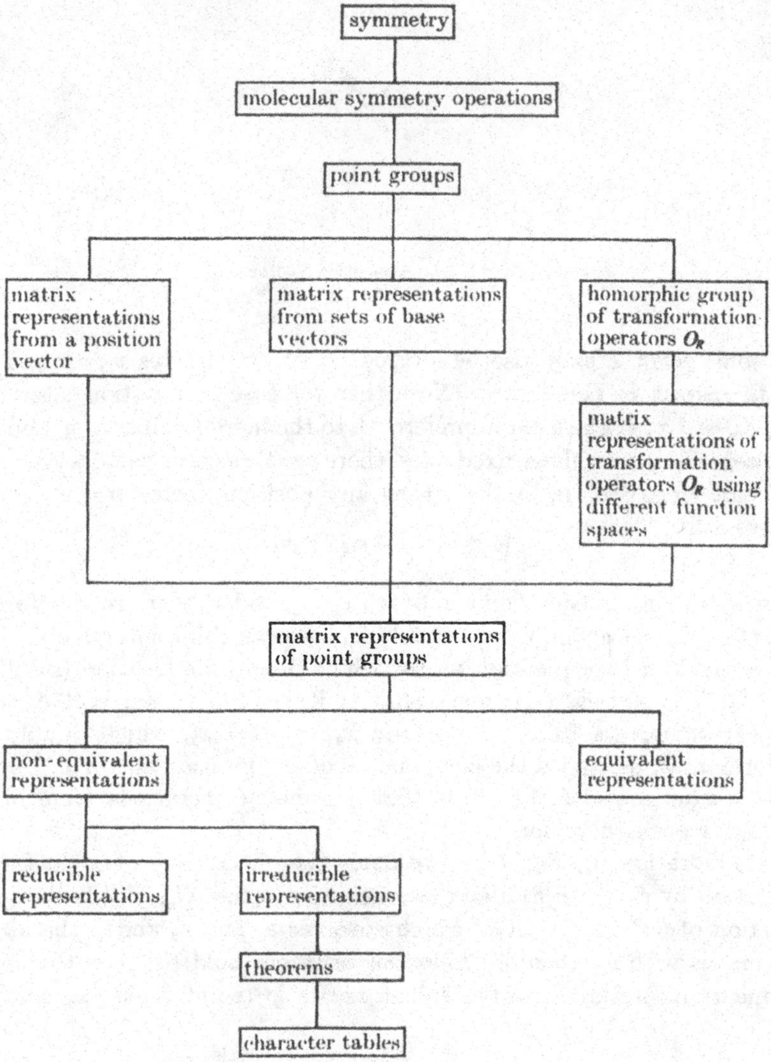

FIG. 5-1.1. Summary.

5-2. Symmetry operations on a position vector

A position vector **p** is a quantity which defines the location of some point P in three-dimensional physical space (see Fig. 5-2.1). If O is the origin of some set of space-fixed axes, the length p of OP and the direction of OP with respect to these axes constitute the position vector. If the set of space-fixed axes are mutually perpendicular, the position

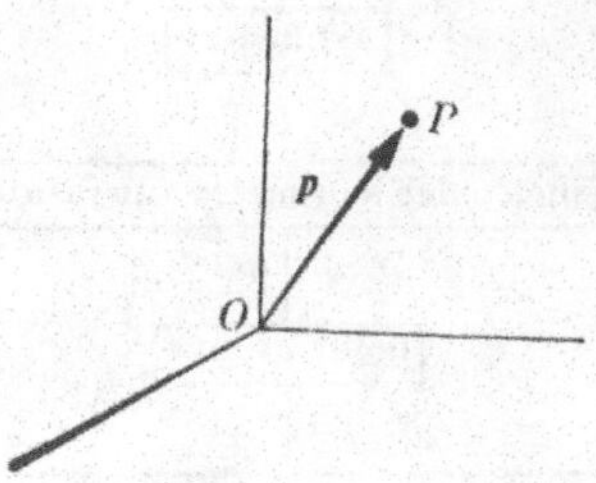

FIG. 5-2.1. A position vector.

of some point P may also be located by its coordinates x_1, x_2, and x_3 with respect to these axes. (Note that for ease of notation later on, x_1, x_2, and x_3 will be used in preference to the more familiar x, y, and z.) If, coinciding with these fixed axes, there are three unit vectors (vectors of unit length) $\mathbf{e}_1$, $\mathbf{e}_2$, and $\mathbf{e}_3$, then any position vector $\mathbf{p}$ can be expressed as

$$\mathbf{p} = x_1\mathbf{e}_1 + x_2\mathbf{e}_2 + x_3\mathbf{e}_3. \tag{5-2.1}$$

Corresponding to each point in space x_1, x_2, and x_3 there is therefore a position vector given by eqn (5-2.1) and we can think interchangeably of a point and the position vector which defines its location (see Fig. 5-2.2). The mutually perpendicular unit vectors $\mathbf{e}_1$, $\mathbf{e}_2$, and $\mathbf{e}_3$, are called orthogonal base vectors and x_1, x_2, and x_3, which double as coordinates, are called the components of the position vector $\mathbf{p}$.

We now consider the effect that symmetry operations have on a point or position vector.

(1) Rotation. In Fig. 5-2.3 we show the effect on $\mathbf{p}$ of a clockwise rotation by θ $(= 2\pi/n)$ about the direction $\mathbf{e}_3$ i.e. C_n. If d is the projection of OP on the plane which contains $\mathbf{e}_1$ and $\mathbf{e}_2$ and ϕ the angle it makes with $\mathbf{e}_1$, then the following relations hold between the components (coordinates) of the initial vector $\mathbf{p}$ (point P) x_1, x_2, and x_3

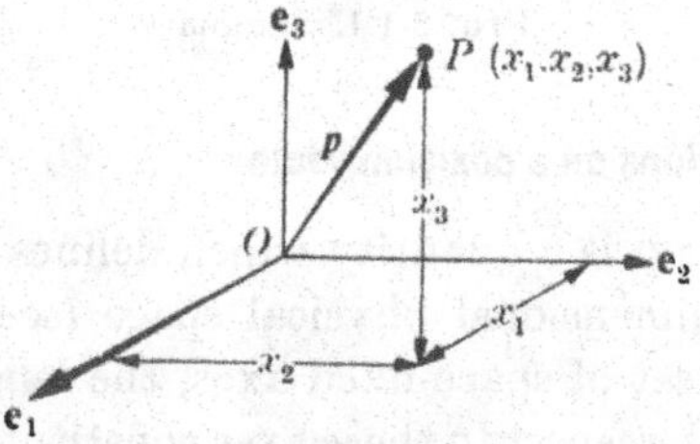

FIG. 5-2.2. Relation between a point and a position vector.

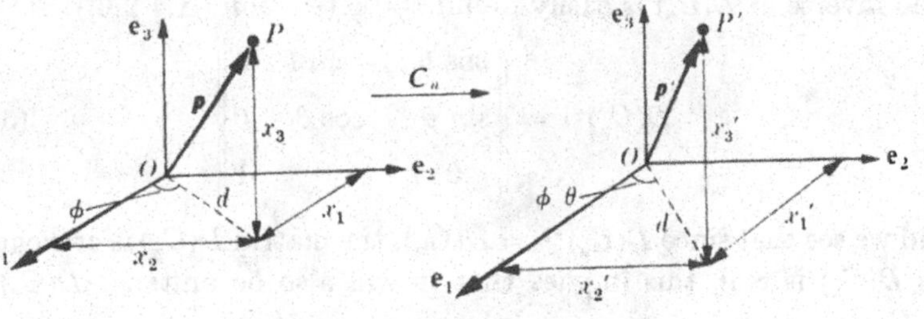

FIG. 5-2.3. Effect of C_n on **p**.

and those of the final vector **p**′ (point P') x_1', x_2', and x_3':

$$x_1' = d\cos(\phi-\theta)$$
$$= d\cos\phi\cos\theta+d\sin\phi\sin\theta$$
$$= d(x_1/d)\cos\theta+d(x_2/d)\sin\theta$$
$$= x_1\cos\theta+x_2\sin\theta \tag{5-2.2}$$
$$x_2' = d\sin(\phi-\theta)$$
$$= d\sin\phi\cos\theta-d\cos\phi\sin\theta$$
$$= d(x_2/d)\cos\theta-d(x_1/d)\sin\theta$$
$$= -x_1\sin\theta+x_2\cos\theta \tag{5-2.3}$$
$$x_3' = x_3. \tag{5-2.4}$$

Eqns (5-2.2) to (5-2.4) can be combined together (see eqn (4-3.8)) to give:

$$\begin{Vmatrix} x_1' \\ x_2' \\ x_3' \end{Vmatrix} = \begin{Vmatrix} \cos\theta & \sin\theta & 0 \\ -\sin\theta & \cos\theta & 0 \\ 0 & 0 & 1 \end{Vmatrix} \begin{Vmatrix} x_1 \\ x_2 \\ x_3 \end{Vmatrix}. \tag{5-2.5}$$

Necessarily, exactly the same set of equations can be obtained from an *anti-clockwise* rotation of θ about $\mathbf{e}_3$ of the base vectors $\mathbf{e}_1$ and $\mathbf{e}_2$, i.e. moving the point clockwise is the same as moving the laboratory axes anti-clockwise.

Eqn (5-2.5) can be used to define a matrix $D(C_n)$ which corresponds to the operation C_n:

$$\begin{Vmatrix} x_1' \\ x_2' \\ x_3' \end{Vmatrix} = D(C_n) \begin{Vmatrix} x_1 \\ x_2 \\ x_3 \end{Vmatrix}, \tag{5-2.6}$$

and

$$D(C_n) = \begin{Vmatrix} \cos\theta & \sin\theta & 0 \\ -\sin\theta & \cos\theta & 0 \\ 0 & 0 & 1 \end{Vmatrix}. \tag{5-2.7}$$

The inverse of $D(C_n)$ is easily found to be (see eqn (A.4-2.6)):

$$D(C_n)^{-1} = \left\| \begin{matrix} \cos\theta & -\sin\theta & 0 \\ \sin\theta & \cos\theta & 0 \\ 0 & 0 & 1 \end{matrix} \right\| \tag{5-2.8}$$

and we see that since $D(C_n)^{-1} = \tilde{D}(C_n)$, the matrix $D(C_n)$ is orthogonal. As $D(C_n)$ is real, this implies that it will also be unitary: $D(C_n)^{-1} = D(C_n)^{\dagger}$.

It is apparent that $D(C_n)^{-1}$ corresponds to a clockwise rotation by $-\theta$ or an anti-clockwise rotation by θ, i.e.

$$D(C_n)^{-1} = \left\| \begin{matrix} \cos\theta & -\sin\theta & 0 \\ \sin\theta & \cos\theta & 0 \\ 0 & 0 & 1 \end{matrix} \right\| = \left\| \begin{matrix} \cos(-\theta) & \sin(-\theta) & 0 \\ -\sin(-\theta) & \cos(-\theta) & 0 \\ 0 & 0 & 1 \end{matrix} \right\|$$
$$= D(C_n^{-1}).$$

(2) Reflection. In Fig. 5-2.4 the effect on **p** of a reflection in the plane containing $\mathbf{e}_2$ and $\mathbf{e}_3(\sigma_{23})$ is shown. Clearly,

$$x_1' = -x_1$$
$$x_2' = x_2$$
$$x_3' = x_3$$

$$\left\| \begin{matrix} x_1' \\ x_2' \\ x_3' \end{matrix} \right\| = \left\| \begin{matrix} -1 & 0 & 0 \\ 0 & 1 & 0 \\ 0 & 0 & 1 \end{matrix} \right\| \left\| \begin{matrix} x_1 \\ x_2 \\ x_3 \end{matrix} \right\|, \tag{5-2.9}$$

and

$$D(\sigma_{23}) = \left\| \begin{matrix} -1 & 0 & 0 \\ 0 & 1 & 0 \\ 0 & 0 & 1 \end{matrix} \right\|. \tag{5-2.10}$$

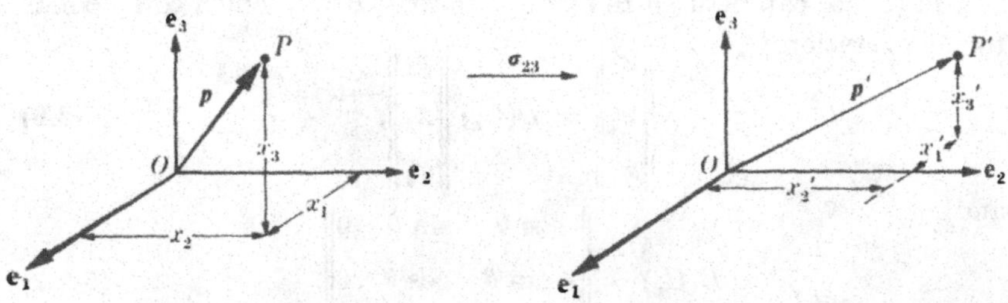

FIG. 5-2.4. Effect of σ_{23} on **p**.

Similarly we can also obtain:

$$D(\boldsymbol{\sigma}_{12}) = \begin{Vmatrix} 1 & 0 & 0 \\ 0 & 1 & 0 \\ 0 & 0 & -1 \end{Vmatrix} \quad \text{and} \quad D(\boldsymbol{\sigma}_{13}) = \begin{Vmatrix} 1 & 0 & 0 \\ 0 & -1 & 0 \\ 0 & 0 & 1 \end{Vmatrix}, \tag{5-2.11}$$

where σ_{12} is the plane containing $\mathbf{e}_1$ and $\mathbf{e}_2$ and σ_{13} is the plane containing $\mathbf{e}_1$ and $\mathbf{e}_3$.

(3) Inversion. The effect of inversion $\boldsymbol{i}$ on a vector will be to invert it, consequently:

$$\begin{aligned} x_1' &= -x_1 \\ x_2' &= -x_2 \\ x_3' &= -x_3 \end{aligned} \quad \text{and} \quad D(\boldsymbol{i}) = \begin{Vmatrix} -1 & 0 & 0 \\ 0 & -1 & 0 \\ 0 & 0 & -1 \end{Vmatrix}. \tag{5-2.12}$$

(4) Rotation–reflection. Consider a rotation by θ $(= 2\pi/n)$ about the $\mathbf{e}_3$ base vector, followed by reflection in the σ_{12} plane. The components of the point vector $\mathbf{p}$ (or, the coordinates of the point P) will be first transformed by the rotation, as in (1), and then these new components (coordinates) will be transformed by the reflection, as in (2). Using matrix notation, these two transformations can be combined into one step (see § 4-3(3)) and we get

$$D(S_n) = \begin{Vmatrix} 1 & 0 & 0 \\ 0 & 1 & 0 \\ 0 & 0 & -1 \end{Vmatrix} \begin{Vmatrix} \cos\theta & \sin\theta & 0 \\ -\sin\theta & \cos\theta & 0 \\ 0 & 0 & 1 \end{Vmatrix} = \begin{Vmatrix} \cos\theta & \sin\theta & 0 \\ -\sin\theta & \cos\theta & 0 \\ 0 & 0 & -1 \end{Vmatrix}. \tag{5-2.13}$$

(5) Identity. This is the 'do nothing' operation, hence:

$$\begin{aligned} x_1' &= x_1 \\ x_2' &= x_2 \\ x_3' &= x_3 \end{aligned} \quad \text{and} \quad D(E) = \begin{Vmatrix} 1 & 0 & 0 \\ 0 & 1 & 0 \\ 0 & 0 & 1 \end{Vmatrix} = E, \tag{5-2.14}$$

where E is the identity matrix.

All of the above matrices are orthogonal.

To summarize, we have found that the effect of any symmetry operation $\boldsymbol{R}$ on a position vector $\mathbf{p} = x_1\mathbf{e}_1 + x_2\mathbf{e}_2 + x_3\mathbf{e}_3$ can be expressed as:

$$\boldsymbol{R}\mathbf{p} = \boldsymbol{R}(x_1\mathbf{e}_1 + x_2\mathbf{e}_2 + x_3\mathbf{e}_3) = x_1'\mathbf{e}_1 + x_2'\mathbf{e}_2 + x_3'\mathbf{e}_3 \tag{5-2.15}$$

where

$$\begin{Vmatrix} x_1' \\ x_2' \\ x_3' \end{Vmatrix} = D(\boldsymbol{R}) \begin{Vmatrix} x_1 \\ x_2 \\ x_3 \end{Vmatrix} \tag{5-2.16}$$

and $D(\boldsymbol{R})$ is a matrix of order three, characteristic of $\boldsymbol{R}$. Eqn (5-2.16) can also be written in the form

$$x'_k = \sum_{j=1}^{3} D_{kj}(\boldsymbol{R})x_j \qquad k = 1, 2, 3 \tag{5-2.17}$$

where $D_{kj}(\boldsymbol{R})$ is the element in the kth row and jth column of matrix $D(\boldsymbol{R})$ and x_j ($j = 1, 2, 3$) are the coordinates of a point or the components of a position vector.

5-3. Matrix representations for $\mathscr{C}_{2h}$ and $\mathscr{C}_{3v}$

Now let us consider two specific point groups: (1) $\mathscr{C}_{2h}$ and (2) $\mathscr{C}_{3v}$.

(1) $\mathscr{C}_{2h}$. A molecule belonging to this point group is planar *trans*-$C_2H_2Cl_2$

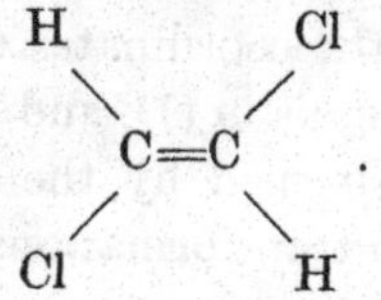

The point group is composed of four symmetry operations: $\boldsymbol{E}$, $\boldsymbol{C}_2$, $\boldsymbol{i}$, and $\boldsymbol{\sigma}_h$ and the group table is given in Table 5-3.1. This table shows the effect of combining one operation with another.

Following the discussion in § 5-2, the matrices which correspond to the four symmetry operations are

$$D(\boldsymbol{E}) = \begin{Vmatrix} 1 & 0 & 0 \\ 0 & 1 & 0 \\ 0 & 0 & 1 \end{Vmatrix}, \quad D(\boldsymbol{C}_2) = \begin{Vmatrix} -1 & 0 & 0 \\ 0 & -1 & 0 \\ 0 & 0 & 1 \end{Vmatrix},$$

$$D(\boldsymbol{i}) = \begin{Vmatrix} -1 & 0 & 0 \\ 0 & -1 & 0 \\ 0 & 0 & -1 \end{Vmatrix}, \quad \text{and} \quad D(\boldsymbol{\sigma}_h) = \begin{Vmatrix} 1 & 0 & 0 \\ 0 & 1 & 0 \\ 0 & 0 & -1 \end{Vmatrix}, \tag{5-3.1}$$

where the base vectors have been chosen such that $\mathbf{e}_3$ coincides with C_2 and $\mathbf{e}_1$ and $\mathbf{e}_2$ lie in the σ_h plane. $D(C_2)$ has been found by replacing θ by π in eqn (5-2.7).

TABLE 5-3.1

Group table for $\mathscr{C}_{2h}$†

	E	C_2	i	σ_h
E	E	C_2	i	σ_h
C_2	C_2	E	σ_h	i
i	i	σ_h	E	C_2
σ_h	σ_h	i	C_2	E

† The order of combining is AB where A is given at the side of the table and B at the top of the table.

Using matrix multiplication as the combining operation, we can construct a group table for these four matrices (Table 5-3.2) e.g.

$$D(C_2)D(i) = \begin{Vmatrix} -1 & 0 & 0 \\ 0 & -1 & 0 \\ 0 & 0 & 1 \end{Vmatrix} \begin{Vmatrix} -1 & 0 & 0 \\ 0 & -1 & 0 \\ 0 & 0 & -1 \end{Vmatrix} = \begin{Vmatrix} 1 & 0 & 0 \\ 0 & 1 & 0 \\ 0 & 0 & -1 \end{Vmatrix}$$

$$= D(\sigma_h).$$

It is apparent from this table that the four matrices form a group, since

(*a*) the product of any two matrices is one of the four,

(*b*) one matrix, $D(E) = E$, is such that when combined with the four, it leaves them unchanged,

(*c*) the associative law holds for matrices,

(*d*) each of the four matrices has an inverse which is one of the four, i.e. $D(E)^{-1} = D(E)$, $D(C_2)^{-1} = D(C_2)$, $D(i)^{-1} = D(i)$, and $D(\sigma_h)^{-1} = D(\sigma_h)$.

Comparison of Tables 5-3.1 and 5-3.2 shows that they are identical in structure (though the elements and combining rules are different) and consequently the matrix group is homomorphic with the point group; we say that the four matrices form a *representation* of $\mathscr{C}_{2h}$.

(2) $\mathscr{C}_{3v}$. Ammonia is an example of a molecule belonging to this point group and it has six symmetry operations which obey the group table introduced in Chapter 3 (Table 3-4.1). If we set up base vectors

TABLE 5-3.2

Group table for the four matrices in eqn (5-3.1)†

	$D(E)$	$D(C_2)$	$D(i)$	$D(\sigma_h)$
$D(E)$	$D(E)$	$D(C_2)$	$D(i)$	$D(\sigma_h)$
$D(C_2)$	$D(C_2)$	$D(E)$	$D(\sigma_h)$	$D(i)$
$D(i)$	$D(i)$	$D(\sigma_h)$	$D(E)$	$D(C_2)$
$D(\sigma_h)$	$D(\sigma_h)$	$D(i)$	$D(C_2)$	$D(E)$

† The order of matrix multiplication is AB where A is given at the side of the table and B at the top of the table.

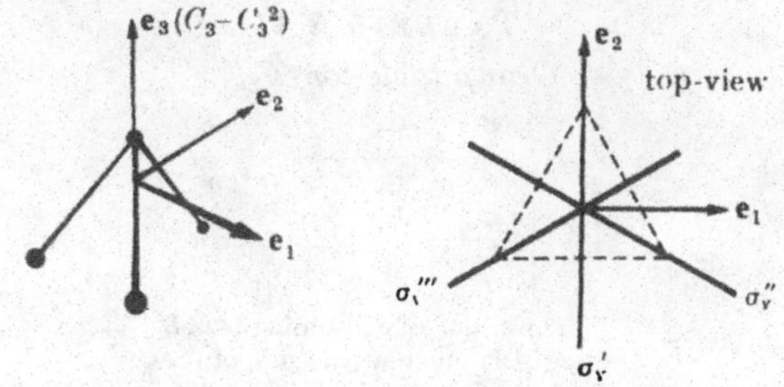

FIG. 5-3.1. Axes for the $\mathscr{C}_{3v}$ point group. The origins of $\mathbf{e}_1$, $\mathbf{e}_2$, and $\mathbf{e}_3$ are at the centre of mass; σ_v', σ_v'', and σ_v''' are perpendicular to the page.

in accordance with Fig. 5-3.1, then the matrices which correspond to E, σ_v' (reflection in the plane containing $\mathbf{e}_2$ and $\mathbf{e}_3$), C_3 (rotation about $\mathbf{e}_3$ with $\theta = 2\pi/3$), and C_3^2 (rotation about $\mathbf{e}_3$ with $\theta = 4\pi/3$) are

$$D(E) = \begin{Vmatrix} 1 & 0 & 0 \\ 0 & 1 & 0 \\ 0 & 0 & 1 \end{Vmatrix}, \qquad D(\sigma_v') = \begin{Vmatrix} -1 & 0 & 0 \\ 0 & 1 & 0 \\ 0 & 0 & 1 \end{Vmatrix},$$

$$D(C_3) = \begin{Vmatrix} -1/2 & \sqrt{3}/2 & 0 \\ -\sqrt{3}/2 & -1/2 & 0 \\ 0 & 0 & 1 \end{Vmatrix}, \quad \text{and} \quad D(C_3^2) = \begin{Vmatrix} -1/2 & -\sqrt{3}/2 & 0 \\ \sqrt{3}/2 & -1/2 & 0 \\ 0 & 0 & 1 \end{Vmatrix}. \tag{5-3.2}$$

Using the same technique as we did for S_n in § 5-2(4), we can also obtain the matrices $D(\sigma_v'')$ and $D(\sigma_v''')$, namely

$$D(\sigma_v'') = D(C_3^2)D(\sigma_v') = \begin{Vmatrix} -1/2 & -\sqrt{3}/2 & 0 \\ \sqrt{3}/2 & -1/2 & 0 \\ 0 & 0 & 1 \end{Vmatrix} \begin{Vmatrix} -1 & 0 & 0 \\ 0 & 1 & 0 \\ 0 & 0 & 1 \end{Vmatrix}$$

$$= \begin{Vmatrix} 1/2 & -\sqrt{3}/2 & 0 \\ -\sqrt{3}/2 & -1/2 & 0 \\ 0 & 0 & 1 \end{Vmatrix} \tag{5-3.3}$$

and

$$D(\sigma_v''') = D(C_3)D(\sigma_v') = \begin{Vmatrix} -1/2 & \sqrt{3}/2 & 0 \\ -\sqrt{3}/2 & -1/2 & 0 \\ 0 & 0 & 1 \end{Vmatrix} \begin{Vmatrix} -1 & 0 & 0 \\ 0 & 1 & 0 \\ 0 & 0 & 1 \end{Vmatrix}$$

$$= \begin{Vmatrix} 1/2 & \sqrt{3}/2 & 0 \\ \sqrt{3}/2 & -1/2 & 0 \\ 0 & 0 & 1 \end{Vmatrix}. \tag{5-3.4}$$

TABLE 5-3.3

Group table for the six matrices in eqns (5-3.2) to (5-3.4)†

	$D(E)$	$D(\sigma_v')$	$D(\sigma_v'')$	$D(\sigma_v''')$	$D(C_3)$	$D(C_3^2)$
$D(E)$	$D(E)$	$D(\sigma_v')$	$D(\sigma_v'')$	$D(\sigma_v''')$	$D(C_3)$	$D(C_3^2)$
$D(\sigma_v')$	$D(\sigma_v')$	$D(E)$	$D(C_3)$	$D(C_3^2)$	$D(\sigma_v'')$	$D(\sigma_v''')$
$D(\sigma_v'')$	$D(\sigma_v'')$	$D(C_3^2)$	$D(E)$	$D(C_3)$	$D(\sigma_v''')$	$D(\sigma_v')$
$D(\sigma_v''')$	$D(\sigma_v''')$	$D(C_3)$	$D(C_3^2)$	$D(E)$	$D(\sigma_v')$	$D(\sigma_v'')$
$D(C_3)$	$D(C_3)$	$D(\sigma_v''')$	$D(\sigma_v')$	$D(\sigma_v'')$	$D(C_3^2)$	$D(E)$
$D(C_3^2)$	$D(C_3^2)$	$D(\sigma_v'')$	$D(\sigma_v''')$	$D(\sigma_v')$	$D(E)$	$D(C_3)$

† The order of matrix multiplication is AB where A is given at the side of the table and B at the top of the table. Also $C_3^2 = C_3^{-1}$.

These six matrices form a group for which the combining rule is matrix multiplication and the group table is that in Table 5-3.3 (the reader is left to confirm this for himself). Since this table is identical in structure to Table 3-4.1, we say that the six matrices form a *representation* of $\mathscr{C}_{3v}$.

It is apparent that we can always get a set of 3×3 matrices, which form a representation of a given point group, by consideration of the effect that the symmetry operations of the point group have on a position vector. Why this works is shown pictorially in Fig. 5-3.2 for the $\sigma_v'' = \sigma_v' C_3$ operation of $\mathscr{C}_{3v}$. The symmetry operation C_3 on the position vector $\mathbf{p}$ followed by σ_v' on $\mathbf{p}'$ produces a vector $\mathbf{p}''$ which is coincidental with the one produced by the operation σ_v'' on $\mathbf{p}$. The matrices $D(C_3)$, $D(\sigma_v')$, $D(\sigma_v'')$ then simply mirror what is being done to the point vector. The general mathematical proof that, if symmetry operations R, S, and T obey the relation $SR = T$, then the matrices $D(R)$, $D(S)$, and $D(T)$, found as above, obey the relation

$$D(S)D(R) = D(T)$$

is given in Appendix A.5-1.

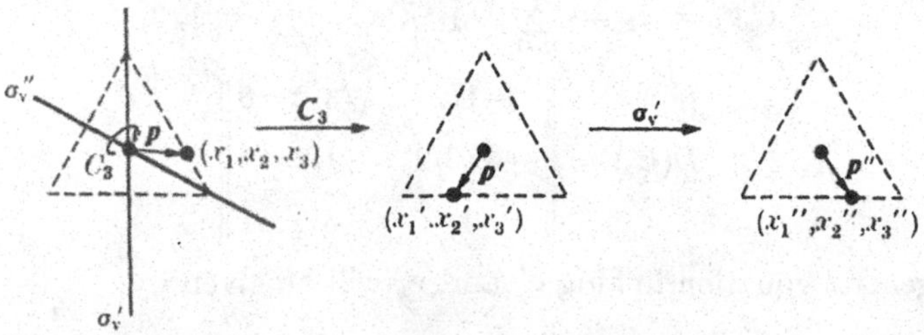

FIG. 5-3.2. The effect of C_3 and σ_v' on a position vector (or point) in the base of a symmetric tripod.

5-4. Matrix representations derived from base vectors

It is also possible to construct matrix representations by considering the effect that the symmetry operations of a point group have on one or more sets of base vectors. We will consider two cases, both using the $\mathscr{C}_{3v}$ point group as an example: (1) the set of base vectors $\mathbf{e}_1$, $\mathbf{e}_2$, and $\mathbf{e}_3$, introduced in § 5-2; (2) three sets of mutually perpendicular base vectors, each located at the foot of a symmetric tripod.

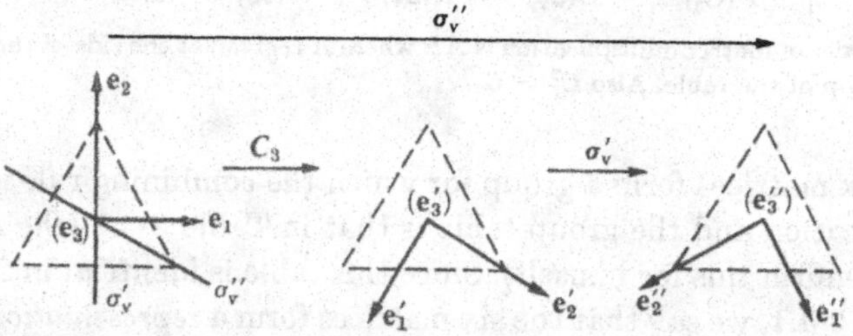

FIG. 5-4.1. The effect of C_3 and σ_v' on a set of base vectors in the base of a symmetric tripod. $\mathbf{e}_3$, $\mathbf{e}_3'$, and $\mathbf{e}_3''$ are perpendicular to the page.

(1) In Fig. 5-4.1 we show, as an example, the effect of the symmetry operations C_3, σ_v', and σ_v'' on the three base vectors $\mathbf{e}_1$, $\mathbf{e}_2$, and $\mathbf{e}_3$. The operation C_3 produces new vectors $\mathbf{e}_1'$, $\mathbf{e}_2'$, and $\mathbf{e}_3'$ and these are transformed to $\mathbf{e}_1''$, $\mathbf{e}_2''$, and $\mathbf{e}_3''$ by σ_v'. The operation σ_v'' will, since $\sigma_v'' = \sigma_v' C_3$, transform $\mathbf{e}_1$, $\mathbf{e}_2$, and $\mathbf{e}_3$ directly into $\mathbf{e}_1''$, $\mathbf{e}_2''$, and $\mathbf{e}_3''$. Therefore, we anticipate that there will be matrices which will link these sets of base vectors and which will behave like the symmetry operations.

If C_3 transforms $\mathbf{e}_1$, $\mathbf{e}_2$, and $\mathbf{e}_3$ to $\mathbf{e}_1'$, $\mathbf{e}_2'$, and $\mathbf{e}_3'$, then we have (see Fig. 5-4.2)

$$C_3\mathbf{e}_1 = \mathbf{e}_1' = -\mathbf{e}_1/2 - (\sqrt{3}/2)\mathbf{e}_2$$
$$C_3\mathbf{e}_2 = \mathbf{e}_2' = (\sqrt{3}/2)\mathbf{e}_1 - \mathbf{e}_2/2$$
$$C_3\mathbf{e}_3 = \mathbf{e}_3' = \mathbf{e}_3$$

which can be written as

$$C_3\mathbf{e}_k = \mathbf{e}_k' = \sum_{j=1}^{3} D_{jk}(C_3)\mathbf{e}_j, \qquad k = 1, 2, 3 \tag{5-4.1}$$

where

$$D(C_3) = \begin{Vmatrix} -1/2 & \sqrt{3}/2 & 0 \\ -\sqrt{3}/2 & -1/2 & 0 \\ 0 & 0 & 1 \end{Vmatrix}.$$

The general equation linking $\mathbf{e}_k'$ and $\mathbf{e}_k$ will be given by:

$$R\mathbf{e}_k = \mathbf{e}_k' = \sum_{j=1}^{3} D_{jk}(R)\mathbf{e}_j, \qquad k = 1, 2, 3. \tag{5-4.2}$$

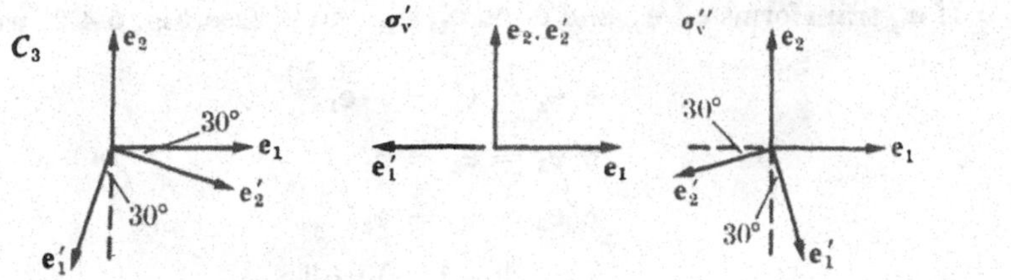

$\mathbf{e}_1' = -(\sin 30°)\mathbf{e}_1 - (\cos 30°)\mathbf{e}_2$ $\quad \mathbf{e}_1' = -\mathbf{e}_1$ $\quad \mathbf{e}_1' = (\sin 30°)\mathbf{e}_1 - (\cos 30°)\mathbf{e}_2$

$\mathbf{e}_2' = (\cos 30°)\mathbf{e}_1 - (\sin 30°)\mathbf{e}_2$ $\quad \mathbf{e}_2' = \mathbf{e}_2$ $\quad \mathbf{e}_2' = -(\cos 30°)\mathbf{e}_1 - (\sin 30°)\mathbf{e}_2$

$\mathbf{e}_3' = \mathbf{e}_3$ $\quad \mathbf{e}_3' = \mathbf{e}_3$ $\quad \mathbf{e}_3' = \mathbf{e}_3$

FIG. 5-4.2. Relation between original and transformed base vectors for C_3, σ_v', and σ_v''. $\mathbf{e}_3$ and $\mathbf{e}_3'$ are perpendicular to the page.

It is important to notice that the way in which we have formed the matrix $D(R)$ is different from the way we did things when we were considering position vectors (cf. eqn (5-2.17)): the components of the new base vectors are used as the *columns* of $D(R)$ whereas, before, the components of the new position vector were used as the *rows* of a matrix. This is what is implied by the order of the subscripts on D in eqn (5-4.2): jk rather than kj. Necessarily

$$\begin{Vmatrix} \mathbf{e}_1' \\ \mathbf{e}_2' \\ \mathbf{e}_3' \end{Vmatrix} \quad \text{does } \textit{not} \text{ equal} \quad D(R) \begin{Vmatrix} \mathbf{e}_1 \\ \mathbf{e}_2 \\ \mathbf{e}_3 \end{Vmatrix}.$$

There is nothing particularly subtle about what we have done, it simply ensures (see the proof in Appendix A.5-2) that the matrices so formed will indeed represent the point group.

As a general rule, equations involving coordinates or components of a position vector will be written in the form:

$$x_i' = \sum_j a_{ij} x_j,$$

and those involving functions (see § 5-7) or base vectors, in the form:

$$f_i' = \sum_j a_{ji} f_j$$

or

$$\mathbf{e}_i' = \sum_j a_{ji} \mathbf{e}_j.$$

In much of the mathematics which follows it is important to bear this rule in mind.

If σ_v' transforms e_1, e_2, and e_3 to e_1', e_2', and e_3' (see Fig. 5-4.2) we find

$$\sigma_v' e_1 = e_1' = -e_1$$
$$\sigma_v' e_2 = e_2' = e_2$$
$$\sigma_v' e_3 = e_3' = e_3$$

and

$$D(\sigma_v') = \begin{Vmatrix} -1 & 0 & 0 \\ 0 & 1 & 0 \\ 0 & 0 & 1 \end{Vmatrix}.$$

If σ_v'' transforms e_1, e_2, and e_3 to e_1', e_2', and e_3' (see Fig. 5-4.2) we find

$$\sigma_v'' e_1 = e_1' = e_1/2 - (\sqrt{3}/2)e_2$$
$$\sigma_v'' e_2 = e_2' = -(\sqrt{3}/2)e_1 - e_2/2$$
$$\sigma_v'' e_3 = e_3' = e_3$$

and

$$D(\sigma_v'') = \begin{Vmatrix} 1/2 & -\sqrt{3}/2 & 0 \\ -\sqrt{3}/2 & -1/2 & 0 \\ 0 & 0 & 1 \end{Vmatrix}.$$

The reader may confirm that $D(\sigma_v')D(C_3) = D(\sigma_v'')$. It is always true that if $SR = T$ and $D(S)$, $D(R)$, and $D(T)$ are defined as in eqn (5-4.2), then $D(S)D(R) = D(T)$; this is proved in Appendix A.5-2. That the matrices found in this way from the base vectors are identical with those found from consideration of the position vector (§ 5-2) is proven in Appendix A.5-3. The base vectors e_1, e_2, and e_3 are said to form a *basis* for a representation of a point group.

(2) If we place sets of base vectors at the feet of a symmetric tripod, these too can produce a matrix representation, as seen in Fig. 5-4.3.

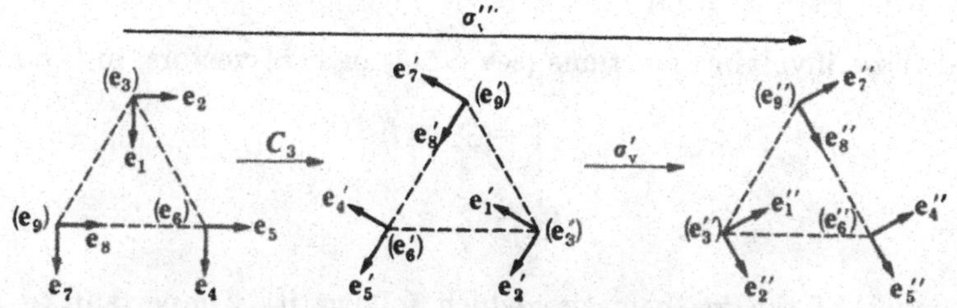

FIG. 5-4.3. Symmetry operations on base vectors located at the feet of a symmetric tripod. e_3, e_6, e_9 etc. are perpendicular to the page.

(By the way, the sets of base vectors need not be parallel.) For example, under C_3, we have

$$\begin{aligned}
C_3\mathbf{e}_1 &= \mathbf{e}_1' = -\mathbf{e}_4/2-(\sqrt{3}/2)\mathbf{e}_5\\
C_3\mathbf{e}_2 &= \mathbf{e}_2' = (\sqrt{3}/2)\mathbf{e}_4-\mathbf{e}_5/2\\
C_3\mathbf{e}_3 &= \mathbf{e}_3' = \mathbf{e}_6\\
C_3\mathbf{e}_4 &= \mathbf{e}_4' = -\mathbf{e}_7/2-(\sqrt{3}/2)\mathbf{e}_8\\
C_3\mathbf{e}_5 &= \mathbf{e}_5' = (\sqrt{3}/2)\mathbf{e}_7-\mathbf{e}_8/2\\
C_3\mathbf{e}_6 &= \mathbf{e}_6' = \mathbf{e}_9\\
C_3\mathbf{e}_7 &= \mathbf{e}_7' = -\mathbf{e}_1/2-(\sqrt{3}/2)\mathbf{e}_2\\
C_3\mathbf{e}_8 &= \mathbf{e}_8' = (\sqrt{3}/2)\mathbf{e}_1-\mathbf{e}_2/2\\
C_3\mathbf{e}_9 &= \mathbf{e}_9' = \mathbf{e}_3
\end{aligned}$$

and we can write

$$C_3\mathbf{e}_k = \mathbf{e}_k' = \sum_{j=1}^{9} D_{jk}(C_3)\mathbf{e}_j, \qquad k = 1, 2 \ldots 9 \tag{5-4.3}$$

and

$$D(C_3) =$$

$$\begin{Vmatrix}
0 & 0 & 0 & 0 & 0 & 0 & -1/2 & \sqrt{3}/2 & 0\\
0 & 0 & 0 & 0 & 0 & 0 & -\sqrt{3}/2 & -1/2 & 0\\
0 & 0 & 0 & 0 & 0 & 0 & 0 & 0 & 1\\
-1/2 & \sqrt{3}/2 & 0 & 0 & 0 & 0 & 0 & 0 & 0\\
-\sqrt{3}/2 & -1/2 & 0 & 0 & 0 & 0 & 0 & 0 & 0\\
0 & 0 & 1 & 0 & 0 & 0 & 0 & 0 & 0\\
0 & 0 & 0 & -1/2 & \sqrt{3}/2 & 0 & 0 & 0 & 0\\
0 & 0 & 0 & -\sqrt{3}/2 & -1/2 & 0 & 0 & 0 & 0\\
0 & 0 & 0 & 0 & 0 & 1 & 0 & 0 & 0
\end{Vmatrix}$$

or, in general,

$$R\mathbf{e}_k = \mathbf{e}_k' = \sum_{j=1}^{n} D_{jk}(R)\mathbf{e}_j, \qquad k = 1, 2 \ldots n. \tag{5-4.4}$$

Again, notice the ordering of the subscripts in eqn (5-4.4). The proof that the $D(R)$ constructed in this way form a representation of the point group is the same as the proof given in Appendix A.5-2. It is obvious that there are a large number of different matrix representations which can be found by choosing different sets of base vectors. The base vectors chosen are said to form a *basis* for the representation of the point group. As with the symmetry operations acting on a position vector, the matrices formed by considering base vectors will be orthogonal.

5-5 Function space

Now we come to a totally different method for producing matrix representations of a point group; a method which involves the concept of a *function space*. The word space is used in this context in a mathematical sense and should not be confused with the more familiar three-dimensional physical space. A function space is a collection or family of mathematical functions which obeys certain rules. These rules are a generalization of those which apply to the family of position vectors in physical space and in order to help in understanding them, the corresponding vector rule will be put in square brackets after each function rule.

Consider the set of functions: $f_1, f_2, \ldots$. If they are to belong to a function space, then the following rules and definitions must hold true.

(1) The addition of any two functions must produce a third function which is also a member of the collection of functions, i.e. belongs to the function space [vector addition of any two position vectors $\mathbf{p}_1$ and $\mathbf{p}_2$ produces another position vector $\mathbf{p}_3$].

(2) The multiplication of any of the functions by a number must produce a function which is also included in the collection [$a\mathbf{p}_1$ is also a position vector].

(3) The rules in (1) and (2) can be combined together: any function which is a linear combination of $f_1, f_2, \ldots$ (i.e. $a_1f_1+a_2f_2+\ldots$) must also be a member of the collection [$a_1\mathbf{p}_1+a_2\mathbf{p}_2+a_3\mathbf{p}_3$ is also a position vector].

(4) The scalar product of any two functions is defined as:

$$(f_i, f_j) = \int f_i^* f_j \, d\tau,$$

where integration is over all the variables of f_i and f_j, [the scalar product of two position vectors $\mathbf{p}$ and $\mathbf{p}'$ is

$$\mathbf{p} \cdot \mathbf{p}' = (\mathbf{p}, \mathbf{p}') = pp' \cos \theta,$$

where θ is the angle between the two vectors, and if $(\mathbf{e}_i, \mathbf{e}_j) = \delta_{ij}$, $i, j = 1, 2, 3$ and

$$\mathbf{p} = x_1\mathbf{e}_1+x_2\mathbf{e}_2+x_3\mathbf{e}_3,$$

$$\mathbf{p}' = x_1'\mathbf{e}_1+x_2'\mathbf{e}_2+x_3'\mathbf{e}_3,$$

then

$$(\mathbf{p}, \mathbf{p}') = x_1x_1'+x_2x_2'+x_3x_3' \,].$$

(5) If n of the functions $f_1, f_2 \ldots$ are linearly independent (i.e. $a_1f_1+a_2f_2+\ldots a_nf_n = 0$ *only* if $a_1 = a_2 = \ldots a_n = 0$), then any of the other functions of the space can be expressed as some linear

combination of these n functions, [in physical space $\mathbf{e}_1$, $\mathbf{e}_2$, $\mathbf{e}_3$ or *any* three non-coplanar vectors, are *linearly independent* and any position vector can be expressed in terms of them]. The n linearly-independent functions are said to *span* the function space and the space is said to be n-dimensional, [$\mathbf{e}_1$, $\mathbf{e}_2$, and $\mathbf{e}_3$ span physical space which is three-dimensional]. Put another way, n is the smallest number of functions from which it is possible to produce all the other functions which belong to the function space. If the n linearly-independent functions are chosen such that the scalar product $(f_i, f_j) = \delta_{ij}$, $i, j = 1, 2 \ldots n$, then they are said to be *orthonormal* (orthogonal and normalized), [$\mathbf{e}_1$, $\mathbf{e}_2$, and $\mathbf{e}_3$ are unit orthogonal vectors]. It is always possible to create n orthonormal functions from n linearly-independent functions. Furthermore, orthonormal functions are always linearly independent and therefore, by definition, the maximum number of orthonormal functions in an n-dimensional space is n, [in a *general* n-dimensional vector space, so far not discussed, the maximum number of orthogonal vectors is n; orthogonality being defined by

$$(\mathbf{p}, \mathbf{p}') = x_1x_1'+x_2x_2'+\ldots x_nx_n' = 0].$$

(6) An n-dimensional function space is defined by specifying n mutually orthogonal, normalized, linearly-independent functions, [$\mathbf{e}_1$, $\mathbf{e}_2$, and $\mathbf{e}_3$ define physical space]; they are called orthonormal *basis functions*.

Now let us consider two examples of a function space. The solutions of the differential equation

$$\frac{d^2f(x)}{dx^2} = -f(x), \qquad 0 < x < 2\pi$$

$f(x) = f_1(x), f_2(x), \ldots$, form a function space, since any combination of them is also a solution of the equation and therefore a member of the collection of functions, i.e.

$$\frac{d^2}{dx^2}[f_i(x)+f_j(x)] = \frac{d^2}{dx^2}f_i(x)+\frac{d^2}{dx^2}f_j(x) = -f_i(x)-f_j(x)$$
$$= -[f_i(x)+f_j(x)],$$
$$\frac{d^2}{dx^2}[af_i(x)] = a\frac{d^2}{dx^2}f_i(x) = a[-f_i(x)] = -[af_i(x)]$$

and

$$\frac{d^2}{dx^2}[a_1f_1(x)+a_2f_2(x)+\ldots] = -[a_1f_1(x)+a_2f_2(x)+\ldots].$$

The general solution of this equation is $f_i(x) = a_i \cos x + b_i \sin x$, where a_i and b_i are arbitrary constants. Since $\cos x$ and $\sin x$ are linearly independent ($\alpha_1 \cos x + \alpha_2 \sin x$ is *only* equal to zero for *all* x if $\alpha_1 = \alpha_2 = 0$) and orthogonal ($\int_0^{2\pi} \cos x \sin x \, dx = 0$), the orthonormal basis functions for this two-dimensional function space are $\cos x/\sqrt{\pi}$ and $\sin x/\sqrt{\pi}$.

Consider the equation:

$$H\psi = E\psi,$$

where H is a linear operator, ψ is some function and E (called the eigenvalue) is a constant. If H is the Hamiltonian operator, this equation will be the Schrödinger equation. If ψ_1 and ψ_2 are two solutions having identical eigenvalues ($E_1 = E_2$), they are said to be degenerate. It is clear that, since H is defined as being linear,

$$H(\psi_1+\psi_2) = H\psi_1+H\psi_2 = E_1\psi_1+E_2\psi_2 = E_1(\psi_1+\psi_2)$$

and

$$H(a\psi_1) = aH\psi_1 = aE_1\psi_1 = E_1(a\psi_1)$$

and *any* linear combination of degenerate solutions will also be a solution of the equation. We can therefore say that any degenerate set of solutions (corresponding to a given eigenvalue) of $H\psi = E\psi$ forms a function space. If n of the solutions are linearly independent, then they can be chosen to be orthonormal and used as basis functions to define an n-dimensional function space. If H is the Hamiltonian operator, then the linearly-independent *wavefunctions* $\psi = \psi_1^v, \psi_2^v, \dots \psi_n^v$ which correspond to $E = E_v$, will be basis functions for an n-dimensional function space. As an example, the three p-orbitals p_x, p_y, and p_z of atomic problems are degenerate and orthogonal and when normalized form an orthonormal basis for the three-dimensional function space of p-orbitals. Any p-orbital can be written, with respect to this basis, as

$$\psi = a_x p_x + a_y p_y + a_z p_z.$$

In § 5-9 we use a function space defined by d-orbitals. The six π-orbitals of benzene are another example of a function space.

5-6. Transformation operators (O_R)

Having defined a function space, we are now in the position of being able to introduce a new group, homomorphic (see § 3-4) with a given point group, in which the elements are transformation operators which operate on the functions of some function space. We will denote the transformation operator which corresponds to the symmetry operation R by O_R. Every O_R will be defined with respect to some particular function space (e.g. the wavefunctions belonging to a given energy level) and they will be such that

(1) they are linear:

$$O_R(af) = a(O_R f)$$

and

$$O_R(f+g) = O_R f + O_R g, \qquad (5\text{-}6.1)$$

where a is any number and f and g are any two functions of the function space,

(2)

$$O_R f = f', \qquad (5\text{-}6.2)$$

where f' also belongs to the function space,

(3) the correspondence between R and O_R retains the multiplication rules, e.g. if $T = SR$, then

$$O_T f = O_{SR} f = O_S(O_R f). \qquad (5\text{-}6.3)$$

Necessarily there will be a unit operator O_E (associated with E) such that

$$O_R O_E = O_E O_R = O_R \qquad (5\text{-}6.4)$$

and each operator O_R will have an inverse O_R^{-1}:

$$O_R^{-1} O_R = O_R O_R^{-1} = O_E \qquad (5\text{-}6.5)$$

and, since $O_{R^{-1}} O_R = O_E$,

$$O_R^{-1} = O_{R^{-1}}. \qquad (5\text{-}6.6)$$

This set of transformation operators O_R associated with the symmetry operations of a given point group will therefore have a group table which is structurally the same as the one for the point group. In the next section we show that, if we introduce a coordinate system into the function space chosen for the O_R, we can define *explicitly* a set of O_R satisfying eqns (5-6.1) to (5-6.6). The reader is warned that since the correspondence between R and O_R is so close, many books (incorrectly) do not distinguish between them.

5-7. A satisfactory set of transformation operators (O_R)

If an n-dimensional function space is defined by the set of linearly-independent basis functions $f_1, f_2, \ldots f_i, \ldots$, and f_n and if these are functions of three Cartesian coordinates x_1, x_2, and x_3, then we can define a transformation operator O_R (corresponding to the symmetry operation R) by the equations:

$$(O_R f_i)(x_1', x_2', x_3') = f_i(x_1, x_2, x_3), \qquad i = 1, 2 \ldots n \qquad (5\text{-}7.1)$$

where a point in physical space with coordinates x_1, x_2, and x_3 is moved by the symmetry operation R to the location x_1', x_2', and x_3'. In other words, the new function $O_R f_i$ assigns the value of the old function f_i at

x_1, x_2, and x_3 to the position x_1', x_2', and x_3'. Knowledge of f_i, x_1, x_2, x_3, x_1', x_2', and x_3' allows one to find the form of $\boldsymbol{O}_R f_i$ (see § 5-9). With this definition, it can be shown that the $\boldsymbol{O}_R$ are linear and that if $\boldsymbol{T} = \boldsymbol{SR}$, then $\boldsymbol{O}_T = \boldsymbol{O}_S \boldsymbol{O}_R$ (see Appendix A.5-4).

The requirement that $\boldsymbol{O}_R f_i$ produces a function belonging to the given function space (see eqn (5-6.2)) will be met by the proper choice of function space (see § 5-8). If this is the case, however, we can write, for an n-dimensional function space defined by the linearly-independent basis functions $f_1, f_2 \ldots$, and f_n,

$$\boldsymbol{O}_R f_k = \sum_{j=1}^{n} D_{jk}(\boldsymbol{R}) f_j, \qquad k = 1, 2 \ldots n \tag{5-7.2}$$

(notice the order of subscripting on D), i.e. the function $\boldsymbol{O}_R f_k$, if it belongs to the function space, must be some linear combination of that space's basis functions.

What is of supreme importance for us, is the fact that the $n \times n$ matrices $D(\boldsymbol{R})$ in eqn (5-7.2) will multiply in the same fashion as the symmetry operations: if $\boldsymbol{T} = \boldsymbol{SR}$, then $D(\boldsymbol{T}) = D(\boldsymbol{S})D(\boldsymbol{R})$, (see Appendix A.5-5). The $D(\boldsymbol{R})$ so found, therefore form an n-dimensional representation of both the point group and the group of transformation operators $\boldsymbol{O}_R$, and the functions $f_1, f_2, \ldots$ and f_n are said to be a basis for the representation.

The operators just described will leave the scalar product of two functions of the function space unchanged: $(\boldsymbol{O}_R f_i, \boldsymbol{O}_R f_j) = (f_i, f_j)$. Such operators are said to be unitary and they can always be represented by unitary matrices (see § 6-4). The proof that the $\boldsymbol{O}_R$ are unitary follows from considering

$$(f_i, f_j) = \int f_i^*(P) f_j(P)\, \mathrm{d}\tau_P,$$

where P is a general point with coordinates x_1, x_2, and x_3 and

$$\mathrm{d}\tau_P = \mathrm{d}x_1\, \mathrm{d}x_2\, \mathrm{d}x_3$$

is the volume element at P. Now a symmetry operation $\boldsymbol{R}$ will move P to a point P' with coordinates (x_1', x_2', and x_3') and the volume element to an equal volume element $\mathrm{d}\tau_{P'} = \mathrm{d}x_1'\, \mathrm{d}x_2'\, \mathrm{d}x_3'$ situated at P'. Furthermore, the operator $\boldsymbol{O}_R$ is defined such that $f_i(P) = (\boldsymbol{O}_R f_i)(P')$ and $f_j(P) = (\boldsymbol{O}_R f_j)(P')$. Hence

$$\begin{aligned}\int f_i^*(P) f_j(P)\, \mathrm{d}\tau_P &= \int [(\boldsymbol{O}_R f_i)(P')]^* [(\boldsymbol{O}_R f_j)(P')]\, \mathrm{d}\tau_{P'} \\ &= \int [(\boldsymbol{O}_R f_i)(P)]^* [(\boldsymbol{O}_R f_j)(P)]\, \mathrm{d}\tau_P\end{aligned}$$

since the range of integration extends over all points P or P'. Therefore,

$$(f_i, f_j) = (O_R f_i, O_R f_j), \tag{5-7.3}$$

that is the scalar product is left unchanged.

Our definition of O_R applied to functions of the coordinates x_1, x_2, and x_3 of a point in physical space, but it can be generalized to apply to functions of any number of variables, as long as we know how those variables change under the symmetry operations. For example, if we let X stand for a complete specification of the coordinates of all the electrons (or all the nuclei) of some molecule, i.e.

$$X = x_1^{(1)}, x_2^{(1)}, x_3^{(1)}, \ldots x_1^{(n)}, x_2^{(n)}, x_3^{(n)}$$

for n electrons (nuclei), and if this specification becomes X' under the symmetry operation R, then we can define O_R by

$$O_R f(X') = f(X), \tag{5-7.4}$$

where f is a function of all the electronic (nuclear) coordinates. The theorems in Appendices A.5-4 and A.5-5 also hold true for this more general definition.

5-8. A caution

We have seen in the previous section that the definition of a set of O_Rs is intimately bound up with some choice of function space. The reader is cautioned, however, that not all function spaces can be used to define O_Rs appropriate for a given point group. For example, the functions $\cos x_1$, $\sin x_1$, $\cos x_2$, and $\sin x_2$ do *not* form a basis for a representation of the $\mathscr{C}_{3v}$ (symmetric tripod) point group; x_1 and x_2 are the coordinates introduced before (see Fig. 5-2.2).

The four functions do define a function space, since the general function is

$$f(x_1, x_2) = a_1 \cos x_1 + a_2 \sin x_1 + a_3 \cos x_2 + a_4 \sin x_2$$

and addition of any two such functions will produce a third which belongs to the space, as does a number times any such function.

However, if we consider the C_3 operation of $\mathscr{C}_{3v}$ (clockwise rotation by $2\pi/3$ about $\mathbf{e}_3$), then

$$x_1' = (-x_1 + \sqrt{3}x_2)/2$$
$$x_2' = (-\sqrt{3}x_1 - x_2)/2$$
$$x_3' = x_3$$

or, inverting,

$$x_1 = (-x_1' - \sqrt{3}x_2')/2$$
$$x_2 = (\sqrt{3}x_1' - x_2')/2$$
$$x_3 = x_3'$$

and selecting the basis function $\cos x_1$, we find

$$O_{C_3}(\cos x_1') = \cos x_1 \qquad \text{(definition of } O_{C_3}\text{)}$$
$$= \cos \tfrac{1}{2}(-x_1' - \sqrt{3}x_2')$$

or, equivalently,

$$O_{C_3}(\cos x_1) = \cos \tfrac{1}{2}(-x_1 - \sqrt{3}x_2).$$

Using the notation $f_1 = \cos x_1, f_2 = \sin x_1, f_3 = \cos x_2$, and $f_4 = \sin x_2$ it is clear that

$$O_{C_3} f_1 \neq \sum_j^4 D_{j1}(C_3) f_j.$$

Consequently, this function space does *not* provide a basis for a representation of the $\mathscr{C}_{3v}$ point group.

5-9. An example of determining O_Rs and $D(R)$s for the $\mathscr{C}_{3v}$ point group using the d-orbital function space

A set of five real d-orbitals defines a function space. In spherical polar coordinates r, θ, and ϕ, they consist of a common radial function times a combination of spherical harmonics $Y_l^m(\theta, \phi)$, $l = 2$ and $m = 0, \pm 1, \pm 2$. The combinations of the five spherical harmonics are chosen such that the orbitals are real.

It is a well known property of spherical harmonics that if we shift the point r, θ, and ϕ to r', θ', and ϕ',† the resulting $Y_l^m(\theta', \phi')$ can be expressed in terms of a linear combination of all the $Y_l^{m'}(\theta, \phi)$ of the same l value ($m' = 0, \pm 1 \ldots \pm l$). The reverse is also true: $Y_l^m(\theta, \phi)$ can be expressed as a linear combination of $Y_l^{m'}(\theta', \phi')$, with $m' = 0, \pm 1, \ldots \pm l$, and therefore,

$$O_R Y_l^m(\theta', \phi') = Y_l^m(\theta, \phi) \qquad \text{(definition of } O_R\text{)}$$
$$= \text{a linear combination of } Y_l^{m'}(\theta', \phi'),\ m' = 0, \pm 1, \ldots \pm l$$

or

$$O_R Y_l^m = \text{a linear combination of } Y_l^{m'},$$

or

$$O_R\, \mathrm{d}_i = \sum_{j=1}^{5} D_{ji}(R)\, \mathrm{d}_j,$$

where $\mathrm{d}_1, \mathrm{d}_2, \ldots \mathrm{d}_5$ are real d-orbitals. The $D(\boldsymbol{R})$ will form a representation of the point group. It is apparent that as the above steps do not specify a particular point group, the d-orbitals (or any set of spherical harmonics of the same l value) can be used as the basis for a representation of *any* point group. With this knowledge, we can now

† Note that in general, symmetry operations do not change the distance r of a point from the origin if the latter is at the centre of mass. We can therefore restrict ourselves to shifts for which r is constant.

use the d-orbital function space to determine a representation of the $\mathscr{C}_{3v}$ point group.

We will define the five d-orbitals by the equations:

$$\begin{aligned}
d_1 &= (x_1^2 - x_2^2)/2,\\
d_2 &= x_1 x_2,\\
d_3 &= x_1 x_3,\\
d_4 &= x_2 x_3,\\
d_5 &= (3x_3^2 - r^2)/(2\sqrt{3}),
\end{aligned}$$

where a common constant and radial function have been omitted (they are not necessary for the purposes of our discussion), and x_1, x_2, and x_3 are the Cartesian coordinates of a point with respect to the base vectors $\mathbf{e}_1$, $\mathbf{e}_2$, and $\mathbf{e}_3$ (see Fig. 5-3.1) and r is the distance of the point x_1, x_2, and x_3 from the origin (centre of mass). The more familiar notation is $d_1 = d_{x^2-y^2}$, $d_2 = d_{xy}$, $d_3 = d_{xz}$, $d_4 = d_{yz}$, $d_5 = d_{z^2}$.

We will first consider the operation of clockwise rotation by an angle θ about the $\mathbf{e}_3$ axis and denote the corresponding transformation operator and matrix by $\boldsymbol{O}_\theta$ and $D(\theta)$, respectively. The relations between the coordinates of a point before (x_1, x_2, and x_3) and after (x_1', x_2', and x_3') rotation are (see eqn (5-2.5))

$$\begin{aligned}
x_1' &= (\cos\theta)x_1 + (\sin\theta)x_2,\\
x_2' &= -(\sin\theta)x_1 + (\cos\theta)x_2,\\
x_3' &= x_3,
\end{aligned}$$

or, taking the inverse:

$$\begin{aligned}
x_1 &= (\cos\theta)x_1' - (\sin\theta)x_2',\\
x_2 &= (\sin\theta)x_1' + (\cos\theta)x_2',\\
x_3 &= x_3'.
\end{aligned}$$

From eqn (5-7.1), we have

$$\begin{aligned}
\boldsymbol{O}_\theta\, d_1(x_1', x_2', x_3') &= d_1(x_1, x_2, x_3)\\
&= (x_1^2 - x_2^2)/2\\
&= [\{(\cos\theta)x_1' - (\sin\theta)x_2'\}^2 - \{(\sin\theta)x_1' + (\cos\theta)x_2'\}^2]/2\\
&= (\cos 2\theta)(x_1'^2 - x_2'^2)/2 - (\sin 2\theta)x_1'x_2'\\
&= (\cos 2\theta)\, d_1(x_1', x_2', x_3') - (\sin 2\theta)\, d_2(x_1', x_2', x_3')
\end{aligned}$$

or, dropping the parameters (they are now the same on both sides of the equation):

$$\boldsymbol{O}_\theta\, d_1 = (\cos 2\theta)\, d_1 - (\sin 2\theta)\, d_2.$$

In the same fashion, we can also obtain

$$\begin{aligned} O_\theta\, d_2 &= (\sin 2\theta)\, d_1 + (\cos 2\theta)\, d_2 \\ O_\theta\, d_3 &= (\cos \theta)\, d_3 - (\sin \theta)\, d_4 \\ O_\theta\, d_4 &= (\sin \theta)\, d_3 + (\cos \theta)\, d_4 \\ O_\theta\, d_5 &= d_5. \end{aligned}$$

These equations define the operator O_θ, that is to say, they specify the new functions created by applying the operator to each of the five basis functions.

Introducing the equation

$$O_\theta\, d_i = \sum_{j=1}^{5} D_{ji}(\theta)\, d_j,$$

we get

$$D(\theta) = \begin{Vmatrix} \cos 2\theta & \sin 2\theta & 0 & 0 & 0 \\ -\sin 2\theta & \cos 2\theta & 0 & 0 & 0 \\ 0 & 0 & \cos\theta & \sin\theta & 0 \\ 0 & 0 & -\sin\theta & \cos\theta & 0 \\ 0 & 0 & 0 & 0 & 1 \end{Vmatrix}$$

(once again, notice the subscripting in $D_{ji}(\theta)$). Hence, if $\theta = 2\pi/3$,

$$D(C_3) = \begin{Vmatrix} -1/2 & -\sqrt{3}/2 & 0 & 0 & 0 \\ \sqrt{3}/2 & -1/2 & 0 & 0 & 0 \\ 0 & 0 & -1/2 & \sqrt{3}/2 & 0 \\ 0 & 0 & -\sqrt{3}/2 & -1/2 & 0 \\ 0 & 0 & 0 & 0 & 1 \end{Vmatrix} \tag{5-9.1}$$

and, if $\theta = 4\pi/3$,

$$D(C_3^2) = \begin{Vmatrix} -1/2 & \sqrt{3}/2 & 0 & 0 & 0 \\ -\sqrt{3}/2 & -1/2 & 0 & 0 & 0 \\ 0 & 0 & -1/2 & -\sqrt{3}/2 & 0 \\ 0 & 0 & \sqrt{3}/2 & -1/2 & 0 \\ 0 & 0 & 0 & 0 & 1 \end{Vmatrix} \tag{5-9.2}$$

Now, let us consider reflection in the σ_v'' plane (see Fig. 5-4.1). We have from eqn (5-3.3)

$$\begin{Vmatrix} x_1' \\ x_2' \\ x_3' \end{Vmatrix} = \begin{Vmatrix} 1/2 & -\sqrt{3}/2 & 0 \\ -\sqrt{3}/2 & -1/2 & 0 \\ 0 & 0 & 1 \end{Vmatrix} \begin{Vmatrix} x_1 \\ x_2 \\ x_3 \end{Vmatrix},$$

or the inverse

$$\begin{Vmatrix} x_1 \\ x_2 \\ x_3 \end{Vmatrix} = \begin{Vmatrix} 1/2 & -\sqrt{3}/2 & 0 \\ -\sqrt{3}/2 & -1/2 & 0 \\ 0 & 0 & 1 \end{Vmatrix} \begin{Vmatrix} x_1' \\ x_2' \\ x_3' \end{Vmatrix}.$$

Hence,

$$\begin{aligned} d_1(x_1, x_2, x_3) &= (x_1^2 - x_2^2)/2 \\ &= \{(\tfrac{1}{2}x_1' - \sqrt{3}/2x_2')^2 - (-\sqrt{3}/2x_1' - \tfrac{1}{2}x_2')^2\}/2 \\ &= -\tfrac{1}{2}(x_1'^2 - x_2'^2)/2 - \sqrt{3}/2x_1'x_2' \\ &= -\tfrac{1}{2}\, d_1(x_1', x_2', x_3') - \sqrt{3}/2\, d_2(x_1', x_2', x_3') \\ d_2(x_1, x_2, x_3) &= x_1x_2 \\ &= (\tfrac{1}{2}x_1' - \sqrt{3}/2x_2')(-\sqrt{3}/2x_1' - \tfrac{1}{2}x_2') \\ &= -\sqrt{3}/2\, d_1(x_1', x_2', x_3') + \tfrac{1}{2}\, d_2(x_1', x_2', x_3'), \text{ etc.} \end{aligned}$$

Therefore,

$$\begin{aligned} O_{\sigma_v''}\, d_1(x_1', x_2', x_3') &= d_1(x_1, x_2, x_3) \\ &= -\tfrac{1}{2}\, d_1(x_1', x_2', x_3') - \sqrt{3}/2\, d_2(x_1', x_2', x_3') \end{aligned}$$

and

$$O_{\sigma_v''}\, d_1 = -\tfrac{1}{2}\, d_1 - \sqrt{3}/2\, d_2$$

and, likewise:

$$\begin{aligned} O_{\sigma_v''}\, d_2 &= -\sqrt{3}/2\, d_1 + \tfrac{1}{2}\, d_2, \\ O_{\sigma_v''}\, d_3 &= \tfrac{1}{2}\, d_3 - \sqrt{3}/2\, d_4, \\ O_{\sigma_v''}\, d_4 &= -\sqrt{3}/2\, d_3 - \tfrac{1}{2}\, d_4, \\ O_{\sigma_v''}\, d_5 &= d_5. \end{aligned}$$

From these equations we obtain

$$D(\sigma_v'') = \begin{Vmatrix} -1/2 & -\sqrt{3}/2 & 0 & 0 & 0 \\ -\sqrt{3}/2 & 1/2 & 0 & 0 & 0 \\ 0 & 0 & 1/2 & -\sqrt{3}/2 & 0 \\ 0 & 0 & -\sqrt{3}/2 & -1/2 & 0 \\ 0 & 0 & 0 & 0 & 1 \end{Vmatrix} \tag{5-9.3}$$

The other symmetry operations of $\mathscr{C}_{3v}$ give

$$D(\sigma_v') = \begin{Vmatrix} 1 & 0 & 0 & 0 & 0 \\ 0 & -1 & 0 & 0 & 0 \\ 0 & 0 & -1 & 0 & 0 \\ 0 & 0 & 0 & 1 & 0 \\ 0 & 0 & 0 & 0 & 1 \end{Vmatrix}, \tag{5-9.4}$$

$$D(\sigma_v''') = \begin{Vmatrix} -1/2 & \sqrt{3}/2 & 0 & 0 & 0 \\ \sqrt{3}/2 & 1/2 & 0 & 0 & 0 \\ 0 & 0 & 1/2 & \sqrt{3}/2 & 0 \\ 0 & 0 & \sqrt{3}/2 & -1/2 & 0 \\ 0 & 0 & 0 & 0 & 1 \end{Vmatrix}, \qquad (5\text{-}9.5)$$

and

$$D(E) = \begin{Vmatrix} 1 & 0 & 0 & 0 & 0 \\ 0 & 1 & 0 & 0 & 0 \\ 0 & 0 & 1 & 0 & 0 \\ 0 & 0 & 0 & 1 & 0 \\ 0 & 0 & 0 & 0 & 1 \end{Vmatrix}. \qquad (5\text{-}9.6)$$

The six 5×5 matrices specified by eqns (5-9.1) to (5-9.6) form a five-dimensional representation of $\mathscr{C}_{3v}$; the basis functions for this representation are the set of five d-orbitals.

It has probably not escaped the reader's notice that *each* of the above matrices has the same structure, that is to say all their non-vanishing elements occur in the same square blocks along the diagonal:

$$\begin{Vmatrix} X & X & 0 & 0 & 0 \\ X & X & 0 & 0 & 0 \\ 0 & 0 & X & X & 0 \\ 0 & 0 & X & X & 0 \\ 0 & 0 & 0 & 0 & X \end{Vmatrix}.$$

What is not so obvious is that since the block structure is *identical* for *every* symmetry operation, the individual blocks themselves, written as matrices, form lower dimensional representations. Take, for example, the block form matrices

$$D(T) = \begin{Vmatrix} D'(T) & [0] \\ [0] & D''(T) \end{Vmatrix},$$

$$D(S) = \begin{Vmatrix} D'(S) & [0] \\ [0] & D''(S) \end{Vmatrix},$$

$$D(R) = \begin{Vmatrix} D'(R) & [0] \\ [0] & D''(R) \end{Vmatrix},$$

where $D'(T)$, $D'(S)$, $D'(R)$ are square arrays of dimension $n \times n$ and $D''(T)$, $D''(S)$, $D''(R)$ are square arrays of dimension $m \times m$ and [0] represents a rectangular array of zeros, if $D(T) = D(S)D(R)$, then the matrix multiplication rule leads to the equations

$$D'(T) = D'(S)D'(R)$$

and

$$D''(T) = D''(S)D''(R) \tag{5-9.7}$$

and consequently, the matrices $D'(R)$ etc. and $D''(R)$ etc. form two lower-dimensional representations.

Another feature of block form matrices is that if a matrix A is in the block form:

$$\begin{Vmatrix} [A_1] & [0] & [0] & \cdots \\ [0] & [A_2] & [0] & \cdots \\ [0] & [0] & [A_3] & \cdots \\ \cdot & \cdot & \cdot & \cdots \end{Vmatrix}$$

then $\det(A) = \det(A_1)\det(A_2)\det(A_3)\ldots$ and if $\det(A) = 0$, then either

$$\det(A_1) = 0$$

or

$$\det(A_2) = 0 \tag{5-9.8}$$

or

$$\det(A_3) = 0 \quad \text{etc.}$$

We will return to the topic of block form matrices in the next chapter; it is the foundation of most of the theorems which follow.

5-10. Determinants as representations

We might note, in passing that, since $\det(AB) = \det(A)\det(B)$ (see Appendix A.4-6), the determinants of any set of matrices which form a representation will themselves act as a one-dimensional representation. That is, if $T = SR$ and $D(T) = D(S)D(R)$ then

$$\det\{D(T)\} = \det\{D(S)D(R)\} = \det\{D(S)\}\det\{D(R)\},$$

and the group of numbers (or 1×1 matrices) det $\{D(R)\}$ etc. is homomorphic with the group of matrices.

5-11. Summary

In this chapter we have shown that there are very many different sets of matrices which behave like the symmetry operations of a given point group. We have constructed these so-called representations by considering the action of the symmetry operations on a position vector or on any number of base vectors. Alternatively, we have found that we can find transformation operators O_R which are homomorphic with the symmetry operations and that from these we can construct representations by considering the actions of the O_R on functions belonging to some function space. In this way, the p-orbitals, etc. form a basis of representation for *any* point group, the six π-orbitals of benzene form a basis of representation for the $\mathscr{D}_{6h}$ point group, (see Chapter 10) etc.

Our next task will be to try and organize, reduce and classify the plethora of representations which we can now create. We will try to eliminate from discussion those which are, in a certain sense, equivalent and those which can, in a certain sense, be 'broken down' into simpler (lower order) representations.

Where we have now got to and where we are going has already been summarized in Fig. 5-1.1.

Appendices

A.5-1. Proof that, if the symmetry operations *R*, *S*, and *T* of a point group obey the relation *T*=*SR*, the matrices *D(R)*, *D(S)*, and *D(T)* found by the consideration of the effect of *R*, *S*, and *T* on a position vector (or point), obey the relation *D (T)*=*D(S)D(R)*

Let

$$\mathbf{p} = x_1\mathbf{e}_1+x_2\mathbf{e}_2+x_3\mathbf{e}_3,$$

$$R\mathbf{p} = \mathbf{p}' = x_1'\mathbf{e}_1+x_2'\mathbf{e}_2+x_3'\mathbf{e}_3,$$

$$S\mathbf{p}' = \mathbf{p}'' = x_1''\mathbf{e}_1+x_2''\mathbf{e}_2+x_3''\mathbf{e}_3$$

and, if $T = SR$,

$$T\mathbf{p} = \mathbf{p}'' = x_1''\mathbf{e}_1+x_2''\mathbf{e}_2+x_3''\mathbf{e}_3$$

(see Fig. 5-3.2), then, from eqn (5-2.17)

$$x_i' = \sum_{j=1}^{3} D_{ij}(R)x_j \qquad i = 1, 2, 3^{\dagger}$$

$$x_k'' = \sum_{j=1}^{3} D_{kj}(T)x_j \qquad k = 1, 2, 3 \tag{A.5-1.1}$$

$$x_k'' = \sum_{i=1}^{3} D_{ki}(S)x_i' \qquad k = 1, 2, 3$$

$$= \sum_{i=1}^{3} D_{ki}(S) \sum_{j=1}^{3} D_{ij}(R)x_j \qquad k = 1, 2, 3$$

$$= \sum_{j=1}^{3} \left[\sum_{i=1}^{3} D_{ki}(S)D_{ij}(R)\right]x_j \quad k = 1, 2, 3. \tag{A.5-1.2}$$

Comparing eqns (A.5-1.1) and (A.5-1.2), we have

$$D_{kj}(T) = \sum_{i=1}^{3} D_{ki}(S)D_{ij}(R) \qquad \begin{matrix} k = 1, 2, 3 \\ j = 1, 2, 3 \end{matrix}$$

† $D_{ij}(R)$ is the matrix element in the ith row and jth column of matrix $D(R)$.

and, because of the matrix multiplication rule, this leads to the desired result $D(T) = D(S)D(R)$.

A.5-2. Proof that the matrices constructed in eqn 5-4.2 (or eqn 5-4.4) form a representation of the point group

$$Re_k = \sum_{j=1}^{n} D_{jk}(R)e_j \qquad k = 1, 2, \ldots, n \quad \text{(N.B. subscript order)}$$

$$Se_j = \sum_{i=1}^{n} D_{ij}(S)e_i \qquad j = 1, 2 \ \ldots, n$$

$$Te_k = \sum_{i=1}^{n} D_{ik}(T)e_i \qquad k = 1, 2, \ldots, n \tag{A.5-2.1}$$

$$\begin{aligned} SRe_k &= S\left[\sum_{j=1}^{n} D_{jk}(R)e_j\right] \\ &= \sum_{j=1}^{n} D_{jk}(R)Se_j \qquad \text{(the symmetry operators are linear, see eqn (2-2.2))} \\ &= \sum_{j=1}^{n} D_{jk}(R) \sum_{i=1}^{n} D_{ij}(S)e_i \\ &= \sum_{i=1}^{n}\left[\sum_{j=1}^{n} D_{ij}(S)D_{jk}(R)\right]e_i \qquad k = 1, 2, \ldots, n. \end{aligned} \tag{A.5-2.2}$$

If $SR = T$, eqn (A.5-2.1) and (A.5-2.2) must be identical and therefore

$$D_{ik}(T) = \sum_{j=1}^{n} D_{ij}(S)D_{jk}(R) \qquad \begin{array}{l} i = 1, 2, \ldots, n \\ k = 1, 2, \ldots, n \end{array}$$

but this is just the matrix multiplication rule, so

$$D(T) = D(S)D(R).$$

A.5-3. Proof that the matrices derived from a position vector are the same as those derived from a single set of base vectors

Consider

$$\mathbf{p} = x_1e_1 + x_2e_2 + x_3e_3,$$

$$\mathbf{p}' = R\mathbf{p} = x_1Re_1 + x_2Re_2 + x_3Re_3,$$

$$e_k' = Re_k = \sum_{j=1}^{3} D_{jk}(R)e_j \qquad k = 1, 2, 3. \tag{A.5-3.1}$$

$$\begin{aligned} \mathbf{p}' &= x_1\sum_{j=1}^{3} D_{j1}(R)e_j + x_2\sum_{j=1}^{3} D_{j2}(R)e_j + x_3\sum_{j=1}^{3} D_{j3}(R)e_j \\ &= \left[\sum_{l=1}^{3} D_{1l}(R)x_l\right]e_1 + \left[\sum_{l=1}^{3} D_{2l}(R)x_l\right]e_2 + \left[\sum_{l=1}^{3} D_{3l}(R)x_l\right]e_3 \\ &= x_1'e_1 + x_2'e_2 + x_3'e_3 \end{aligned}$$

and hence

$$x_i' = \sum_{l=1}^{3} D_{il}(R)x_l \qquad i = 1, 2, 3$$

or changing subscripts

$$x'_k = \sum_{j=1}^{3} D_{kj}(R)x_j \qquad k = 1, 2, 3. \tag{A.5-3.2}$$

Consequently, the same matrix $D(R)$ defines x'_k through eqn (A-5-3.2) (or eqn (5-2.17)) and $\mathbf{e}'_k$ through eqn (A.5-3.1) (or eqn (5-4.2)).

A.5-4. Proof that the operators O_R are (1) linear, (2) homomorphic with the symmetry operations R

(1) (a) If f is a function, a a number and $g = af$, then

$$\begin{aligned}(O_R g)(x'_1, x'_2, x'_3) &= g(x_1, x_2, x_3)\\ &= af(x_1, x_2, x_3)\\ &= a(O_R f)(x'_1, x'_2, x'_3).\end{aligned}$$

This equation must hold for any point (x'_1, x'_2, x'_3), hence

$$O_R(af) = a(O_R f). \tag{A.5-4.1}$$

(b) Consider two functions f and g and let $h = f+g$, then

$$(O_R f)(x'_1, x'_2, x'_3) = f(x_1, x_2, x_3)$$

$$(O_R g)(x'_1, x'_2, x'_3) = g(x_1, x_2, x_3)$$

and

$$\begin{aligned}&O_R[f(x'_1, x'_2, x'_3)+g(x'_1, x'_2, x'_3)]\\ &= (O_R h)(x'_1, x'_2, x'_3) = h(x_1, x_2, x_3)\\ &= f(x_1, x_2, x_3)+g(x_1, x_2, x_3)\\ &= (O_R f)(x'_1, x'_2, x'_3)+(O_R g)(x'_1, x'_2, x'_3).\end{aligned}$$

Since this equation holds for any point x'_1, x'_2, x'_3, we find

$$O_R(f+g) = (O_R f)+(O_R g). \tag{A.5-4.2}$$

(2) Let R transform the point (x_1, x_2, x_3) to (x'_1, x'_2, x'_3), S the point (x'_1, x'_2, x'_3) to (x''_1, x''_2, x''_3) and $T(= SR)$ the point (x_1, x_2, x_3) to (x''_1, x''_2, x''_3). Then, if f and g are functions belonging to the function space,

$$(O_T f)(x''_1, x''_2, x''_3) = f(x_1, x_2, x_3) \tag{A.5-4.3}$$

and

$$(O_S g)(x''_1, x''_2, x''_3) = g(x'_1, x'_2, x'_3). \tag{A.5-4.4}$$

Now let $g = O_R f$, then

$$g(x'_1, x'_2, x'_3) = (O_R f)(x'_1, x'_2, x'_3) = f(x_1, x_2, x_3)$$

and therefore, going back to eqn (A.5-4.4),

$$(O_S O_R f)(x''_1, x''_2, x''_3) = f(x_1, x_2, x_3).$$

Comparing this last equation with eqn (A.5-4.3), we have the desired result

$$O_S O_R f = O_T f. \tag{A.5-4.5}$$

A. 5-5. Proof that the matrices derived from O_R form a representation of the point group

We must show that, if $T = SR$ (or, equivalently, $O_T = O_S O_R$, see the previous appendix), then $D(T) = D(S)D(R)$.

$$O_R f_k = \sum_{j=1}^{n} D_{jk}(R) f_j$$

$$O_S f_j = \sum_{i=1}^{n} D_{ij}(S) f_i$$

$$O_T f_k = \sum_{i=1}^{n} D_{ik}(T) f_i$$

$$O_T f_k = O_{SR} f_k = O_S(O_R f_k) = O_S\left\{\sum_{j=1}^{n} D_{jk}(R) f_j\right\}$$

$$= \sum_{j=1}^{n} D_{jk}(R)(O_S f_j) \qquad (O_S \text{ is linear, see the previous appendix})$$

$$= \sum_{j=1}^{n} D_{jk}(R)\left\{\sum_{i=1}^{n} D_{ij}(S) f_i\right\}.$$

That is,

$$O_T f_k = \sum_{i=1}^{n}\left\{\sum_{j=1}^{n} D_{ij}(S) D_{jk}(R)\right\} f_i$$

therefore,

$$D_{ik}(T) = \sum_{j=1}^{n} D_{ij}(S) D_{jk}(R)$$

which is the matrix multiplication rule, therefore $D(T) = D(S)D(R)$, the desired result.

PROBLEMS

5.1. Consider the following planar symmetric figure.

(a) Determine the distinct symmetry operations which take it into itself; construct the group multiplication table for these operations, and identify the point group to which this figure belongs.

(b) Find a set of two-dimensional matrices which are in one-to-one correspondence with the above symmetry operations, and verify that they have the same group multiplication table as the symmetry operations.

5.2. The table below gives the effects of the transformation operators O_R for the symmetry operations R of the point group $\mathscr{D}_4$ on four functions f_1, f_2, f_3, and f_4. Construct a four-dimensional representation of $\mathscr{D}_4$.

$R =$	E	C_4	C_4^3	C_2	C'_{2a}	C'_{2b}	C''_{2a}	C''_{2b}
f_1	f_1	f_2	f_4	f_3	$-f_4$	$-f_2$	$-f_1$	$-f_3$
f_2	f_2	f_3	f_1	f_4	$-f_3$	$-f_1$	$-f_4$	$-f_2$
f_3	f_3	f_4	f_2	f_1	$-f_2$	$-f_4$	$-f_3$	$-f_1$
f_4	f_4	f_1	f_3	f_2	$-f_1$	$-f_3$	$-f_2$	$-f_4$

5.3. Consider a set of base vectors located on the nuclei of the molecule SO_2 as in the figure below ($\mathbf{e}_3$, $\mathbf{e}_6$, $\mathbf{e}_9$ are perpendicular to the page).

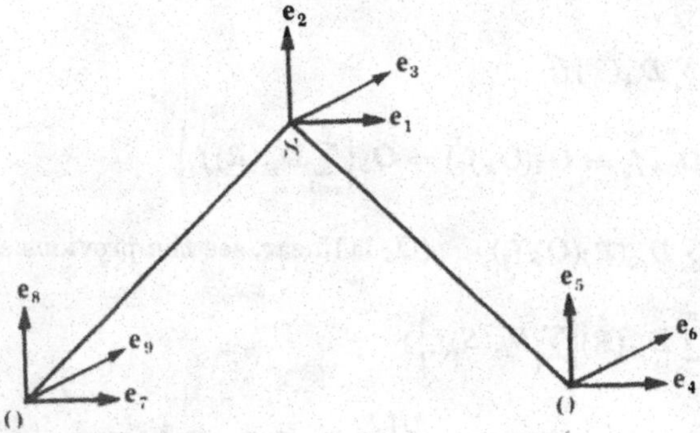

Construct a nine-dimensional matrix representation for the point group to which SO_2 belongs.

5.4. For the point group $\mathscr{D}_{2h}$:
(a) construct a three-dimensional matrix representation using three real p-orbitals as basis functions,
(b) construct a five-dimensional matrix representation using five real d-orbitals as basis functions.

5.5. Consider the planar trivinylmethyl radical with seven π-orbitals located as shown below:

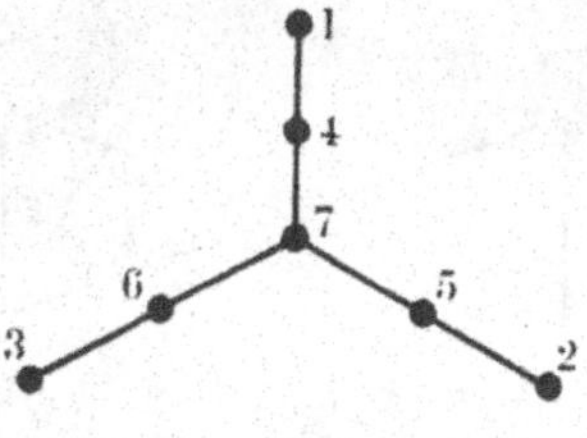

Using these π-orbitals as basis functions, construct a seven-dimensional representation of the $\mathscr{C}_3$ point group.

6. Equivalent and reducible representations

6-1. Introduction

HAVING spent a considerable effort in creating many different matrix representations for the point groups we now, ironic as it may seem, devote an equal effort to eliminating many of them from further consideration. We do this in two ways.

In the first place, we consider those representations which are produced by the same transformation operators O_R and the same function space but with different choices of basis functions describing that space to be *equivalent*. We will see that any pair of such *equivalent representations* have corresponding matrices which are linked by a similarity transformation (see § 4-5). As it will always be possible to find a set of basis functions which produce unitary matrices (a *unitary representation*), convenience dictates that we choose such a set for producing a representation which is typical of the other equivalent ones.

In the second place, we restrict our discussion to those representations which are composed of matrices which cannot be simultaneously broken down (reduced) by a similarity transformation into block form (e.g. the matrices in § 5-9). It will be for these *irreducible representations* that (in the next chapter) we will be able to prove a number of far reaching theorems, one of the most important of which is the theorem that the number of non-equivalent irreducible representations is equal to the number of classes in the point group. So that, for example, for the point group $\mathscr{C}_{3v}$, which has three classes, rather than dealing with an infinite number of representations we will have only the three which are non-equivalent and irreducible to worry about. Also in the next chapter we will show how to obtain, with the least amount of work, the essential information concerning the non-equivalent irreducible representations which exist for any point group.

Irreducible representations are important to the chemist since they provide directly a great deal of information about the nature of vibrational and electronic wavefunctions.

6-2. Equivalent representations

Let us consider a particular n-dimensional function space, that is one which requires n linearly-independent *basis functions* to specify any function belonging to it. There will be many possible different sets of linearly-independent functions which can act as a basis for this space but we will take just the following two:

$$f_1, f_2, \dots f_n$$
$$g_1, g_2, \dots g_n$$

e.g. if we were considering the d-orbital function space, the basis functions could be a set of five real d-orbitals or a set of five complex d-orbitals, the latter being just combinations of the former (and vice versa) and both could be used to define any function belonging to the d-orbital space.

Since the g functions are a basis for the space, any f function can be written as a linear combination of g functions:

$$f_k = \sum_{j=1}^{n} A_{jk} g_j \qquad k = 1, 2, \dots n \tag{6-2.1}$$

(note the subscripting on the coefficient A_{jk}), and any g function can be written as a linear combination of f functions:

$$g_j = \sum_{i=1}^{n} B_{ij} f_i \qquad j = 1, 2, \dots n. \tag{6-2.2}$$

We can combine eqns (6-2.1) and (6-2.2) together to give

$$f_k = \sum_{j=1}^{n} A_{jk} \left(\sum_{i=1}^{n} B_{ij} f_i \right) = \sum_{i=1}^{n} \left(\sum_{j=1}^{n} B_{ij} A_{jk} \right) f_i$$

and since the f's are linearly independent, this equation implies that

$$\sum_{j=1}^{n} B_{ij} A_{jk} = \delta_{ik} \qquad \begin{matrix} i = 1, 2, \dots n \\ k = 1, 2, \dots n \end{matrix} \tag{6-2.3}$$

since then:

$$f_k = \sum_{i=1}^{n} \delta_{ik} f_i = f_k.$$

The left hand side of eqn (6-2.3) is the element in the ith row and kth column of the matrix formed by multiplying matrix B (composed of the elements B_{ij}) by matrix A (composed of the elements A_{ij}), the right hand side is the i, k element of the identity matrix E, hence

$$BA = E.$$

We can also write

$$g_j = \sum_{i=1}^{n} B_{ij} f_i = \sum_{i=1}^{n} B_{ij} \left(\sum_{k=1}^{n} A_{ki} g_k \right) = \sum_{k=1}^{n} \left(\sum_{i=1}^{n} A_{ki} B_{ij} \right) g_k$$

and

$$\sum_{i=1}^{n} A_{ki}B_{ij} = \delta_{kj} \qquad \begin{matrix} k = 1, 2, \ldots n \\ j = 1, 2, \ldots n \end{matrix}$$

and obtain

$$AB = E.$$

Therefore the matrices A and B are related to each other in the following way:

$$B = A^{-1} \tag{6-2.4}$$

and

$$A = B^{-1}$$

Now let us consider the transformation operator $\boldsymbol{O_R}$ (corresponding to the symmetry operation $\boldsymbol{R}$) which is defined by the equation which we have had before:

$$\boldsymbol{O_R} f(x_1', x_2', x_3') = f(x_1, x_2, x_3) \tag{6-2.5}$$

where x_1, x_2, x_3 and x_1', x_2', x_3' are the Cartesian coordinates of a point before and after the application of $\boldsymbol{R}$, respectively. Let us assume that $\boldsymbol{O_R}$ in conjunction with our chosen function space produces matrices which represent $\boldsymbol{O_R}$ and $\boldsymbol{R}$, i.e.

$$\boldsymbol{O_R} f_i = \sum_{k=1}^{n} D_{ki}^{f}(\boldsymbol{R}) f_k \tag{6-2.6}$$

or

$$\boldsymbol{O_R} g_j = \sum_{l=1}^{n} D_{lj}^{g}(\boldsymbol{R}) g_l. \tag{6-2.7}$$

The coefficients $D_{ki}^{f}(\boldsymbol{R})$ will form the matrix $D^{f}(\boldsymbol{R})$ which represents $\boldsymbol{R}$ in the f basis and the coefficients $D_{lj}^{g}(\boldsymbol{R})$ will form the matrix $D^{g}(\boldsymbol{R})$ which represents $\boldsymbol{R}$ in the g basis.

We can readily obtain an alternative equation to eqn (6-2.7) in the following way:

$$\begin{aligned}
\boldsymbol{O_R} g_j &= \boldsymbol{O_R}\left(\sum_{i=1}^{n} B_{ij} f_i\right) && \text{(from eqn (6-2.2))} \\
&= \sum_{i=1}^{n} B_{ij}(\boldsymbol{O_R} f_i) && (\boldsymbol{O_R} \text{ is linear}) \\
&= \sum_{i=1}^{n} B_{ij}\left(\sum_{k=1}^{n} D_{ki}^{f}(\boldsymbol{R}) f_k\right) && \text{(from eqn (6-2.6))} \\
&= \sum_{i=1}^{n} B_{ij} \sum_{k=1}^{n} D_{ki}^{f}(\boldsymbol{R})\left(\sum_{l=1}^{n} A_{lk} g_l\right) && \text{(from eqn (6-2.1))} \\
&= \sum_{l=1}^{n}\left(\sum_{k=1}^{n}\sum_{i=1}^{n} A_{lk} D_{ki}^{f}(\boldsymbol{R}) B_{ij}\right) g_l.
\end{aligned}$$

Comparing this result with eqn (6-2.7) and recalling the rule for the product of three matrices eqn (4-3.6), we have:

$$D^g_{lj}(R) = \sum_{k=1}^{n} \sum_{i=1}^{n} A_{lk} D^f_{ki}(R) B_{ij}$$

or

$$D^g(R) = AD^f(R)B$$

or, since $A = B^{-1}$,

$$D^g(R) = B^{-1}D^f(R)B \tag{6-2.8}$$

or

$$D^f(R) = A^{-1}D^g(R)A. \tag{6-2.9}$$

Consequently, the matrices in the g basis representation are related to those in the f basis representation simply by a similarity transformation (see § 4-5); the two representations are said to be *equivalent*.

It is clear that a change of basis for a given function space does not affect the multiplication rules, i.e. if

$$D^f(SR) = D^f(S)D^f(R)$$

then

$$\begin{aligned} D^g(SR) &= B^{-1}D^f(SR)B \\ &= B^{-1}D^f(S)D^f(R)B \\ &= B^{-1}D^f(S)BB^{-1}D^f(R)B \\ &= D^g(S)D^g(R). \end{aligned}$$

To summarize, the different sets of matrices that can be used to represent the operators O_R in a given space are different realizations of what is really the same representation. Two representations of a point group are *equivalent* if matrices B and B^{-1} exist for which

$$D^g(R) = B^{-1}D^f(R)B$$

for *every* operation R of the point group. Conversely, making the same similarity transformation on all matrices of a given representation is equivalent to simply changing the basis of the chosen function space.

6-3. An example of equivalent representations

As an example of equivalent representations, we will consider the p-orbital function space. This space may be described by three real p-orbitals: p_1, p_2, p_3 (commonly written as p_x, p_y, p_z) and we will call this the f basis. Alternatively, we may take three complex p-orbitals: p_1', p_2', p_3' (commonly written as p_1, p_{-1}, p_0) and we will call this the g basis. These two sets of functions are related by the equations:

$$p_1 = (p_1' + p_2')/\sqrt{2} \qquad p_1' = (p_1 + ip_2)/\sqrt{2}$$

$$p_2 = -i(p_1' - p_2')/\sqrt{2} \qquad p_2' = (p_1 - ip_2)/\sqrt{2}$$

$$p_3 = p_3' \qquad p_3' = p_3$$

hence, using the notation of the previous section

$$A = \begin{Vmatrix} 1/\sqrt{2} & -i/\sqrt{2} & 0 \\ 1/\sqrt{2} & i/\sqrt{2} & 0 \\ 0 & 0 & 1 \end{Vmatrix}, \qquad B = \begin{Vmatrix} 1/\sqrt{2} & 1/\sqrt{2} & 0 \\ i/\sqrt{2} & -i/\sqrt{2} & 0 \\ 0 & 0 & 1 \end{Vmatrix}$$

and it can be verified that the equation $AB = E$ is obeyed.

The three real p-orbitals may be written in terms of a set of Cartesian coordinates x_1, x_2, and x_3 as

$$\begin{aligned} p_1 &= F(r)x_1 \\ p_2 &= F(r)x_2 \\ p_3 &= F(r)x_3 \end{aligned}$$

where $F(r)$ is a function of the radial distance r, and is common to all three orbitals. If O_θ is the transformation operator which corresponds to a clockwise rotation by an angle θ about the $\mathbf{e}_3$ axis (which defines the coordinate x_3) and if this rotation takes the point at x_1, x_2, x_3 to x_1', x_2', x_3', then (see eqn (5-2.5))

$$\begin{aligned} x_1' &= (\cos\theta)x_1 + (\sin\theta)x_2 \\ x_2' &= -(\sin\theta)x_1 + (\cos\theta)x_2 \\ x_3' &= x_3 \end{aligned}$$

or, taking the inverse

$$\begin{aligned} x_1 &= (\cos\theta)x_1' - (\sin\theta)x_2' \\ x_2 &= (\sin\theta)x_1' + (\cos\theta)x_2' \\ x_3 &= x_3'. \end{aligned}$$

Hence

$$\begin{aligned} O_\theta p_1(x_1', x_2', x_3') &= p_1(x_1, x_2, x_3) \qquad \text{(definition of } O_\theta\text{)} \\ &= F(r)x_1 \\ &= F(r)\{(\cos\theta)x_1' - (\sin\theta)x_2'\} \\ &= (\cos\theta)p_1(x_1', x_2', x_3') - (\sin\theta)p_2(x_1', x_2', x_3') \end{aligned}$$

or,

$$O_\theta p_1 = (\cos\theta)p_1 - (\sin\theta)p_2$$

and by carrying out similar steps for p_2 and p_3, we obtain

$$O_\theta p_k = \sum_{j=1}^{3} D^f_{jk}(\theta)p_j \qquad k = 1, 2, 3$$

where

$$D^f(\theta) = \begin{Vmatrix} \cos\theta & \sin\theta & 0 \\ -\sin\theta & \cos\theta & 0 \\ 0 & 0 & 1 \end{Vmatrix}. \tag{6-3.1}$$

The matrix which corresponds to $D^f(\theta)$ in the equivalent representation using the g basis p_1', p_2', p_3', $D^g(\theta)$, is given by

$$D^g(\theta) = B^{-1}D^f(\theta)B = AD^f(\theta)B$$

$$= \begin{Vmatrix} 1/\sqrt{2} & -i/\sqrt{2} & 0 \\ 1/\sqrt{2} & i/\sqrt{2} & 0 \\ 0 & 0 & 1 \end{Vmatrix} \begin{Vmatrix} \cos\theta & \sin\theta & 0 \\ -\sin\theta & \cos\theta & 0 \\ 0 & 0 & 1 \end{Vmatrix} \begin{Vmatrix} 1/\sqrt{2} & 1/\sqrt{2} & 0 \\ i/\sqrt{2} & -i/\sqrt{2} & 0 \\ 0 & 0 & 1 \end{Vmatrix}$$

$$= \begin{Vmatrix} 1/\sqrt{2} & -i/\sqrt{2} & 0 \\ 1/\sqrt{2} & i/\sqrt{2} & 0 \\ 0 & 0 & 1 \end{Vmatrix} \begin{Vmatrix} e^{i\theta}/\sqrt{2} & e^{-i\theta}/\sqrt{2} & 0 \\ ie^{i\theta}/\sqrt{2} & -ie^{-i\theta}/\sqrt{2} & 0 \\ 0 & 0 & 1 \end{Vmatrix}$$

$$= \begin{Vmatrix} e^{i\theta} & 0 & 0 \\ 0 & e^{-i\theta} & 0 \\ 0 & 0 & 1 \end{Vmatrix}. \qquad (6\text{-}3.2)$$

Or, alternatively, we may obtain this same matrix by considering

$$O_\theta p_k'(x_1', x_2', x_3') = p_k'(x_1, x_2, x_3) \qquad k = 1, 2, 3$$

and carrying out the same steps for the complex basis functions p_1', p_2', p_3' as we did for the real basis functions p_1, p_2, p_3.

In Table 6-3.1 we show the matrices for all of the operations of the $\mathscr{C}_{3v}$ point group using both real and complex p-orbitals as basis functions. For the operations C_3 and C_3^2 we have simply replaced θ by $2\pi/3$ and $4\pi/3$ respectively in both eqn (6-3.1) and eqn (6-3.2). The matrices for the reflection operations have been obtained in a fashion similar to that used for the rotations. In carrying out these steps it has been assumed that p_1, p_2, and p_3 lie along the vectors $\mathbf{e}_1$, $\mathbf{e}_2$, and $\mathbf{e}_3$, respectively (see Fig. 5-3.1). For obvious reasons the matrix representation in the real basis is identical to the one given in § 5-3(2) and, further, the reader may verify for himself that the matrices using the complex basis obey the $\mathscr{C}_{3v}$ group table (Table 3-4.1).

6-4. Unitary representations

If we have a number of equivalent representations of a particular point group, it is useful to choose just one of them as a prototype for all the others. It makes sense that the one we choose for this role has matrices which are unitary, since unitary matrices are much easier to handle and manipulate than non-unitary matrices. The reader will recall (Appendix A.4-1(g)) that a unitary matrix is defined by $A^{-1} = A^\dagger$. Just as there are two ways of interpreting equivalent representations

TABLE 6-3.1

Equivalent representations of $\mathscr{C}_{3v}$ using the p-orbital function space†

Operation	*Real basis*	*Complex basis*
E	$\begin{Vmatrix} 1 & 0 & 0 \\ 0 & 1 & 0 \\ 0 & 0 & 1 \end{Vmatrix}$	$\begin{Vmatrix} 1 & 0 & 0 \\ 0 & 1 & 0 \\ 0 & 0 & 1 \end{Vmatrix}$
C_3	$\begin{Vmatrix} -c & s & 0 \\ -s & -c & 0 \\ 0 & 0 & 1 \end{Vmatrix}$	$\begin{Vmatrix} \varepsilon & 0 & 0 \\ 0 & \varepsilon^* & 0 \\ 0 & 0 & 1 \end{Vmatrix}$
C_3^2	$\begin{Vmatrix} -c & -s & 0 \\ s & -c & 0 \\ 0 & 0 & 1 \end{Vmatrix}$	$\begin{Vmatrix} \varepsilon^* & 0 & 0 \\ 0 & \varepsilon & 0 \\ 0 & 0 & 1 \end{Vmatrix}$
σ_v'	$\begin{Vmatrix} -1 & 0 & 0 \\ 0 & 1 & 0 \\ 0 & 0 & 1 \end{Vmatrix}$	$\begin{Vmatrix} 0 & -1 & 0 \\ -1 & 0 & 0 \\ 0 & 0 & 1 \end{Vmatrix}$
σ_v''	$\begin{Vmatrix} c & -s & 0 \\ -s & -c & 0 \\ 0 & 0 & 1 \end{Vmatrix}$	$\begin{Vmatrix} 0 & -\varepsilon^* & 0 \\ -\varepsilon & 0 & 0 \\ 0 & 0 & 1 \end{Vmatrix}$
σ_v'''	$\begin{Vmatrix} c & s & 0 \\ s & -c & 0 \\ 0 & 0 & 1 \end{Vmatrix}$	$\begin{Vmatrix} 0 & -\varepsilon & 0 \\ -\varepsilon^* & 0 & 0 \\ 0 & 0 & 1 \end{Vmatrix}$

† $c = \cos(\pi/3) = \frac{1}{2}$, $s = \sin(\pi/3) = \sqrt{3}/2$, $\varepsilon = \exp(2\pi i/3)$.

(change of basis functions or a similarity transformation on the matrices), so there are two ways of proving that it is always possible to find a *unitary representation* which is equivalent to any given representation.

If we choose our basis functions for a particular function space to be orthonormal (orthogonal and normalized) i.e. $(f_i, f_j) = \int f_i^* f_j \, d\tau = \delta_{ij}$, then, since the transformation operators are unitary (§ 5-7), the representation created will consist of unitary matrices. This is proved in Appendix A.6-1. It should be stated that it *is* always possible to find an orthonormal basis and one way, the Schmidt orthogonalization process, is given in Appendix A.6-2.

Alternatively, we can prove that there is always a similarity transformation which will transform simultaneously all of the matrices of a representation into unitary matrices. This is proved in Appendix A.6-3.

From now on therefore it will be no restriction to consider, if we wish to, only unitary representations.

6-5. Reducible representations

It is convenient at this stage to introduce the symbol commonly used for a matrix representation, namely Γ. Different representations for a point group can then be distinguished by a superscript on this symbol, for example the representation in the f basis in § 6-2 could be symbolized by Γ^f and that in the g basis by Γ^g. It is important to understand that Γ is not a symbol for a single matrix but for the whole set of matrices which constitute the representation.

Suppose that Γ is an n-dimensional representation of a group of transformation operators $\boldsymbol{O}_R$ acting on the functions of an n-dimensional function space and that we have basis functions $f_1, f_2, \ldots, f_n$ with the property that the first m $(m \leqslant n)$ are transformed among themselves for *all* $\boldsymbol{O}_R$ (e.g. in § 6-3, the p-orbitals p_1 and p_2 were transformed among themselves by all $\boldsymbol{O}_R$ and so $m = 2$ for this case):

$$\begin{aligned}
\boldsymbol{O}_R f_1 &= D_{11}(\boldsymbol{R})f_1 + \ldots D_{m1}f_m + 0 . f_{m+1} + \ldots 0 . f_n \\
&\quad \vdots \\
\boldsymbol{O}_R f_m &= D_{1m}(\boldsymbol{R})f_1 + \ldots D_{mm}f_m + 0 . f_{m+1} + \ldots 0 . f_n .
\end{aligned}$$

The matrices will then *all* take the form:

$$D(\boldsymbol{R}) = \begin{Vmatrix} D^1(\boldsymbol{R}) & [Q] \\ [0] & D^2(\boldsymbol{R}) \end{Vmatrix} \tag{6-5.1}$$

where $D^1(\boldsymbol{R})$ is a $m \times m$ block of elements, $D^2(\boldsymbol{R})$ is a $(n-m)\times(n-m)$ block of elements, [Q] is a $m \times (n-m)$ block of elements and [0] stands for a $(n-m)\times m$ block of zeros.

If the basis $f_1, f_2 \ldots f_n$ is chosen to be orthonormal, then the matrices $D(\boldsymbol{R})$ will be unitary $[D(\boldsymbol{R})^{-1} = D(\boldsymbol{R})^\dagger]$ and since

$$D(\boldsymbol{R}^{-1})D(\boldsymbol{R}) = D(\boldsymbol{R}^{-1}\boldsymbol{R}) = D(\boldsymbol{E}) = E$$

we have

$$\begin{aligned}
D(\boldsymbol{R}^{-1}) &= D(\boldsymbol{R})^{-1} && (6\text{-}5.2) \\
&= D(\boldsymbol{R})^\dagger \\
&= \begin{Vmatrix} D^1(\boldsymbol{R})^\dagger & [0] \\ [Q]^\dagger & D^2(\boldsymbol{R})^\dagger \end{Vmatrix}. && (6\text{-}5.3)
\end{aligned}$$

As $\boldsymbol{R}^{-1}$ is one of the operations of the point group, $D(\boldsymbol{R}^{-1})$ must also have the form of eqn (6-5.1) and consequently, comparing eqns (6-5.1) and (6-5.3), [Q] must be zero and

$$D(\boldsymbol{R}) = \begin{Vmatrix} D^1(\boldsymbol{R}) & [0] \\ [0] & D^2(\boldsymbol{R}) \end{Vmatrix}. \tag{6-5.4}$$

When $D(\boldsymbol{R})$ is of block form like this, it is said to be fully reducible.

If there is a similarity transformation (or, what is the same thing, a change of basis) such that all the matrices in some representation Γ are brought into identical block form, then Γ is said to be a *reducible representation*. If there is not, then Γ is said to be an *irreducible representation.*

As we have stated before (eqn (5-9.7)), the lower dimensional matrices formed from the blocks can themselves form a representation of the point group. If, for example, $D(\boldsymbol{R})$ are matrices for the representation Γ and if a matrix A exists such that

$$A^{-1}D(\boldsymbol{R})A = \begin{Vmatrix} D^1(\boldsymbol{R}) & [0] & [0] \\ [0] & D^2(\boldsymbol{R}) & [0] \\ [0] & [0] & D^3(\boldsymbol{R}) \end{Vmatrix} \text{ for all } \boldsymbol{R},$$

where [0] stands for rectangular arrays of zeros, then the matrices $D^1(\boldsymbol{R})$, $D^2(\boldsymbol{R})$, and $D^3(\boldsymbol{R})$ form, if they are different and non-equivalent, three new and different representations Γ^1, Γ^2, and Γ^3 for the point group. We write this symbolically as:

$$\Gamma = \Gamma^1 \oplus \Gamma^2 \oplus \Gamma^3.$$

This equation is a highly abbreviated version of what we have just done and must be interpreted with care. The symbol $\oplus$ does *not* mean addition and the equation should be read as: 'the representation Γ can be reduced through a similarity transformation to three representations Γ^1, Γ^2, and Γ^3.'

It is usual to take any reduction that can be carried out as far as possible, that is to reduce Γ to irreducible representations. Quite often the same or an equivalent irreducible representation will occur more than once, we will then write

$$\begin{aligned} \Gamma &= a_1\Gamma^1 \oplus a_2\Gamma^2 \ldots \\ &= \sum_{\nu} a_\nu \Gamma^\nu \end{aligned} \tag{6-5.5}$$

where a_ν is the number of times Γ^ν or its equivalent occurs in Γ and the Γ^ν are non-equivalent and irreducible representations.

Consider the $\mathscr{C}_{3v}$ point group and the transformation operators $\boldsymbol{O_R}$ for the d-orbital function space. If we do not choose our five linearly-independent basis functions for this space with any particular care, we will produce a five-dimensional representation Γ for $\mathscr{C}_{3v}$, the matrices of which are not in block form. If, however, we carry out the similarity transformation on each matrix which corresponds to changing the basis functions to those of § 5-9, we will obtain the matrices of eqns (5-9.1) to (5-9.6), which are in the same block form and Γ will have been

TABLE 6-5.1

The non-equivalent irreducible representations for $\mathscr{C}_{3v}$ using the d-orbital function space

R	Γ^1	Γ^2
E	1	$\begin{Vmatrix} 1 & 0 \\ 0 & 1 \end{Vmatrix}$
C_3	1	$\begin{Vmatrix} -\frac{1}{2} & -\sqrt{3}/2 \\ \sqrt{3}/2 & -\frac{1}{2} \end{Vmatrix}$
C_3^2	1	$\begin{Vmatrix} -\frac{1}{2} & \sqrt{3}/2 \\ -\sqrt{3}/2 & -\frac{1}{2} \end{Vmatrix}$
σ_v'	1	$\begin{Vmatrix} 1 & 0 \\ 0 & -1 \end{Vmatrix}$
σ_v''	1	$\begin{Vmatrix} -\frac{1}{2} & -\sqrt{3}/2 \\ -\sqrt{3}/2 & \frac{1}{2} \end{Vmatrix}$
σ_v'''	1	$\begin{Vmatrix} -\frac{1}{2} & \sqrt{3}/2 \\ \sqrt{3}/2 & \frac{1}{2} \end{Vmatrix}$

reduced to two two-dimensional representations and one one-dimensional representation. Another change of basis functions (in fact, simply interchanging d_3 and d_4, i.e. writing $d_3 = x_2x_3$ and $d_4 = x_1x_3$) shows that these two two-dimensional representations are equivalent. Clearly the one-dimensional representation is irreducible and, though we have not proved it, so are the two-dimensional ones. We can therefore write

$$\Gamma = \Gamma^1 \oplus 2\Gamma^2$$

where Γ^1 and Γ^2 are given in Table 6-5.1.

Our next task is to discover the relationship between the matrix elements of non-equivalent irreducible representations, the restrictions on the number of such representations, simple criteria for testing for irreducibility and a method for readily carrying out the reduction of a reducible representation.

Appendices

A.6-1. Proof that the transformation operators O_R will produce a unitary representation if orthonormal basis functions are used

If an n-dimensional space is characterized by the n orthonormal basis functions $f_1, f_2, \ldots f_n$, then, by definition, the scalar product is

$$(f_i, f_j) = \int f_i^* f_j \, d\tau = \delta_{ij} \qquad \begin{array}{l} i = 1, 2, \ldots, n \\ j = 1, 2, \ldots, n. \end{array}$$

In § 5-7 we showed that the transformation operators O_R are unitary therefore

$$(O_R f_i, O_R f_j) = \int (O_R f_i)^*(O_R f_j) \, d\tau = (f_i, f_j)$$

and hence

$$\int (O_R f_i)^*(O_R f_j) \, d\tau = \delta_{ij} \qquad \begin{array}{l} i = 1, 2, \ldots, n \\ j = 1, 2, \ldots, n. \end{array}$$

Coupling this equation with eqn (6-2.6) and omitting the superscript f on the matrices, we obtain

$$\begin{aligned} \delta_{ij} &= \int \left(\sum_{k=1}^{n} D_{ki}(R) f_k \right)^* \left(\sum_{l=1}^{n} D_{lj}(R) f_l \right) d\tau \\ &= \sum_{k=1}^{n} \sum_{l=1}^{n} D_{ki}(R)^* D_{lj}(R) \int f_k^* f_l \, d\tau \\ &= \sum_{k=1}^{n} \sum_{l=1}^{n} D_{ki}(R)^* D_{lj}(R) \, \delta_{kl} \\ &= \sum_{k=1}^{n} D_{ki}(R)^* D_{kj}(R). \end{aligned}$$

From consideration of the formula for the product of two matrices, it is apparent that the above relationship leads to

$$D(R)^\dagger D(R) = E$$

or

$$D(R)^{-1} = D(R)^\dagger.$$

Hence, in an orthonormal basis the matrices $D(R)$ which represent unitary operators O_R are all unitary.

A.6-2. The Schmidt orthogonalization process

Consider the set of linearly-independent functions $\psi_1, \psi_2, \ldots \psi_n$ where

$$\alpha_1 \psi_1 + \alpha_2 \psi_2 + \ldots \alpha_n \psi_n = 0$$

only if $\alpha_1 = \alpha_2 = \ldots \alpha_n = 0$. The scalar product is defined by

$$(\psi_i, \psi_j) = \int \psi_i^* \psi_j \, d\tau$$

[see § 5-5(4)]. Then the functions $\phi_1, \phi_2, \dots \phi_n$ defined by the following equations will be orthogonal:

$$\phi_1 = \psi_1$$

$$\phi_2 = \psi_2 - \frac{(\phi_1, \psi_2)}{(\phi_1, \phi_1)}\phi_1$$

$$\phi_3 = \psi_3 - \frac{(\phi_1, \psi_3)}{(\phi_1, \phi_1)}\phi_1 - \frac{(\phi_2, \psi_3)}{(\phi_2, \phi_2)}\phi_2$$

$$\vdots$$

$$\phi_n = \psi_n - \frac{(\phi_1, \psi_n)}{(\phi_1, \phi_1)}\phi_1 \dots - \frac{(\phi_{n-1}, \psi_n)}{(\phi_{n-1}, \phi_{n-1})}\phi_{n-1}$$

$$= \psi_n - \sum_{i=1}^{n-1} \frac{(\phi_i, \psi_n)}{(\phi_i, \phi_i)}\phi_i.$$

To prove that these functions are orthogonal we must first show that these definitions have meaning, i.e. we must verify that each $\phi_k \neq 0$. This is done by induction. We clearly have $\phi_1 \neq 0$. Assume that, for some $k \geqslant 2$, $\phi_1, \phi_2 \dots \phi_{k-1}$ are all non-zero, then the definition of ϕ_k is significant since we recognize that ϕ_k can be written as a linear combination of $\psi_1, \psi_2, \dots \psi_k$ in which ψ_k has the coefficient 1 and the $\psi_1, \psi_2, \dots \psi_k$ are linearly independent. It follows that $\phi_k \neq 0$. Thus $\phi_1 \neq 0, \dots, \phi_n \neq 0$.

We next show, again by induction, that $\phi_1, \dots \phi_n$ form an orthogonal set. Assume that for some $k \geqslant 2$, $\phi_1, \dots \phi_{k-1}$ form an orthogonal set. Now

$$\phi_k = \psi_k - \sum_{i=1}^{k-1} \frac{(\phi_i, \psi_k)}{(\phi_i, \phi_i)}\phi_i$$

and therefore, forming the scalar product (ϕ_j, ϕ_k) where $1 \leqslant j < k$, we obtain

$$(\phi_j, \phi_k) = (\phi_j, \psi_k) - \sum_{i=1}^{k-1} \frac{(\phi_i, \psi_k)}{(\phi_i, \phi_i)}(\phi_j, \phi_i)$$

$$= (\phi_j, \psi_k) - \frac{(\phi_j, \psi_k)}{(\phi_j, \phi_j)}(\phi_j, \phi_j) = 0.$$

Thus $\phi_1, \dots \phi_k$ form an orthogonal set and by induction so do $\phi_1, \dots \phi_n$.

A normalized set of functions can be formed from the ϕ's by taking:

$$\chi_i = \frac{\phi_i}{\sqrt{(\phi_i, \phi_i)}},$$

so that finally $(\chi_i, \chi_j) = \delta_{ij}$ and $\chi_1, \dots \chi_n$ form an orthonormal set.

It is also true that for a general vector space any set of linearly-independent vectors can be combined in analogous fashion to give a set of orthonormal vectors. In this case the scalar product is defined by

$$(\mathbf{r}, \mathbf{s}) = rs \cos\theta \text{ (see § 5-5(4), (5))}.$$

A.6-3. Proof that any representation is equivalent, through a similarity transformation, to a unitary representation

To prove this, we first construct the matrix

$$H = \sum_{R} D(R)D(R)^{\dagger}, \tag{A.6-3.1}$$

that is we form a new matrix by adding together the matrices obtained by multiplying each $D(R)$ by its adjoint $D(R)^{\dagger}$ where $D(R)$ is the matrix representing the symmetry operation R or the transformation operator O_R. The sum in eqn (A.6-3.1) is over all the operations R of the point group. Each product $D(R)D(R)^{\dagger}$ is Hermitian (see Problem 4.5) and hence the sum of them H is also Hermitian. Because of this fact H can be diagonalized by a unitary transformation (see § 4-5)

$$X = U^{-1}HU \tag{A.6-3.2}$$

where X is diagonal and U is unitary.

If we define new matrices, equivalent to $D(R)$, as

$$D'(R) = U^{-1}D(R)U, \tag{A.6-3.3}$$

then it can be shown by the usual matrix manipulations, that the adjoints of the $D'(R)$ are

$$D'(R)^{\dagger} = U^{-1}D(R)^{\dagger}U.$$

We can now obtain

$$\begin{aligned} X &= U^{-1}HU \\ &= \sum_{R} U^{-1}D(R)D(R)^{\dagger}U \\ &= \sum_{R} U^{-1}D(R)UU^{-1}D(R)^{\dagger}U \\ &= \sum_{R} D'(R)D'(R)^{\dagger}. \end{aligned}$$

If we consider a typical diagonal element of X, we have:

$$X_{ii} = \sum_{R}\sum_{j} D'_{ij}(R)D'_{ij}(R)^{*}$$

and we see that not only is X diagonal but that it has only real positive elements.

We can therefore find the square root of X, $X^{\frac{1}{2}}$, simply by taking the square root of each diagonal element: $(X^{\frac{1}{2}})_{ii} = (X_{ii})^{\frac{1}{2}}$, the off-diagonal elements of $X^{\frac{1}{2}}$ will, of course, be zero. Also, since $X^{\frac{1}{2}}$ is diagonal and real

$$(X^{\frac{1}{2}})^{\dagger} = X^{\frac{1}{2}}$$

and

$$(X^{-\frac{1}{2}})^{\dagger} = X^{-\frac{1}{2}}.$$

Now we define the matrices $D''(R)$, equivalent to both $D(R)$ and $D'(R)$, by the equations

$$D''(R) = X^{-\frac{1}{2}}D'(R)X^{\frac{1}{2}}. \tag{A.6-3.4}$$

It can be shown that the adjoints of these matrices are

$$D''(R)^{\dagger} = X^{\frac{1}{2}}D'(R)^{\dagger}X^{-\frac{1}{2}}.$$

We can prove that the matrices $D''(\boldsymbol{R})$ are all unitary. Consider the particular operation $\boldsymbol{S}$, then

$$\begin{aligned} D''(\boldsymbol{S})D''(\boldsymbol{S})^{\dagger} &= X^{-\frac{1}{2}}D'(\boldsymbol{S})X^{\frac{1}{2}}X^{\frac{1}{2}}D'(\boldsymbol{S})^{\dagger}X^{-\frac{1}{2}} \\ &= X^{-\frac{1}{2}}D'(\boldsymbol{S})U^{-1}HUD'(\boldsymbol{S})^{\dagger}X^{-\frac{1}{2}} \\ &= X^{-\frac{1}{2}}D'(\boldsymbol{S})U^{-1}\sum_{\boldsymbol{R}} D(\boldsymbol{R})D(\boldsymbol{R})^{\dagger}UD'(\boldsymbol{S})^{\dagger}X^{-\frac{1}{2}} \\ &= X^{-\frac{1}{2}}D'(\boldsymbol{S})\sum_{\boldsymbol{R}} U^{-1}D(\boldsymbol{R})UU^{-1}D(\boldsymbol{R})^{\dagger}UD'(\boldsymbol{S})^{\dagger}X^{-\frac{1}{2}} \\ &= X^{-\frac{1}{2}}D'(\boldsymbol{S})\sum_{\boldsymbol{R}} D'(\boldsymbol{R})D'(\boldsymbol{R})^{\dagger}D'(\boldsymbol{S})^{\dagger}X^{-\frac{1}{2}} \\ &= X^{-\frac{1}{2}}\sum_{\boldsymbol{R}} D'(\boldsymbol{S})D'(\boldsymbol{R})[D'(\boldsymbol{S})D'(\boldsymbol{R})]^{\dagger}X^{-\frac{1}{2}}. \end{aligned}$$

If $\boldsymbol{SR} = \boldsymbol{T}$, then $D(\boldsymbol{S})D(\boldsymbol{R}) = D(\boldsymbol{T})$, see Appendix A.5-5, and since the primed matrices are obtained through a similarity transformation, they are equivalent to the unprimed ones and we have $D'(\boldsymbol{S})D'(\boldsymbol{R}) = D'(\boldsymbol{T})$. Furthermore, by the Rearrangement Theorem (Appendix A.3-1), as $\boldsymbol{R}$ runs over the symmetry operations of the point group, so does $\boldsymbol{SR}(= \boldsymbol{T})$. Hence

$$\begin{aligned} D''(\boldsymbol{S})D''(\boldsymbol{S})^{\dagger} &= X^{-\frac{1}{2}}\sum_{\boldsymbol{T}} D'(\boldsymbol{T})D'(\boldsymbol{T})^{\dagger}X^{-\frac{1}{2}} \\ &= X^{-\frac{1}{2}}\sum_{\boldsymbol{T}} U^{-1}D(\boldsymbol{T})UU^{-1}D(\boldsymbol{T})^{\dagger}UX^{-\frac{1}{2}} \\ &= X^{-\frac{1}{2}}U^{-1}\sum_{\boldsymbol{T}} D(\boldsymbol{T})D(\boldsymbol{T})^{\dagger}UX^{-\frac{1}{2}} \\ &= X^{-\frac{1}{2}}U^{-1}HUX^{-\frac{1}{2}} \\ &= X^{-\frac{1}{2}}XX^{-\frac{1}{2}} \\ &= \boldsymbol{E}. \end{aligned}$$

The matrices $D''(\boldsymbol{R})$ are therefore unitary and the similarity transformation which produces them is (combining eqns (A.6-3.3) and (A.6-3.4))

$$D''(\boldsymbol{R}) = Z^{-1}D(\boldsymbol{R})Z \qquad \text{(A.6-3.5)}$$

where $Z = UX^{\frac{1}{2}}$ and U and X are defined by eqn (A.6-3.2).†

† The reader will have noticed that in determining $X^{\frac{1}{2}}$ we have written $(X^{\frac{1}{2}})_{ii} = (X_{ii})^{\frac{1}{2}}$ rather than $(X^{\frac{1}{2}})_{ii} = -(X_{ii})^{\frac{1}{2}}$ which is equally valid. By doing so we have merely chosen from the several square roots of X the one which is called the *positive square root*.

7. Irreducible representations and character tables

7-1. Introduction

SINCE it is the non-equivalent irreducible representations which reflect the essence of a point group, it is upon these which we now focus our attention. Others which are equivalent or can through a similarity transformation be reduced to these are, in a certain sense, superfluous: they contain no new information about the point group and we can safely ignore them.

In this chapter we introduce a theorem which is central to the use we make of irreducible representations in solving quantum mechanical problems. This theorem is called the Great Orthogonality Theorem or the Key Theorem in Representation Theory; both of these names give an indication of the theorem's importance. What it does is to show the relationships which exist between the matrix elements of the non-equivalent irreducible representations. Though the proof of this theorem is fairly complicated, it has an elegance and beauty, the discovery of which is the reward for those who master it. The implications of the theorem are many and include such things as the fact that the number of irreducible representations is less than or equal to the number of classes in the point group, and that for a given point group the possible dimensions of its irreducible representations are restricted. It also leads to a formula for the number of times a given irreducible representation occurs in a reducible representation.

We introduce in this chapter the word 'character'. A character is the trace (sum of the diagonal elements) of any matrix which is a part of a representation of a point group. Many of the properties of a point group can be deduced from the characters of its irreducible representations alone rather than from the matrices themselves; this greatly simplifies things. Furthermore, the Great Orthogonality Theorem leads to rules which allow us to construct tables of characters of the irreducible representations without explicitly knowing the matrices.

We end this chapter with an example of the determination of the irreducible representations produced by certain basis functions using the rules and theorems which we have developed. It is at this point that we are ready, at last, to produce results of genuine chemical interest from the sole knowledge of the point group to which a molecule belongs.

7-2. The Great Orthogonality Theorem

This theorem states that if Γ^μ and Γ^ν are two non-equivalent irreducible representations with matrices $D^\mu(R)$ and $D^\nu(R)$ (of dimensions n_μ and n_ν, respectively) for each operation R of the group $\mathscr{G}$ then the matrix elements are related by the equation

$$\sum_R D^\mu_{ik}(R) D^\nu_{mj}(R^{-1}) = (g/n_\mu)\delta_{\mu\nu}\delta_{ij}\delta_{km}, \qquad (7\text{-}2.1)$$

where g is the order of the group and the sum is over all of the operations R. The proof is given in Appendix A.7-1.

If we assume that the irreducible representations are unitary, and we can do this without any loss of generality (see § 6-4), then

$$D^\nu(R^{-1}) = D^\nu(R)^{-1} = D^\nu(R)^\dagger$$

and eqn (7-2.1) becomes:

$$\sum_R D^\mu_{ik}(R) D^\nu_{jm}(R)^* = (g/n_\mu)\delta_{\mu\nu}\delta_{ij}\delta_{km}. \qquad (7\text{-}2.2)$$

For a given unitary irreducible representation Γ^μ there will be n_μ^2 matrix elements corresponding to each R. If the operations are $R = R_1, R_2, \ldots R_g$, then the g matrix elements of a chosen i and j value,

$$D^\mu_{ij}(R_1),\ D^\mu_{ij}(R_2),\ \ldots\ D^\mu_{ij}(R_g)$$

can be considered to be the components of a g-dimensional vector. Since there are n_μ^2 such sets of matrix elements (corresponding to $i = 1, 2, \ldots n_\mu$ and $j = 1, 2, \ldots n_\mu$), there will be n_μ^2 such g-dimensional vectors for each irreducible representation. Eqn (7-2.2) shows that the vectors from a given unitary irreducible representation are orthogonal to one another and to the vectors formed in a similar way from any other non-equivalent unitary irreducible representation. This is because, as the reader will recall (Section 5-5(5)), two vectors

$$\mathbf{p} = x_1\mathbf{e}_1 + x_2\mathbf{e}_2 \ldots + x_n\mathbf{e}_n$$
$$\mathbf{p}' = x_1'\mathbf{e}_1 + x_2'\mathbf{e}_2 \ldots + x_n'\mathbf{e}_n$$

are said to be orthogonal if

$$(\mathbf{p},\mathbf{p}') = x_1x_1' + x_2x_2' \ldots + x_nx_n' = 0$$

or, if the vectors are complex, if

$$(\mathbf{p},\mathbf{p}') = x_1x_1'^* + x_2x_2'^* \ldots + x_nx_n'^* = 0.$$

Eqn (7-2.2) with $\mu \neq \nu$ or $i \neq j$ or $k \neq m$ is of the same form as this last equation.

Now, the maximum number of g-dimensional vectors which can be orthogonal is g. Consider, for example, a two-dimensional vector space containing the two orthogonal vectors:

$$\mathbf{p} = x_1\mathbf{e}_1 + x_2\mathbf{e}_2$$
$$\mathbf{p}' = x_1'\mathbf{e}_1 + x_2'\mathbf{e}_2$$

where

$$x_1 x_1'^* + x_2 x_2'^* = 0$$

or

$$x_1^* x_1' + x_2^* x_2' = 0. \tag{7-2.3}$$

If there were a third vector

$$\mathbf{p}'' = x_1''\mathbf{e}_1 + x_2''\mathbf{e}_2$$

orthogonal to $\mathbf{p}$ and $\mathbf{p}'$, then

$$x_1 x_1''^* + x_2 x_2''^* = 0$$

and

$$x_1' x_1''^* + x_2' x_2''^* = 0.$$

Solving these last two equations we get:

$$\left(\frac{x_1}{x_2} - \frac{x_1'}{x_2'}\right) x_1''^* = 0$$

and

$$\left(\frac{x_2}{x_1} - \frac{x_2'}{x_1'}\right) x_2''^* = 0$$

or, since by eqn (7-2.3)

$$\frac{x_2'}{x_1'} = -\frac{x_1^*}{x_2^*}$$

and

$$\frac{x_1'}{x_2'} = -\frac{x_2^*}{x_1^*},$$

we have

$$\left(\frac{x_1}{x_2} + \frac{x_2^*}{x_1^*}\right) x_1''^* = \left(\frac{x_1^2 + x_2^2}{x_1^* x_2}\right) x_1''^* = 0$$

and

$$\left(\frac{x_2}{x_1} + \frac{x_1^*}{x_2^*}\right) x_2''^* = \left(\frac{x_2^2 + x_1^2}{x_1 x_2^*}\right) x_2''^* = 0.$$

Since $x_1^2 + x_2^2 \neq 0$, we have $x_1''^* = x_2''^* = 0$ or $x_1'' = x_2'' = 0$ and hence a two dimensional vector orthogonal to $\mathbf{p}$ and $\mathbf{p}'$ cannot exist. By extension, this is generally true i.e. the maximum number of orthogonal g-dimensional vectors is g.

We therefore have the following result:

$$\sum_\mu n_\mu^2 \leqslant g, \tag{7-2.4}$$

that is: the sum over all of the non-equivalent irreducible representations of the square of the dimensions of the representations is less than or equal to the order of the group g. This result places restrictions on both the number and size of the irreducible representations of a group. In Appendix A.7-2 we show that, in fact, the equality holds i.e.

$$\sum_{\mu=1}^{r} n_\mu^2 = g,$$

where r is the number of non-equivalent irreducible representations.

7-3. Characters

The trace of a matrix is the sum of its diagonal elements:

$$\text{Trace}(A) = \sum_i A_{ii}.$$

The trace of a matrix which represents an element of a group (or an operation of a point group) is called a *character* and is usually given the symbol χ. $\chi(\boldsymbol{R})$ is thus the character of the operation $\boldsymbol{R}$ in the representation which has matrices $D(\boldsymbol{R})$, i.e.

$$\chi(\boldsymbol{R}) = \text{Trace}\{D(\boldsymbol{R})\},$$

and if we are considering the representation Γ^μ, then we write

$$\chi^\mu(\boldsymbol{R}) = \text{Trace}\{D^\mu(\boldsymbol{R})\}.$$

The complete set of characters for a given representation for the elements of a group is called the *character of the representation.* The characters of the two representations introduced in Table 6-3.1 are shown in Table 7-3.1.

We have already seen (eqn (4-5.4)) that the traces of two matrices which are related by a similarity transformation are identical. Since equivalent representations have matrices which are linked by a common similarity transformation, the characters of two equivalent

TABLE 7-3.1

Characters of the representations given in Table 6-3.1

$\boldsymbol{R}$	$\chi(\boldsymbol{R})$ (*real basis*)	$\chi(\boldsymbol{R})$† (*complex basis*)
$\boldsymbol{E}$	3	3
$\boldsymbol{C}_3$	0	0
$\boldsymbol{C}_3^2$	0	0
$\boldsymbol{\sigma}_v$	1	1
$\boldsymbol{\sigma}_v'$	1	1
$\boldsymbol{\sigma}_v''$	1	1

† In the construction of this set of characters, the fact that $\varepsilon+\varepsilon^* = -1$ has been used, $\varepsilon = \exp(2\pi i/3)$.

representations will be identical; this is borne out by Table 7-3.1. The reverse proposition: that if the characters of two representations are identical, then the representations are *necessarily* equivalent, will be proved in the next section.

If two operations P and Q of a point group are linked by a third operation X of the point group by the relation

$$P = X^{-1}QX, \tag{7-3.1}$$

then P and Q are conjugate to each other and are said to belong to the same *class* (see § 3-5). If $D(P)$, $D(Q)$, $D(X)$, and $D(X^{-1})$ are the matrices in some representation of the point group representing the operations P, Q, X, and X^{-1}, then necessarily these matrices must mirror eqn (7-3.1), i.e.:

$$D(P) = D(X^{-1})D(Q)D(X). \tag{7-3.2}$$

Since

$$X^{-1}X = E,$$

$$D(X^{-1})D(X) = D(E) = E$$

and

$$D(X^{-1}) = D(X)^{-1},$$

then:

$$D(P) = D(X)^{-1}D(Q)D(X)$$

and $D(P)$ and $D(Q)$ are related by a similarity transformation. From this we see that the characters of operations belonging to the same class are identical, i.e. if

$$P = X^{-1}QX$$

then

$$\chi(P) = \chi(Q).$$

This result, too, is borne out by Table 7-3.1, where, for the $\mathscr{C}_{3v}$ point group E forms one class, C_3 and C_3^2 form another, and σ_v', σ_v'', and σ_v''' form a third.

From eqn (7-2.1) (the Great Orthogonality Theorem) we can obtain for the non-equivalent irreducible representations Γ^μ and Γ^ν:

$$\sum_R D^\mu_{ii}(R)D^\nu_{jj}(R^{-1}) = (g/n_\mu)\delta_{\mu\nu}\delta_{ij}\delta_{ij} = (g/n_\mu)\delta_{\mu\nu}\delta_{ij}$$

and if we sum over i and j, we get:

$$\sum_R \sum_{i=1}^{n_\mu} D^\mu_{ii}(R) \sum_{j=1}^{n_\nu} D^\nu_{jj}(R^{-1}) = (g/n_\mu)\delta_{\mu\nu} \sum_{i=1}^{n_\mu} \sum_{j=1}^{n_\nu} \delta_{ij}$$

and

$$\sum_R \chi^\mu(R)\chi^\nu(R^{-1}) = (g/n_\mu)\delta_{\mu\nu}n_\mu = g\delta_{\mu\nu}. \tag{7-3.3}$$

If the irreducible representations Γ^μ and Γ^ν are chosen to be unitary (no loss of generality), then eqn (7-3.3) becomes

$$\sum_R \chi^\mu(R)\chi^\nu(R)^* = g\delta_{\mu\nu} \tag{7-3.4}$$

since

$$\chi^{\nu}(R^{-1}) = \sum_{i=1}^{n_\nu} D^{\nu}_{ii}(R^{-1}) = \sum_{i=1}^{n_\nu} D^{\nu}_{ii}(R)^* = \chi^{\nu}(R)^*$$

for unitary matrices. If the operations of the group are

$$R_1, R_2, \ldots R_g$$

we can interpret the characters

$$\chi^{\mu}(R_1), \chi^{\mu}(R_2), \ldots \chi^{\mu}(R_g)$$

as the components of a g-dimensional vector and eqn (7-3.4) then implies that such vectors for different non-equivalent unitary irreducible representations are orthogonal.

We will collectively symbolize the operations of the ith class of a point group $\mathscr{G}$ of order g by C_i and we will symbolize the number of operations in the ith class by g_i and the number of classes in the group by k, so that

$$\sum_{i=1}^{k} g_i = g.$$

For example, for the $\mathscr{C}_{3v}$ point group we will write:

$$\begin{aligned} C_1 &= E, \\ C_2 &= C_3 \text{ or } C_3^2, \\ C_3 &= \sigma_v' \text{ or } \sigma_v'' \text{ or } \sigma_v''', \\ g_1 &= 1, \\ g_2 &= 2, \\ g_3 &= 3, \\ g &= 6, \\ k &= 3. \end{aligned}$$

Since the characters of the operations of the same class are identical, we can write eqn (7-3.4) as:

$$\sum_{i=1}^{k} g_i \chi^{\mu}(C_i)\chi^{\nu}(C_i)^* = g\delta_{\mu\nu} \qquad (7\text{-}3.5)$$

where the sum now runs over the different classes and $\chi^{\mu}(C_i)$ is the character of any operation in the ith class in the Γ^{μ} irreducible representation.

Rearranging eqn (7-3.5) as

$$\sum_{i=1}^{k} \{g_i^{\frac{1}{2}}\chi^{\mu}(C_i)\}\{g_i^{\frac{1}{2}}\chi^{\nu}(C_i)\}^* = g\delta_{\mu\nu}, \qquad (7\text{-}3.6)$$

we see that the numbers

$$g_1^{\frac{1}{2}}\chi^\mu(C_1),\, g_2^{\frac{1}{2}}\chi^\mu(C_2), \ldots\, g_k^{\frac{1}{2}}\chi^\mu(C_k)$$

can be interpreted as the components of a k-dimensional vector and that similar vectors from the other irreducible representations will be orthogonal to it. Since the maximum number of orthogonal k-dimensional vectors is k, the maximum number of non-equivalent irreducible representations must also be k, so that if r is the total number of non-equivalent irreducible representations, then

$$r \leqslant k. \tag{7-3.7}$$

In fact we can show (Appendix A.7-3) that it is the equality which holds.

7-4. Number of times an irreducible representation occurs in a reducible one

Consider the reducible representation Γ^{red}: we can write (see eqn (6-5.5))

$$\Gamma^{\text{red}} = a_1\Gamma^1 \oplus a_2\Gamma^2 \oplus \ldots a_\nu\Gamma^\nu \oplus \ldots a_k\Gamma^k = \sum_{\nu=1}^{k} a_\nu\Gamma^\nu$$

where Γ^ν is the νth irreducible representation, a_ν is the number of times Γ^ν appears in Γ^{red} and k is the number of classes in the point group (the number of irreducible representations equals k, see Appendix A.7-3). The characters of the matrices belonging to Γ^{red}, $\chi^{\text{red}}(\boldsymbol{R})$, are the same as the characters of the matrices in their reduced form since only a similarity transformation has been carried out for each one. But the sum of the diagonal elements of the matrices in their reduced form is simply the sum of the diagonal elements of the irreducible matrices $\chi^\nu(\boldsymbol{R})$ which occur, multiplied by the number of times that they occur. Consequently, we have

$$\chi^{\text{red}}(\boldsymbol{R}) = \sum_{\nu=1}^{k} a_\nu\chi^\nu(\boldsymbol{R}) \tag{7-4.1}$$

for each $\boldsymbol{R}$, where $\chi^\nu(\boldsymbol{R})$ is the character of $\boldsymbol{R}$ in the Γ^ν irreducible representation.

If we multiply eqn (7-4.1) on both sides by $\chi^\mu(\boldsymbol{R})^*$, the conjugate complex of the character of $\boldsymbol{R}$ in the Γ^μ irreducible representation, and sum over all operations $\boldsymbol{R}$ of the point group, we obtain, by using eqn (7-3.4),

$$\sum_{\boldsymbol{R}} \chi^{\text{red}}(\boldsymbol{R})\chi^\mu(\boldsymbol{R})^* = \sum_{\boldsymbol{R}}\sum_{\nu=1}^{k} a_\nu\chi^\nu(\boldsymbol{R})\chi^\mu(\boldsymbol{R})^* = \sum_{\nu=1}^{k} a_\nu g\delta_{\mu\nu} = ga_\mu.$$

Therefore,

$$a_\mu = g^{-1} \sum_R \chi^{\text{red}}(R)\chi^\mu(R)^* = g^{-1} \sum_{i=1}^{k} g_i \chi^{\text{red}}(C_i)\chi^\mu(C_i)^* \qquad (7\text{-}4.2)$$

where C_i denotes any operation of the ith class and g_i is the number of operations in the ith class. Eqn (7-4.2) is an extremely useful formula, which can be easily applied (provided the appropriate characters are available), for determining the number of times a_μ that the Γ^μ irreducible representation occurs in the Γ^{red} reducible representation.

We are now in a position to show that two representations with a one-to-one correspondence in characters for each operation, are *necessarily* equivalent (see § 7-3). If we consider two different non-equivalent irreducible representations then, since the characters are orthogonal (eqn (7-3.4)), there cannot be a one-to-one correspondence. If we consider two different reducible representations Γ^a and Γ^b then, by eqn (7-4.2), if the characters are the same, the reduction will also be the same, that is the number of times Γ^μ occurs in Γ^a (a_μ) will, by the formula, be the same as the number of times Γ^μ occurs in Γ^b. The reduced matrices can therefore be brought to the same form by re-ordering the basis functions of either Γ^a or Γ^b. The reduced matrices are therefore equivalent and necessarily Γ^a and Γ^b from whence the reduced matrices came (via a similarity transformation) must also be equivalent. Hence, we have proved our proposition.

7-5. Criterion for irreducibility

From eqns (7-4.1) and (7-3.4) we can discover a simple condition for a representation to be irreducible. Consider the representation Γ^a, we can write its characters as

$$\chi^a(R) = \sum_{\mu=1}^{k} a_\mu \chi^\mu(R) \qquad (7\text{-}5.1)$$

and multiplying both sides by $\chi^a(R)^*$ and summing over all the operations R of the point group, we obtain:

$$\begin{aligned}
\sum_R \chi^a(R)\chi^a(R)^* &= \sum_R \left\{\sum_{\mu=1}^{k} a_\mu \chi^\mu(R)\right\}\left\{\sum_{\nu=1}^{k} a_\nu \chi^\nu(R)\right\}^* \\
&= \sum_{\mu=1}^{k}\sum_{\nu=1}^{k} a_\mu a_\nu \left\{\sum_R \chi^\mu(R)\chi^\nu(R)^*\right\} \\
&= \sum_{\mu=1}^{k}\sum_{\nu=1}^{k} a_\mu a_\nu g \delta_{\mu\nu} \\
&= g \sum_{\mu=1}^{k} a_\mu^2.
\end{aligned}$$

(Note that the numbers a_μ are necessarily real, i.e. $a_\mu = a_\mu^*$.)

Now, if Γ^a is irreducible, inspection of eqn (7-5.1) shows that all the a_μ are zero except for one which is unity, this one corresponding to the particular irreducible representation which is identical with Γ^a. So, if Γ^a is irreducible, the characters must satisfy:

$$\sum_R \chi^a(R)\chi^a(R)^* = g$$

or

$$\sum_{i=1}^{k} g_i\chi^a(C_i)\chi^a(C_i)^* = g \tag{7-5.2}$$

which gives a simple test of whether a representation is irreducible or not.

7-6. The reduction of a reducible representation

When we come to apply the results we have so far discovered to quantum mechanical situations, we will find that the application usually revolves around the reduction of some reducible representation for the point group concerned. We have already seen how to find out which irreducible representations appear in the reduction of a reducible representation, namely if we write

$$\Gamma^{\text{red}} = \sum_{\nu=1}^{k} a_\nu \Gamma^\nu,$$

then

$$a_\nu = g^{-1} \sum_R \chi^{\text{red}}(R)\chi^\nu(R)^* = g^{-1} \sum_{i=1}^{k} g_i\chi^{\text{red}}(C_i)\chi^\nu(C_i)^*.$$

Now we ask the parallel question—what is the new choice of basis functions for the function space (the one which produced Γ^{red}) which will produce matrices in their fully reduced form? Once again we are looking at the opposite side of the coin whose two faces are a similarity transformation and a change of basis functions. To answer the question we have posed, we will invoke the Great Orthogonality Theorem and carry out a certain amount of straightforward algebra.

Suppose that we have found k different function spaces for a given point group, where k is the number of classes or irreducible representations for the point group, and suppose that each function space provides the basis functions for one of the k irreducible representations. If the dimension of the νth irreducible representation is n_ν, there will be n_ν orthonormal basis functions describing the νth function space. We will write these sets of basis functions as

$$f_1^\nu, f_2^\nu, \ldots f_{n_\nu}^\nu \qquad \nu = 1, 2, \ldots k$$

where the superscript on f shows the function space to which it belongs. By definition, these functions must obey an equation of the same form as eqn (5-7.2), that is,

$$O_R f_q^\nu = \sum_{p=1}^{n_\nu} D_{pq}^\nu(R) f_p^\nu \qquad \begin{aligned} q &= 1, 2, \dots n_\nu \\ \nu &= 1, 2, \dots k \end{aligned} \tag{7-6.1}$$

and this equation must be satisfied for every operation R of the point group. Since we will choose the basis functions to be orthonormal, the matrices $D^\nu(R)$ will be unitary (see § 6-4).

Now let us multiply eqn (7-6.1) by $D_{ij}^\mu(R)^*$ and sum over all R. From eqn (7-2.2) we have

$$\begin{aligned} \sum_R D_{ij}^\mu(R)^* O_R f_q^\nu &= \sum_{p=1}^{n_\nu} \sum_R D_{ij}^\mu(R)^* D_{pq}^\nu(R) f_p^\nu \\ &= \sum_{p=1}^{n_\nu} (g/n_\mu)\delta_{\mu\nu}\delta_{ip}\delta_{jq} f_p^\nu \\ &= (g/n_\mu)\delta_{\mu\nu}\delta_{jq} f_i^\nu \end{aligned} \tag{7-6.2}$$

and, if we define a new operator P_{ij}^μ by

$$P_{ij}^\mu = \sum_R D_{ij}^\mu(R)^* O_R \tag{7-6.3}$$

then

$$P_{ij}^\mu f_q^\nu = (g/n_\mu)\delta_{\mu\nu}\delta_{jq} f_i^\nu. \tag{7-6.4}$$

P_{ij}^μ is an operator which is a definite linear combination of the transformation operators O_R with coefficients which are related to the matrices of Γ^μ; it is (for reasons which will be clear later) called a *projection operator*. If $\mu = \nu$ and $q = j$, eqn (7-6.4) becomes

$$P_{ij}^\nu f_j^\nu = (g/n_\nu) f_i^\nu \tag{7-6.5}$$

and Van Vleck has called this equation the basis function generating machine, since from one basis function f_j^ν the others can be generated.

Furthermore, we can create another projection operator P^μ by the equation

$$P^\mu = \sum_{i=1}^{n_\mu} P_{ii}^\mu = \sum_{i=1}^{n_\mu} \sum_R D_{ii}^\mu(R)^* O_R = \sum_R \chi^\mu(R)^* O_R \tag{7-6.6}$$

and hence, using eqn (7-6.4),

$$P^\mu f_q^\nu = \sum_{i=1}^{n_\mu} (g/n_\mu)\delta_{\mu\nu}\delta_{iq} f_i^\nu$$

or, if $\mu \neq \nu$,

$$P^\mu f_q^\nu = 0 \qquad \begin{aligned} q &= 1, 2, \dots n_\nu \\ \mu &= 1, 2, \dots k \\ \nu &= 1, 2, \dots k \end{aligned} \tag{7-6.7}$$

and, if $\mu = \nu$,

$$P^\mu f_q^\mu = (g/n_\mu) f_q^\mu \qquad \begin{aligned} q &= 1, 2, \dots n_\mu \\ \mu &= 1, 2, \dots k. \end{aligned} \tag{7-6.8}$$

Any function belonging to the function space which has been used to produce Γ^ν can necessarily be written as some linear combination of

$f_1^\nu, f_2^\nu, \ldots f_{n_\nu}^\nu$ and we will denote such a general function by f_{gen}^ν. Then from eqn (7-6.7) if $\mu \neq \nu$ we obtain,

$$P^\mu f_{\text{gen}}^\nu = P^\mu(\alpha_1 f_1^\nu + \alpha_2 f_2^\nu + \ldots \alpha_{n_\nu} f_{n_\nu}^\nu)$$

$$= 0 \qquad \begin{matrix} \mu = 1, 2, \ldots k \\ \nu = 1, 2, \ldots k \end{matrix} \qquad (7\text{-}6.9)$$

and from eqn (7-6.8), if $\mu = \nu$

$$P^\mu f_{\text{gen}}^\mu = (g/n_\mu) f_{\text{gen}}^\mu \qquad \mu = 1, 2, \ldots k. \qquad (7\text{-}6.10)$$

We now see why P^μ is called a projection operator; it annihilates any function which does not belong to the μth space and projects out (and multiplies by g/n_μ) any function which does.

Let us now consider the n-dimensional reducible representation Γ^{red} which is produced from the function space whose basis functions are $g_1, g_2, \ldots g_n$, and let us assume that in the reduction of Γ^{red} no irreducible representation of the point group occurs more than once. One way of looking at the reduction is to see it as a change of basis functions from $g_1, g_2, \ldots g_n$ to

$$f_1^1, f_2^1, \ldots f_{n_1}^1; \ldots ; f_1^\nu, f_2^\nu, \ldots f_{n_\nu}^\nu; \ldots ; f_1^k, f_2^k, \ldots f_{n_k}^k;$$

where k is the number of irreducible representations. From this it follows that it must be possible to express the g functions as linear combinations of all of the f functions:

$$g_s = \sum_{\nu=1}^{k} \sum_{i=1}^{n_\nu} c_{\nu i}^s f_i^\nu \qquad s = 1, 2, \ldots n$$

or as

$$g_s = \sum_{\nu=1}^{k} f_s^\nu \qquad s = 1, 2, \ldots n$$

where

$$f_s^\nu = \sum_{i=1}^{n_\nu} c_{\nu i}^s f_i^\nu \qquad s = 1, 2, \ldots n.$$

The functions f_s^ν ($s = 1, 2, \ldots n$) must be functions which belong to the space which produces Γ^ν, since they are simply linear combinations of the basis functions which define that space. If we choose one of the irreducible representations, say Γ^μ, and apply the corresponding projection operator P^μ to g_s, we obtain from eqns (7-6.9) and (7-6.10),

$$P^\mu g_s = \sum_{\nu=1}^{k} P^\mu f_s^\nu = \frac{g}{n_\mu} f_s^\mu \qquad s = 1, 2, \ldots n \qquad \mu = 1, 2, \ldots k. \qquad (7\text{-}6.11)$$

Since P^μ is a linear combination of the operators O_R, and $O_R g_s$ is a known linear combination of $g_1, g_2, \ldots g_n$, $P^\mu g_s$ must also be a linear combination of $g_1, g_2, \dots g_n$. So, from eqn (7-6.11) we can obtain a linear combination of g's which are proportional to functions f_s^μ which belong to Γ^μ and if we apply P^μ to each g function in turn, we get n linear combinations of g's which belong to Γ^μ and from which we can find n_μ which are linearly independent and, if we wish, orthonormal.

Hence we have a method of finding basis functions which belong to a given irreducible representation, if we are given some function space which produces a reducible representation. Notice that in addition to the O_R, the construction of P^μ (eqn (7-6.6)) requires only the knowledge of the characters of the Γ^μ representation.

If Γ^μ occurs, for example, twice in Γ^{red} then

$$g_s = f_s^1 + \dots f_s^\mu + f_s^{\mu'} \dots + f_s^k$$

and f_s^μ and $f_s^{\mu'}$ are both functions belonging to Γ^μ and are both linear combinations of $f_1^\mu, f_2^\mu, \dots, f_n^\mu$, they differ solely in the coefficients $c_{\mu i}^s$, i.e.

$$f_s^\mu = \sum_{i=1}^{n_\mu} c_{\mu i}^s f_i^\mu \qquad s = 1, 2, \dots n$$

and

$$f_s^{\mu'} = \sum_{i=1}^{n_\mu} c_{\mu i}^{s'} f_i^\mu \qquad s = 1, 2, \dots n.$$

Our method is not capable of separating these two functions; we will always get a combination:

$$P^\mu g_s = (g/n_\mu)(f_s^\mu + f_s^{\mu'}).$$

So that in a case like this we will obtain a mixture of two sets of functions each of which *alone* would be sufficient to define a function space leading to Γ^μ.

The usefulness of the results of this section will be exemplified by the problem in § 7-9.

7-7 Character tables and their construction

Since we will continually be requiring the characters of the irreducible representations of the point groups, it is convenient to put them together in tables known as *character tables*. In the character table of a point group each row refers to a particular irreducible representation and, since the characters of operations of the same class are identical, only a single entry $\chi^\mu(C_i)$ is made for all the operations of a given class. The columns are headed by a representative element from each class preceded by the number of elements or operations in that class g_i.

For example, the $\mathscr{C}_{3v}$ point group has three classes (and necessarily three irreducible representations) and its character table is shown in

TABLE 7-7.1

The character table for the $\mathscr{C}_{3v}$ point group†

$\mathscr{C}_{3v}$	E	$2C_3$	$3\sigma_v$
A_1	1	1	1
A_2	1	1	−1
E	2	−1	0

† The first column shows the labels (see §7-8) of the three non-equivalent irreducible representations Γ^{A_1}, Γ^{A_2} and Γ^{E}.

Table 7-7.1. The first row corresponds to Γ^1 in Table 6-5.1 and the last row to Γ^2 in that table. (We have not previously discussed the middle representation.) The headings of the columns in Table 7-7.1 are E, $2C_3$ and $3\sigma_v$ and they imply the identity operation E (one class), the two rotations C_3 and C_3^2 (another class), and the three reflections σ_v', σ_v'', and σ_v''' (a third class). The names of the three representations (A_1, A_2, and E) will be discussed later.

It is easy to check that the characters in Table 7-7.1 satisfy the orthogonality relationship (eqn (7-3.5)):

$$\sum_{i=1}^{k} g_i \chi^{\mu}(C_i)\chi^{\nu}(C_i)^* = g\delta_{\mu\nu},$$

for example, the characters of Γ^{A_1} are orthogonal to those of Γ^{A_2}:

$$(1\times 1\times 1)+(2\times 1\times 1)+(3\times 1\times -1) = 0,$$

those of Γ^{A_1} are orthogonal to those of Γ^{E}:

$$(1\times 1\times 2)+(2\times 1\times -1)+(3\times 1\times 0) = 0,$$

and those of Γ^{A_2} are orthogonal to those of Γ^{E}:

$$(1\times 1\times 2)+(2\times 1\times -1)+(3\times -1\times 0) = 0.$$

There also exists an orthogonality relationship between the columns of the character table:

$$\sum_{\nu=1}^{k} \chi^{\nu}(C_i)\chi^{\nu}(C_j)^* = (g/g_j)\delta_{ij} \qquad (7\text{-}7.1)$$

and the reader may confirm for himself that this equation too is satisfied. The proof of eqn (7-7.1) is part of the proof that the number of irreducible representations is equal to the number of classes of a point group (see eqn (A.7-3.10)).

Though the character tables for all the important point groups are readily available (see, for example, Appendix I at the end of this book), it makes a convenient summary of our results to see how the tables can normally be deduced without explicit knowledge of the matrices themselves. The following four rules can be used:

(1) The sum of the squares of the dimensions of the irreducible representations is equal to the order of the point group,

$$\sum_{\mu=1}^{k} n_\mu^2 = g,$$

(the proof of this is given in Appendix A.7-2). Since the identity operation is always represented by the unit or identity matrix, the first column of a character table is $\chi^\mu(E) = n_\mu$. Also we have

$$\sum_{\mu=1}^{k} \{\chi^\mu(E)\}^2 = g.$$

Since the matrices $\|1\|, \|1\|, \ldots$ form a one-dimensional totally symmetric irreducible representation of any point group, it is customary to put the corresponding characters in the first row and so $\chi^1(C_i) = 1$.

(2) The number of irreducible representations r is equal to the number of classes k; the proof of this is given in Appendix A.7-3.

(3) The rows must satisfy

$$\sum_{i=1}^{k} g_i \chi^\mu(C_i)\chi^\nu(C_i)^* = g\delta_{\mu\nu}.$$

(4) The columns must satisfy

$$\sum_{\nu=1}^{k} \chi^\nu(C_i)\chi^\nu(C_j)^* = (g/g_i)\delta_{ij}.$$

From these four rules it is easy, for example, to construct Table 7-7.1. There are three classes for $\mathscr{C}_{3v}$ and therefore three irreducible representations. The only three numbers whose squares add up to six (the order of the group) are 1, 1, and 2. We therefore immediately have:

$\mathscr{C}_{3v}$	E	$2C_3$	$3\sigma_v$
	1	1	1
	1	a	b
	2	c	d

and have only to determine a, b, c, and d. From rule (3) we have

$$1+2a+3b = 0$$
$$1+2a^2+3b^2 = 6$$

and

$$2+2c+3d = 0$$
$$4+2c^2+3d^2 = 6$$

hence $a = 1$, $b = -1$, $c = -1$, and $d = 0$.

There are several general methods for calculating the characters of the irreducible representations which are more systematic than the method we have given. Their drawback, however, is that they involve long and complex calculations and are only feasible when use is made of high speed computers (see, for example, John D. Dixon, *Numerische Mathematik* **10,** 446 (1967)). Furthermore, Esko Blokker has described a theory for the construction of the irreducible representations of the finite groups from their characters and though complicated, it can be conveniently programmed for a computer (see, *International journal of quantum chemistry* VI, 925, (1972)). The reader who is interested in the part that computers can play in group theory is recommended to read the article by J. J. Cannon in the *Communications of the association for computing machinery* **12,** 3 (1969).

7-8. Notation for irreducible representations

The symbols formulated by R. S. Mulliken are used to distinguish the irreducible representations of the various point groups. In this section we will outline the general points of the notation and the reader is referred to Mulliken's report† for the details.

One-dimensional irreducible representations are labeled either A or B according to whether the character of a $2\pi/n$ (proper or improper) rotation about the symmetry axis of highest order n is $+1$ or -1, respectively. For the point groups $\mathscr{C}_1$, $\mathscr{C}_s$, and $\mathscr{C}_i$ which have no symmetry axis, all one-dimensional representations are labeled A. For $\mathscr{D}_2$ and $\mathscr{D}_{2h}$ there are three C_2 axes and the three C_2 operations fall in different classes; those one-dimensional representations for which the

† This report was published in *The Journal of chemical physics*, 23, 1997 (1955). The reader should note that on page 2003 of this report the third line below Table VI should read: for $\mathscr{D}_{4h}$, $\sigma_v = iC_2$, $\sigma_d = iC_2'$; for $\mathscr{D}_{6h}$, $\sigma_v = iC_2'$, $\sigma_d = iC_2$. Also, in the diagram for $\mathscr{D}_{4h}$ in Fig. 1, the σ_v and σ_d planes should be interchanged. When these corrections are made, the definitions are the same as those in Fig. 3-6.1 of this book, with the proviso that our C_2' axis is Mulliken's C_2 axis and our C_2'' axis is Mulliken's C_2' axis.

Further information on notation is contained in G. Herzberg's *Molecular spectra and molecular structure*, vol. II, Van Nostrand Reinhold.

characters of all three C_2 operations are $+1$ are labeled A, while the other one-dimensional representations are labeled B. For $\mathscr{D}_{nd}$, the character of S_{2n} determines the label of the one-dimensional representations.

Two-dimensional irreducible representations are labeled E, which should not be confused with the identity element or the identity matrix.

Three-dimensional irreducible representations can be labeled either T or F; usually T is used in electronic problems and F in vibrational problems.

If a point group contains the operation of inversion, a subscript g (from the German word *gerade*) or u (from the German word *ungerade*) is added to the label according to whether the character of i is positive or negative respectively. The inversion operation is always represented by $+1$ or -1 times the identity matrix; hence the character is either $+n_\mu$ or $-n_\mu$ where n_μ is the dimension of the representation. Point groups which contain i are $\mathscr{C}_{nh}$ (n even), $\mathscr{D}_{nh}$ (n even), $\mathscr{D}_{nd}$ (n odd), $\mathscr{O}_h$ and $\mathscr{D}_{\infty h}$ and these point groups are often written as $\mathscr{C}_n \otimes \mathscr{C}_i$ (n even), $\mathscr{D}_n \otimes \mathscr{C}_i$ (n even), $\mathscr{D}_n \otimes \mathscr{C}_i$ (n odd), $\mathscr{O} \otimes \mathscr{C}_i$ and $\mathscr{C}_{\infty v} \otimes \mathscr{C}_i$ respectively (i.e. $\mathscr{G} \otimes \mathscr{C}_i$),† since they contain all the operations (R) of $\mathscr{G}$ plus all those one can obtain by combining each R with i (i.e. Ri). There are twice as many classes in $\mathscr{G} \otimes \mathscr{C}_i$ as in $\mathscr{G}$ and therefore twice as many irreducible representations. Thus for each irreducible representation of $\mathscr{G}$ there will be represented by $+1$ or -1 times an identity matrix and that there are twice as many irreducible representations in $\mathscr{G} \otimes \mathscr{C}_s$ as in $\mathscr{G}$. For $\Gamma^{\mu'}$

$$\chi^{\mu_g}(R) = \chi^{\mu}(R),$$
$$\chi^{\mu_g}(iR) = \chi^{\mu}(R)$$

and for Γ^{μ_u}

$$\chi^{\mu_u}(R) = \chi^{\mu}(R),$$
$$\chi^{\mu_u}(iR) = -\chi^{\mu}(R)$$

(for a proof of these equations, see Appendix C of Schonland's book *Molecular symmetry*).

If the point group has a σ_h operation but no i operation (groups $\mathscr{C}_{nh}$ and $\mathscr{D}_{nh}$ with n odd) the labels are primed or double primed according to whether the character of σ_h is positive or negative, respectively. The situation is similar to the one in the previous paragraph in that σ_h will be represented by $+1$ or -1 times an identity matrix and that there are twice as many irreducible representations in $\mathscr{G} \otimes \mathscr{C}_s$ as in $\mathscr{G}$. For $\Gamma^{\mu'}$

$$\chi^{\mu'}(R) = \chi^{\mu}(R),$$
$$\chi^{\mu'}(\sigma_h R) = \chi^{\mu}(R)$$

† This notation is referred to again at the end of § 8-3.

and for $\Gamma^{\mu''}$

$$\chi^{\mu''}(R) = \chi^{\mu}(R),$$
$$\chi^{\mu''}(\sigma_h R) = -\chi^{\mu}(R).$$

If one can write a point group either as $\mathscr{G}\otimes\mathscr{C}_i$ or $\mathscr{G}\otimes\mathscr{C}_s$ (e.g. $\mathscr{D}_{4h}$) the former takes precedence.

If necessary, numerical subscripts are added to the labels to distinguish the non-equivalent irreducible representations which are not distinguished by the foregoing rules. Except for the fact that the totally symmetric representation (one-dimensional unit matrices) is numbered and listed first, the numbering is arbitrary and the reader is referred to Appendix I or Mulliken's report for the internationally accepted conventions.

If a one-dimensional representation has complex characters $a, b, c, \ldots$ then there must be another equally acceptable representation with the characters: $a^*, b^*, c^*, \ldots$ since for a one-dimensional representation the character of an operation equals the single matrix element representing the operation. These pairs are usually bracketed together and labeled E. In fact quite often the reduction which produces the pair of irreducible representations is not carried out, since no useful information is gained by it and anyway the two always occur together.

The two infinite point groups $\mathscr{C}_{\infty v}$ and $\mathscr{D}_{\infty h}$ ($= \mathscr{C}_{\infty v}\otimes\mathscr{C}_i$) have their own notation. Because these groups have an infinite number of elements the theorems we have given do not apply and other more elaborate methods are needed to find their irreducible representations. In $\mathscr{C}_{\infty v}$ the pairs of rotations $C(\phi)$, $C(-\phi)$ through equal and opposite angles, belong to a class of two operations, one class for each ϕ value. All the reflections σ_v belong to one class. The point group has two one-dimensional irreducible representations and an infinite number of two-dimensional ones. In Mulliken's notation these would be labeled $A_1, A_2, E_1, E_2, \ldots$ but a different notation, using Greek letters and which was developed in early spectroscopic work, is usually employed. The character table for $\mathscr{C}_{\infty v}$ is shown in Table 7-8.1.

TABLE 7-8.1

Character table for $\mathscr{C}_{\infty v}$

$\mathscr{C}_{\infty v}$	E	$2C(\phi)$	$\infty\sigma_v$
$A_1 = \Sigma^+$	1	1	1
$A_2 = \Sigma^-$	1	1	-1
$E_1 = \Pi$	2	$2\cos\phi$	0
$E_2 = \Delta$	2	$2\cos 2\phi$	0
$E_3 = \Phi$	2	$2\cos 3\phi$	0
$\cdots$	.	.	.

7-9. An example of the determination of the irreducible representations to which certain functions belong

In order to demonstrate some of the conclusions of this chapter, we pose the question: for the $\mathscr{D}_{4h}$ point group, for which irreducible representations do three real p-orbitals and five real d-orbitals or their combinations form a basis of representation?

In Fig. 7-9.1 the symmetry elements for the $\mathscr{D}_{4h}$ point group are shown (see also Fig. 3-6.1) as well as our choice of x, y and z axes (this choice establishes the orientation of the p- and d-orbitals). In Table 7-9.1 the corresponding character table is given in full.

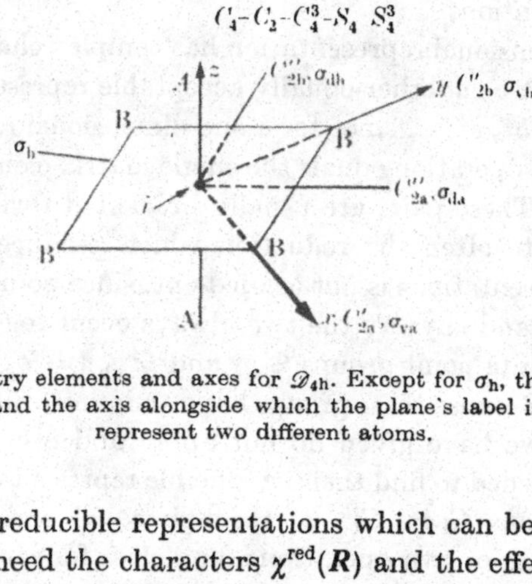

FIG. 7-9.1. Symmetry elements and axes for $\mathscr{D}_{4h}$. Except for σ_h, the symmetry planes contain the z axis and the axis alongside which the plane's label is written. A and B represent two different atoms.

To find the irreducible representations which can be produced from the orbitals we need the characters $\chi^{red}(\boldsymbol{R})$ and the effect of $\boldsymbol{O_R}$ on the orbitals. Taking the last point first, we find $\boldsymbol{O_R}g$ for all $\boldsymbol{R}$ of $\mathscr{D}_{4h}$ and $g = p_1, p_2, p_3, d_1, d_2, d_3, d_4, d_5$ and the results are given in Table 7-9.2. We have carried out this kind of step before (see § 5-9) and for this particular point group and axis choice, the process is particularly simple, for example we have

$$p_1 \propto x$$
$$p_2 \propto y$$
$$p_3 \propto z$$
$$d_1 \propto x^2-y^2$$
$$d_2 \propto xy$$
$$d_3 \propto xz$$
$$d_4 \propto yz$$
$$d_5 \propto 3z^2-r^2$$

TABLE 7-9.1

Character table for $\mathscr{D}_{4h}$†

$\mathscr{D}_{4h}$	E	C_4	C_4^3	C_2	C'_{2a}	C'_{2b}	C''_{2a}	C''_{2b}	i	S_4	S_4^3	σ_h	σ_{va}	σ_{vb}	σ_{da}	σ_{db}
A_{1g}	1	1	1	1	1	1	1	1	1	1	1	1	1	1	1	1
A_{2g}	1	1	1	1	−1	−1	−1	−1	1	1	1	1	−1	−1	−1	−1
B_{1g}	1	−1	−1	1	1	1	−1	−1	1	−1	−1	1	1	1	−1	−1
B_{2g}	1	−1	−1	1	−1	−1	1	1	1	−1	−1	1	−1	−1	1	1
E_g	2	0	0	−2	0	0	0	0	2	0	0	−2	0	0	0	0
A_{1u}	1	1	1	1	1	1	1	1	−1	−1	−1	−1	−1	−1	−1	1
A_{2u}	1	1	1	1	−1	−1	−1	−1	−1	−1	1	−1	1	1	1	1
B_{1u}	1	−1	−1	1	1	1	−1	−1	−1	1	1	−1	−1	1	1	1
B_{2u}	1	−1	−1	1	−1	−1	1	1	−1	1	1	−1	1	1	−1	−1
E_u	2	0	0	−2	0	0	0	0	−2	0	0	2	0	0	0	0

† For convenience, the characters are given for each operation rather than, as is customary, for each class.

TABLE 7-9.2

The functions $O_R g$ for the transformation operations of $\mathscr{D}_{4h}$

g	E	C_4	C_4^3	C_2	C'_{2a}	C'_{2b}	C''_{2a}	C''_{2b}	i	S_4	S_4^3	σ_h	σ_{va}	σ_{vb}	σ_{da}	σ_{db}
p_1	p_1	$-p_2$	p_2	$-p_1$	p_1	$-p_1$	p_2	$-p_2$	$-p_1$	$-p_2$	p_2	p_1	p_1	$-p_1$	p_2	$-p_2$
p_2	p_2	p_1	$-p_1$	$-p_2$	$-p_2$	p_2	p_1	$-p_1$	$-p_2$	p_1	$-p_1$	p_2	$-p_2$	p_2	p_1	$-p_1$
p_3	p_3	p_3	p_3	p_3	$-p_3$	$-p_3$	$-p_3$	$-p_3$	$-p_3$	$-p_3$	$-p_3$	$-p_3$	p_3	p_3	p_3	p_3
d_1	d_1	$-d_1$	$-d_1$	d_1	d_1	d_1	$-d_1$	$-d_1$	d_1	$-d_1$	$-d_1$	d_1	d_1	d_1	$-d_1$	$-d_1$
d_2	d_2	$-d_2$	$-d_2$	d_2	$-d_2$	$-d_2$	d_2	d_2	d_2	$-d_2$	$-d_2$	d_2	$-d_2$	$-d_2$	d_2	d_2
d_3	d_3	$-d_4$	d_4	$-d_3$	$-d_3$	d_3	$-d_4$	d_4	d_3	d_4	$-d_4$	$-d_3$	d_3	$-d_3$	d_4	$-d_4$
d_4	d_4	d_3	$-d_3$	$-d_4$	d_4	$-d_4$	$-d_3$	d_3	d_4	$-d_3$	d_3	$-d_4$	$-d_4$	d_4	d_3	$-d_3$
d_5	d_5	d_5	d_5	d_5	d_5	d_5	d_5	d_5	d_5	d_5	d_5	d_5	d_5	d_5	d_5	d_5

and if we consider O_{C_4}, under the C_4 operation we have (see eqn (5-2.5))

$$\begin{Vmatrix} x' \\ y' \\ z' \end{Vmatrix} = \begin{Vmatrix} 0 & 1 & 0 \\ -1 & 0 & 0 \\ 0 & 0 & 1 \end{Vmatrix} \begin{Vmatrix} x \\ y \\ z \end{Vmatrix} \quad \text{or} \quad \begin{Vmatrix} x \\ y \\ z \end{Vmatrix} = \begin{Vmatrix} 0 & -1 & 0 \\ 1 & 0 & 0 \\ 0 & 0 & 1 \end{Vmatrix} \begin{Vmatrix} x' \\ y' \\ z' \end{Vmatrix}$$

and thus

$$\begin{aligned} O_{C_4}\mathrm{p}_1(x', y', z') &= \mathrm{p}_1(x, y, z) \qquad \text{(from the definition of } O_R) \\ &\propto x \\ &\propto -y' \\ &= -\mathrm{p}_2(x', y', z') \end{aligned}$$

or, since the coordinates are now the same on both sides of the equation,

$$O_{C_4}\mathrm{p}_1 = -\mathrm{p}_2.$$

Also, using a slightly different but nonetheless straight-forward way, we have

$$\begin{aligned} O_{C_4}\mathrm{d}_1(x', y', z') &\propto O_{C_4}(x'^2 - y'^2) \\ &\propto O_{C_4}\mathrm{p}_1(x', y', z')O_{C_4}\mathrm{p}_1(x', y', z') \\ &\quad -O_{C_4}\mathrm{p}_2(x', y', z')O_{C_4}\mathrm{p}_2(x', y', z') \\ &\propto \{-\mathrm{p}_2(x', y', z')\}^2 - \{\mathrm{p}_1(x', y', z')\}^2 \\ &= -\mathrm{d}_1(x', y', z') \end{aligned}$$

or

$$O_{C_4}\mathrm{d}_1 = -\mathrm{d}_1.$$

From Table 7-9.2 and using eqn (5-7.2) we can find the diagonal elements of the matrices which represent the $\mathscr{D}_{4h}$ point group in the p-orbital basis and in the d-orbital basis. From these elements we get the characters of two reducible representations; they are shown in Table 7-9.3. By applying eqn (7-4.2)

$$a_\mu = g^{-1} \sum_R \chi^{\text{red}}(R)\chi^\mu(R)^*$$

we have

$$\Gamma^{\text{red}}\ (\text{p-basis}) = \Gamma^{A_{2u}} \oplus \Gamma^{E_u}$$

and

$$\Gamma^{\text{red}}\ (\text{d-basis}) = \Gamma^{A_{1g}} \oplus \Gamma^{B_{1g}} \oplus \Gamma^{B_{2g}} \oplus \Gamma^{E_g}.$$

So that we know that there are p-orbitals or combinations of p-orbitals which form a basis for the irreducible representations $\Gamma^{A_{2u}}$ and Γ^{E_u} and d-orbitals or combinations of d-orbitals which form a basis for the irreducible representations $\Gamma^{A_{1g}}$, $\Gamma^{B_{1g}}$, $\Gamma^{B_{2g}}$ and Γ^{E_g}.

TABLE 7-9.3

The characters of the reducible representations using p-orbitals and d-orbitals

	E	C_4	C_4^3	C_2	C_{2a}'	C_{2b}'	C_{2a}''	C_{2b}''	i	S_4	S_4^3	σ_h	σ_{va}	σ_{vb}	σ_{da}	σ_{db}
p-basis	3	1	1	−1	−1	−1	−1	−1	−3	−1	−1	1	1	1	1	1
d-basis	5	−1	−1	1	1	1	1	1	5	−1	−1	1	1	1	1	1

To find out which orbitals or combinations of orbitals produce which representations, we make use of the projection operator P^{μ} defined in eqn (7-6.6) as

$$P^{\mu} = \sum_{R} \chi^{\mu}(R)^{*} O_{R}.$$

By applying this operator to each of the orbitals in turn we project out functions belonging to Γ^{μ}. In Table 7-9.4 we collect together the results

TABLE 7-9.4

The results of applying P^{μ} for each irreducible representation of $\mathscr{D}_{4h}$ to the three p- and five d-orbitals

μ	p_1	p_2	p_3	d_1	d_2	d_3	d_4	d_5
A_{1g}	0	0	0	0	0	0	0	$16d_5$
A_{2g}	0	0	0	0	0	0	0	0
B_{1g}	0	0	0	$16d_1$	0	0	0	0
B_{2g}	0	0	0	0	$16d_2$	0	0	0
E_g	0	0	0	0	0	$8d_3$	$8d_4$	0
A_{1u}	0	0	0	0	0	0	0	0
A_{2u}	0	0	$16p_3$	0	0	0	0	0
B_{1u}	0	0	0	0	0	0	0	0
B_{2u}	0	0	0	0	0	0	0	0
E_u	$8p_1$	$8p_2$	0	0	0	0	0	0

of applying the ten projection operators for the $\mathscr{D}_{4h}$ point group to the eight functions. These results are found by simply combining Tables 7-9.1 and 7-9.2. Recalling that (eqn (7-6.11))

$$P^{\mu}g_s = (g/n_{\mu})f_s^{\mu},$$

we see that p_1 and p_2 (or p_x and p_y) belong to Γ^{E_u} and form a basis for that two-dimensional irreducible representation; p_3 (or p_z) belongs to $\Gamma^{A_{2u}}$; d_1 to $\Gamma^{B_{1g}}$; d_2 to $\Gamma^{B_{2g}}$; d_3 and d_4 form a basis for the two-dimensional representation Γ^{E_g}; and d_5 belongs to $\Gamma^{A_{1g}}$.

Appendices

A.7-1. The Great Orthogonality Theorem

To prove this theorem we first have to prove two theorems (sometimes called Schur's lemmas) concerning those matrices which commute with all the matrices of an irreducible representation. We will call these theorems Theorem I and Theorem II.

(1) Theorem I

This theorem states that the only matrix which commutes (see § 4-3) with all the matrices of an irreducible representation is a constant matrix. We have to show therefore that if

$$AD(R) = D(R)A \qquad \text{for all } R \qquad \text{(A.7-1.1)}$$

and the $D(\boldsymbol{R})$ are irreducible, *then* $A = \lambda E$, where λ is a constant, i.e.

$$A = \begin{Vmatrix} \lambda & 0 & 0 & . \\ 0 & \lambda & 0 & . \\ 0 & 0 & \lambda & . \\ . & . & . & . \end{Vmatrix}.$$

First we diagonalize the commuting matrix A by a matrix X

$$Z = X^{-1}AX$$

and define the new matrices

$$D'(\boldsymbol{R}) = X^{-1}D(\boldsymbol{R})X \qquad (\text{all } \boldsymbol{R}).$$

Then

$$\begin{aligned} D'(\boldsymbol{R})Z - ZD'(\boldsymbol{R}) &= X^{-1}D(\boldsymbol{R})XX^{-1}AX - X^{-1}AXX^{-1}D(\boldsymbol{R})X \\ &= X^{-1}D(\boldsymbol{R})AX - X^{-1}AD(\boldsymbol{R})X \\ &= X^{-1}[D(\boldsymbol{R})A - AD(\boldsymbol{R})]X. \end{aligned}$$

Hence, if $D(\boldsymbol{R})A = AD(\boldsymbol{R})$,

$$D'(\boldsymbol{R})Z - ZD'(\boldsymbol{R}) = 0$$

and Z and $D'(\boldsymbol{R})$ commute:

$$D'(\boldsymbol{R})Z = ZD'(\boldsymbol{R})$$

and comparing matrix elements on both sides of this last equation, we have:

$$\sum_{k=1}^{n} D'_{ik}(\boldsymbol{R})Z_{kj} = \sum_{k=1}^{n} Z_{ik}D'_{kj}(\boldsymbol{R}) \qquad \begin{matrix} i = 1, 2, \ldots, n \\ j = 1, 2, \ldots, n \end{matrix}$$

where n is the dimension of the representation. But, since Z is diagonal,

$$Z_{kj} = 0 \quad \text{for} \quad k \neq j$$

and

$$Z_{ik} = 0 \quad \text{for} \quad k \neq i,$$

therefore,

$$D'_{ij}(\boldsymbol{R})Z_{jj} = Z_{ii}D'_{ij}(\boldsymbol{R})$$

or

$$D'_{ij}(\boldsymbol{R})(Z_{jj} - Z_{ii}) = 0 \qquad \begin{matrix} i = 1, 2, \ldots, n \\ j = 1, 2, \ldots, n. \end{matrix} \qquad \text{(A.7-1.2)}$$

Consider one specific diagonal element of Z, say the first Z_{11}: if it is different from all the others, then eqn (A.7-1.2) shows that if $i = 1$

$$D'_{1j}(\boldsymbol{R}) = 0 \qquad j = 2, 3, \ldots n$$

and that if $j = 1$

$$D'_{i1}(\boldsymbol{R}) = 0 \qquad i = 2, 3, \ldots n$$

so that the first row and column except the diagonal element are zero and hence the $D'(\boldsymbol{R})$ are in block form and all the $D(\boldsymbol{R})$ have been reduced by the transformation $X^{-1}D(\boldsymbol{R})X$. But we stated that the $D(\boldsymbol{R})$ are irreducible and therefore Z_{11} cannot be different from the other diagonal elements. By extension we find, therefore, that all the diagonal elements of the diagonal matrix Z are the same:

$$Z = \begin{Vmatrix} \lambda & 0 & 0 & . \\ 0 & \lambda & 0 & . \\ 0 & 0 & \lambda & . \\ . & . & . & . \end{Vmatrix} = \lambda E$$

where λ = constant, and hence

$$A = X(\lambda E)X^{-1} = \lambda E.$$

(2) *Theorem II*

This theorem states that if for some group there are two ***different*** irreducible representations Γ^{μ} and Γ^{ν} with matrices $D^{\mu}(\boldsymbol{R})$ and $D^{\nu}(\boldsymbol{R})$ of dimension n_{μ} and n_{ν} respectively and if a rectangular matrix A exists such that

$$AD^{\nu}(\boldsymbol{R}) = D^{\mu}(\boldsymbol{R})A \qquad \text{for all } \boldsymbol{R},$$

then

Case (a) if $n_{\mu} = n_{\nu}$, either $\det(A) \neq 0$ and the two representations are equivalent or else $A = 0$ (the null matrix),

Case (b) if $n_{\mu} \neq n_{\nu}$, $A = 0$.

Without loss of generality we can assume that the $D^{\mu}(\boldsymbol{R})$ and $D^{\nu}(\boldsymbol{R})$ are unitary and $n_{\nu} \leqslant n_{\mu}$. Then:

$$AD^{\nu}(\boldsymbol{R}) = D^{\mu}(\boldsymbol{R})A \qquad \text{for all } \boldsymbol{R} \tag{A.7-1.3}$$

and taking adjoints (see eqn (4-3.15))

$$D^{\nu}(\boldsymbol{R})^{\dagger}A^{\dagger} = A^{\dagger}D^{\mu}(\boldsymbol{R})^{\dagger}$$

and since $D^{\mu}(\boldsymbol{R})$ and $D^{\nu}(\boldsymbol{R})$ are unitary

$$D^{\nu}(\boldsymbol{R}^{-1})A^{\dagger} = A^{\dagger}D^{\mu}(\boldsymbol{R}^{-1}) \qquad \text{(see eqn. (6-5.2))}$$

and multiplying both sides of this equation by A, we get

$$AD^{\nu}(\boldsymbol{R}^{-1})A^{\dagger} = AA^{\dagger}D^{\mu}(\boldsymbol{R}^{-1}). \tag{A.7-1.4}$$

Since the inverses $\boldsymbol{R}^{-1}$ are all, by definition, operations of the group, eqn (A.7-1.3) implies

$$AD^{\nu}(\boldsymbol{R}^{-1}) = D^{\mu}(\boldsymbol{R}^{-1})A$$

and multiplying by $A^{\dagger}$ we get:

$$AD^{\nu}(\boldsymbol{R}^{-1})A^{\dagger} = D^{\mu}(\boldsymbol{R}^{-1})AA^{\dagger}. \tag{A.7-1.5}$$

Comparing eqns (A.7-1.4) and (A.7-1.5) we obtain

$$AA^{\dagger}D^{\mu}(\boldsymbol{R}^{-1}) = D^{\mu}(\boldsymbol{R}^{-1})AA^{\dagger}.$$

Thus, by Theorem I,

$$AA^{\dagger} = \lambda E \tag{A.7-1.6}$$

where λ is a constant.

Case (*a*). If $n_\mu = n_\nu$, then A and $AA^\dagger$ are square matrices and the determinant of $AA^\dagger$ is

$$\det(AA^\dagger) = \det(\lambda E)$$
$$\det(A)\det(A^\dagger) = \lambda^{n_\mu} \qquad \text{(see Appendix A.4-6)}$$
$$\{\det(A)\}^2 = \lambda^{n_\mu}.$$

If $\lambda \neq 0$, then $\det(A) \neq 0$ and A has an inverse and we can write

$$D^\nu(R) = A^{-1}D^\mu(R)A \qquad \text{(all } R\text{)}$$

and the two representations are equivalent. If $\lambda = 0$, then $AA^\dagger = 0$ and

$$\sum_j A_{ij}A^*_{kj} = 0 \qquad \begin{matrix} i = 1, 2, \ldots n_\mu \\ k = 1, 2, \ldots n_\mu \end{matrix}$$

if $i = k$, this becomes

$$\sum_j A^2_{ij} = 0 \qquad i = 1, 2, \ldots n_\mu$$

which is only possible if all $A_{ij} = 0$, hence $A = 0$.

Case (*b*). If $n_\nu < n_\mu$, then A has n_ν columns and n_μ rows and we can fill A out to a $n_\mu \times n_\mu$ square matrix B by adding $(n_\mu - n_\nu)$ columns of zeros. Clearly,

$$BB^\dagger = AA^\dagger$$

and since

$$\det(BB^\dagger) = \{\det(B)\}^2 = 0$$

so $\det(AA^\dagger) = 0$, but by eqn (A.7-1.6)

$$\det(AA^\dagger) = \det(\lambda E) = \lambda^{n_\mu}$$

hence $\lambda = 0$ and, recalling Case (a), $A = 0$.

(3) *Proof of the Great Orthogonality Theorem*

This theorem states that

$$\sum_R D^\mu_{ik}(R)D^\nu_{mj}(R^{-1}) = (g/n_\mu)\,\delta_{\mu\nu}\,\delta_{ij}\,\delta_{km}$$

where $D^\mu(R)$ and $D^\nu(R)$ are the matrices for two non-equivalent irreducible representations of dimension n_μ and n_ν for the group $\mathscr{G}$ of order g. There are two parts to this proof.

(i) Take the irreducible representation with matrices $D(S)$ of dimensions $n \times n$, where S runs over the operations of the group $\mathscr{G}$ and construct the matrix

$$A = \sum_S D(S)XD(S^{-1}) \qquad \text{(A.7-1.7)}$$

where X is any arbitrary matrix and the summation is over all the operations of the group. We can show that

$$AD(R) = D(R)A \qquad \text{(all } R\text{)}$$

and hence $A = \lambda E$ (see Theorem I). We do this through the following steps

$$\begin{aligned} D(\boldsymbol{R})A &= D(\boldsymbol{R}) \sum_{\boldsymbol{S}} D(\boldsymbol{S})XD(\boldsymbol{S}^{-1}) \\ &= \sum_{\boldsymbol{S}} D(\boldsymbol{R})D(\boldsymbol{S})XD(\boldsymbol{S}^{-1})D(\boldsymbol{R}^{-1})D(\boldsymbol{R})\ddagger \\ &= \left[\sum_{\boldsymbol{S}} D(\boldsymbol{RS})XD\{(\boldsymbol{RS})^{-1}\}\right]D(\boldsymbol{R}) \\ &= \left[\sum_{\boldsymbol{T}} D(\boldsymbol{T})XD(\boldsymbol{T}^{-1})\right]D(\boldsymbol{R}) \\ &= AD(\boldsymbol{R}). \end{aligned}$$

The penultimate step is true since as $\boldsymbol{S}$ runs through the operations of the group so, by the Rearrangement Theorem (Appendix A.3-1), does $\boldsymbol{RS}(= \boldsymbol{T})$. Replacing $\boldsymbol{T}$ by $\boldsymbol{S}$ gives the final step.

Therefore $A = \lambda E$, the value of λ depending on the choice of X. Let us choose X to have all zero elements except in the kth row and mth column and let this exception be $X_{km} = 1$. The value of λ under these circumstances, we will symbolize as λ_{km}.

From eqn (A.7-1.7) and using the rule for matrix multiplication, we can obtain

$$A_{ij} = \sum_{\boldsymbol{S}} \sum_{p=1}^{n} \sum_{q=1}^{n} D_{ip}(\boldsymbol{S})X_{pq}D_{qj}(\boldsymbol{S}^{-1}) = \lambda_{km}\,\delta_{ij} \qquad \begin{aligned} i &= 1, 2, \ldots n \\ j &= 1, 2, \ldots n. \end{aligned}$$

Or, since $X_{pq} = 0$ unless $p = k$ and $q = m$ and $X_{km} = 1$,

$$\sum_{\boldsymbol{S}} D_{ik}(\boldsymbol{S})D_{mj}(\boldsymbol{S}^{-1}) = \lambda_{km}\,\delta_{ij} \qquad \begin{aligned} i &= 1, 2, \ldots n \\ j &= 1, 2, \ldots n. \end{aligned} \qquad \text{(A.7-1.8)}$$

Setting $i = j$ and summing over i, we obtain

$$\sum_{\boldsymbol{S}} \sum_{i=1}^{n} D_{ik}(\boldsymbol{S})D_{mi}(\boldsymbol{S}^{-1}) = n\lambda_{km}$$

or,

$$\begin{aligned} n\lambda_{km} &= \sum_{\boldsymbol{S}} \left\{\sum_{i=1}^{n} D_{mi}(\boldsymbol{S}^{-1})D_{ik}(\boldsymbol{S})\right\} \\ &= \sum_{\boldsymbol{S}} D_{mk}(\boldsymbol{S}^{-1}\boldsymbol{S}) \\ &= \sum_{\boldsymbol{S}} D_{mk}(\boldsymbol{E}) \\ &= \sum_{\boldsymbol{S}} \delta_{mk} \\ &= g\,\delta_{mk} \qquad (g = \text{number of elements in } \mathscr{G}) \end{aligned}$$

and hence

$$\lambda_{km} = (g/n)\,\delta_{mk} = (g/n)\,\delta_{km}$$

‡ Note that $D(\boldsymbol{R}^{-1})D(\boldsymbol{R}) = D(\boldsymbol{R}^{-1}\boldsymbol{R}) = D(\boldsymbol{E}) = E$.

and eqn (A.7-1.8) becomes

$$\sum_S D_{ik}(S)D_{mj}(S^{-1}) = (g/n)\,\delta_{km}\,\delta_{ij},$$

or for the μth irreducible representation Γ^μ:

$$\sum_S D^\mu_{ik}(S)D^\mu_{mj}(S^{-1}) = (g/n_\mu)\,\delta_{km}\,\delta_{ij}. \tag{A.7-1.9}$$

(ii) Construct the matrix

$$A = \sum_S D^\mu(S)XD^\nu(S^{-1})$$

where Γ^μ and Γ^ν are two non-equivalent irreducible representations of the group $\mathscr{G}$ with dimensions n_μ and n_ν and X is any arbitrary matrix. Matrix A satisfies Theorem II since

$$\begin{aligned} D^\mu(R)A &= \sum_S D^\mu(R)D^\mu(S)XD^\nu(S^{-1}) \\ &= \sum_S D^\mu(R)D^\mu(S)XD^\nu(S^{-1})D^\nu(R^{-1})D^\nu(R) \\ &= \left[\sum_S D^\mu(RS)XD^\nu\{(RS)^{-1}\}\right]D^\nu(R) \\ &= \left[\sum_T D^\mu(T)XD^\nu(T^{-1})\right]D^\nu(R) \\ &= AD^\nu(R) \end{aligned}$$

[these steps are similar to the ones we carried out in Part (i)].

As $D^\mu(R)$ and $D^\nu(R)$ are chosen to be non-equivalent, Theorem II requires that $A = 0$, hence, choosing the matrix X as before, we get

$$\sum_S D^\mu_{ik}(S)D^\nu_{mj}(S^{-1}) = 0 \qquad \begin{aligned} i &= 1, 2, \ldots n_\mu \\ k &= 1, 2, \ldots n_\mu \\ m &= 1, 2, \ldots n_\nu \\ j &= 1, 2, \ldots n_\nu. \end{aligned} \tag{A.7-1.10}$$

Combining eqns (A.7-1.9) and (A.7-1.10) and replacing the symbol S by the symbol R, we have:

$$\sum_R D^\mu_{ik}(R)D^\nu_{mj}(R^{-1}) = (g/n_\mu)\,\delta_{\mu\nu}\,\delta_{ij}\,\delta_{km}. \tag{A.7-1.11}$$

When $\mu = \nu$, or $\Gamma^\mu = \Gamma^\nu$ (the same representation) we have eqn (A.7-1.9) and when $\mu \neq \nu$ we have eqn (A.7-1.10). Eqn (A.7-1.11) represents the Great Orthogonality Theorem.

A.7-2. Proof that $\sum_\mu n_\mu^2 = g$

The proof that the sum of the squares of the dimensions of the irreducible representations is equal to the order of the group has three parts: (1) introduction of the regular representation, (2) the Celebrated Theorem, (3) the final steps.

(1) The regular representation

The regular representation is a reducible representation composed of matrices constructed as follows: first write down the group multiplication table in such a way that the order of the rows corresponds to the inverses of the operations heading the columns; in this way E will appear only along the diagonal of the table. For example, from Table 3-4.2 we would have

	E	A	B	C	D	F
$E^{-1}(=E)$	E	A	B	C	D	F
$A^{-1}(=A)$	A	E	D	F	B	C
$B^{-1}(=B)$	B	F	E	D	C	A
$C^{-1}(=C)$	C	D	F	E	A	B
$D^{-1}(=F)$	F	B	C	A	E	D
$F^{-1}(=D)$	D	C	A	B	F	E

Next, the matrix of the regular representation of the operation R is formed from the resulting table by replacing R by unity and *all* the other operations by zero. For example, in the above case we have

$$D^{\text{reg}}(A) = \begin{Vmatrix} 0 & 1 & 0 & 0 & 0 & 0 \\ 1 & 0 & 0 & 0 & 0 & 0 \\ 0 & 0 & 0 & 0 & 0 & 1 \\ 0 & 0 & 0 & 0 & 1 & 0 \\ 0 & 0 & 0 & 1 & 0 & 0 \\ 0 & 0 & 1 & 0 & 0 & 0 \end{Vmatrix}.$$

It is clear that $\chi^{\text{reg}}(E) = g$ and that if $R \neq E$ then $\chi^{\text{reg}}(R) = 0$, since only $D^{\text{reg}}(E)$ has non-zero elements on the diagonal and it has unity g times (g = order of the group).

We must now confirm that the matrices formed in the above way do indeed form a representation of the group. We have to prove that if the operations of the group are $R_1, R_2, \ldots R_g$, then

$$D^{\text{reg}}(R_aR_b) = D^{\text{reg}}(R_a)D^{\text{reg}}(R_b) \qquad \begin{matrix} a = 1,2,\ldots g \\ b = 1,2,\ldots g \end{matrix} \tag{A.7-2.1}$$

or, in terms of the matrix elements,

$$D^{\text{reg}}_{ki}(R_aR_b) = \sum_{j=1}^{g} D^{\text{reg}}_{kj}(R_a)D^{\text{reg}}_{ji}(R_b). \tag{A.7-2.2}$$

From the way we have set up the matrices, we know that

$$D^{\text{reg}}_{ki}(R_aR_b) = \begin{cases} 1 & \text{if } R_k^{-1}R_i = R_aR_b \\ 0 & \text{if } R_k^{-1}R_i \neq R_aR_b \end{cases} \tag{A.7-2.3}$$

$$D^{\text{reg}}_{kj}(R_a) = \begin{cases} 1 & \text{if } R_k^{-1}R_j = R_a \\ 0 & \text{if } R_k^{-1}R_j \neq R_a \end{cases}$$

and

$$D^{\text{reg}}_{ji}(R_b) = \begin{cases} 1 & \text{if} \quad R_j^{-1}R_i = R_b \\ 0 & \text{if} \quad R_j^{-1}R_i \neq R_b. \end{cases}$$

By the Rearrangement Theorem (Appendix A.3-1) there will only be one value of j for which $R_k^{-1}R_j = R_a$ and only one value of j for which $R_j^{-1}R_i = R_b$. The sum over j in eqn (A.7-2.2) will vanish unless for some single j value *both* $D^{\text{reg}}_{kj}(R_a)$ and $D^{\text{reg}}_{ji}(R_b)$ are simultaneously non-zero and this will occur, if at all, if

$$(R_k^{-1}R_j)(R_j^{-1}R_i) = R_aR_b$$

or

$$R_k^{-1}R_i = R_aR_b$$

in which case the sum will equal unity. So we have

$$\sum_{j=1}^{g} D^{\text{reg}}_{kj}(R_a)D^{\text{reg}}_{ji}(R_b) = \begin{cases} 1 & \text{if} \quad R_k^{-1}R_i = R_aR_b \\ 0 & \text{if} \quad R_k^{-1}R_i \neq R_aR_b \end{cases}$$

which when compared with eqn (A.7-2.3) shows that eqn (A.7-2.2) is true and hence the matrices do form a representation of the group.

(2) *The Celebrated Theorem*

This theorem states that the number of times each irreducible representation Γ^{μ} occurs in the regular representation Γ^{reg} is equal to the dimension of Γ^{μ} (n_{μ}). This is easily proved by using eqn (7-4.2). We find, since $\chi^{\text{reg}}(E) = g$ and if $R \neq E$, $\chi^{\text{reg}}(R) = 0$, that

$$\begin{aligned} a_{\mu} &= g^{-1} \sum_{R} \chi^{\text{reg}}(R)\chi^{\mu}(R)^* \\ &= g^{-1}\chi^{\text{reg}}(E)\chi^{\mu}(E)^* \\ &= g^{-1}gn_{\mu} \\ &= n_{\mu}. \end{aligned} \tag{A.7-2.4}$$

(3) *Proof that* $\sum_{\mu} n_{\mu}^2 = g$

By its construction the dimension of Γ^{reg} is equal to the order of the group g but it must also be equal to the sum of the dimensions of all of the irreducible representations to which it can be reduced, that is

$$\sum_{\mu} a_{\mu}n_{\mu} = g$$

but since for Γ^{reg} $a_{\mu} = n_{\mu}$, we obtain

$$\sum n_{\mu}^2 = g.$$

A. 7-3. Proof that the number of irreducible representations r equals the number of classes k

This proof is quite complex and we will first summarize the definitions (some of which have appeared before, see § 3-5) which are to be used.

C_i = any operation in the ith class

$C_1 = E$, class 1 is the identity operation

$C_{i'}$ = any operation in the collection of inverses of the operations in the ith class
R_j^m = jth operation of the mth class
R = any operation of the group, irrespective of its class
K_i = the sum of the operations of the ith class
$K_1 = E$
$K_{i'}$ = the sum of the inverses of the operations of the ith class
g_i = the number of elements in the ith class
$g_1 = 1$.

The proof of $k = r$ has two parts of which the first is to prove that K_iK_j is solely composed of *complete* classes.

(1) *Proof that* $K_iK_j = \sum_{p=1}^{k} c_{ijp}K_p$

To understand what it is we wish to prove, consider the symmetric tripod point group $\mathscr{C}_{3v}$ for which there are three classes and define

$$K_1 = E$$
$$K_2 = \sigma_v' + \sigma_v'' + \sigma_v'''$$
$$K_3 = C_3 + C_3^2.$$

Then

$$K_1K_1 = K_1$$
$$K_1K_2 = K_2$$
$$K_1K_3 = K_3$$
$$K_2K_2 = 3K_1 + 3K_3$$
$$K_2K_3 = 2K_2$$
$$K_3K_3 = 2K_1 + K_3$$

since, for example,

$$\begin{aligned} K_2K_2 &= (\sigma_v'+\sigma_v''+\sigma_v''')(\sigma_v'+\sigma_v''+\sigma_v''') \\ &= \sigma_v'\sigma_v'+\sigma_v'\sigma_v''+\sigma_v'\sigma_v'''+\sigma_v''\sigma_v'+\sigma_v''\sigma_v''+\sigma_v''\sigma_v''' \\ &\quad +\sigma_v'''\sigma_v'+\sigma_v'''\sigma_v''+\sigma_v'''\sigma_v''' \\ &= E+C_3+C_3^2+C_3^2+E+C_3+C_3+C_3^2+E \\ &= 3E+3(C_3+C_3^2) \\ &= 3K_1+3K_3 \end{aligned}$$

and we see that the product of any two classes produces a linear combination of *complete* classes.

Consider the transformation $R^{-1}C_iR$ for all operations of the ith class with some operation R of the group; the operations produced by this transformation are

(*a*) equal in number to the number in the ith class;
(*b*) all different (from the uniqueness of group multiplication which is itself a result of the Rearrangement Theorem);
(*c*) members of the ith class (by definition).

As there is no opportunity for duplication, they are therefore the same as the ith class, hence

$$R^{-1}K_iR = K_i.$$

Now, we know that K_iK_j must, at least, be *some* linear combination of the operations of the group, so:

$$K_iK_j = \ldots+aR_1^m+bR_2^m+\ldots \qquad \text{(A.7-3.1)}$$

where R_1^m and R_2^m are two operations of the mth class, conjugate to each other by definition

$$R^{-1}R_1^mR = R_2^m, \qquad \text{(for at least one } R \text{ of the group)} \qquad \text{(A.7-3.2)}$$

and a and b are positive integers.

If we use the same R as in eqn (A.7-3.2) to transform the left-hand side of eqn (A.7-3.1), we get

$$\begin{aligned} R^{-1}K_iK_jR &= R^{-1}K_iRR^{-1}K_jR \\ &= K_iK_j \\ &= \ldots+aR_1^m+bR_2^m+\ldots \qquad (\text{since } R^{-1}K_iR = K_i, \text{ etc.}) \end{aligned}$$

and transforming the right-hand side of eqn (A.7-3.1), we get

$$\ldots+aR^{-1}R_1^mR+bR^{-1}R_2^mR = \ldots+aR_2^m+\ldots.$$

Therefore

$$\ldots aR_1^m+bR_2^m+\ldots = \ldots+aR_2^m+\ldots$$

and hence $a = b$. By generalizing this result we see that every operation in the mth class occurs equally often in K_iK_j, hence, K_iK_j consists only of whole classes

$$K_iK_j = \sum_{p=1}^{k} c_{ijp}K_p \qquad \text{(A.7-3.3)}$$

where c_{ijp} are positive integers.

(2) *Proof that* $k = r$

We will define the matrix Δ_i^ν by:

$$\Delta_i^\nu = \sum_{m=1}^{g_i} D^\nu(R_m^i) \qquad \text{(A.7-3.4)}$$

that is, Δ_i^ν is the sum of all the matrices representing the operations of the ith class in the Γ^ν irreducible representation. Then Δ_i^ν commutes with $D^\nu(R)$ for all R, since

$$\begin{aligned} \Delta_i^\nu &= \sum_{m=1}^{g_i} D^\nu(R_m^i) \\ &= \sum_{m=1}^{g_i} D^\nu(R^{-1}R_m^iR) \qquad \text{(see Part } (1)) \\ &= \sum_{m=1}^{g_i} D^\nu(R^{-1})D^\nu(R_m^i)D^\nu(R) \\ &= D^\nu(R^{-1})\left\{\sum_{m=1}^{g_i} D^\nu(R_m^i)\right\}D^\nu(R) \\ &= D^\nu(R)^{-1}\,\Delta_i^\nu D^\nu(R) \end{aligned}$$

[note that $D^\nu(R^{-1}) = D^\nu(R)^{-1}$, see eqn (6-5.2)] and therefore

$$D^\nu(R)\,\Delta_i^\nu = \Delta_i^\nu D^\nu(R).$$

Because of this commutative property, we can use Theorem I (Appendix A.7-1(1)) and write:

$$\Delta_i^\nu = \lambda_i^\nu E \tag{A.7-3.5}$$

where $\lambda_i^\nu E$ is a constant matrix. Eqn (A.7-3.5) implies that the characters satisfy

$$\begin{aligned} g_i\chi^\nu(C_i) &= \lambda_i^\nu n_\nu \\ &= \lambda_i^\nu \chi^\nu(E) \end{aligned}$$

so:

$$\lambda_i^\nu = g_i\,\frac{\chi^\nu(C_i)}{\chi^\nu(E)}. \tag{A.7-3.6}$$

From eqn (A.7-3.3) (see Part (*1*)) we get, by using the Γ^ν representation,

$$\begin{aligned} \Delta_i^\nu \Delta_j^\nu = \sum_{m=1}^{g_i} D^\nu(R_m^i) \sum_{l=1}^{g_j} D^\nu(R_l^j) &= \sum_{m=1}^{g_i}\sum_{l=1}^{g_j} D^\nu(R_m^i R_l^j) \\ &= \sum_{p=1}^{k} c_{ijp} \sum_{q=1}^{g_p} D^\nu(R_q^p) \\ &= \sum_{p=1}^{k} c_{ijp}\,\Delta_p^\nu \end{aligned}$$

or

$$\lambda_i^\nu \lambda_j^\nu = \sum_{p=1}^{k} c_{ijp}\lambda_p^\nu$$

and from eqn (A.7-3.6)

$$\frac{g_i\chi^\nu(C_i)}{\chi^\nu(E)}\,\frac{g_j\chi^\nu(C_j)}{\chi^\nu(E)} = \sum_{p=1}^{k} c_{ijp}\frac{g_p\chi^\nu(C_p)}{\chi^\nu(E)}$$

$$g_i g_j \chi^\nu(C_i)\chi^\nu(C_j) = \chi^\nu(E)\sum_{p=1}^{k} c_{ijp} g_p \chi^\nu(C_p). \tag{A.7-3.7}$$

In eqn (A.7-3.3) E appears in K_iK_j only if $j = i'$, (that is the inverses of the operations in K_i all appear in one class and we denote the sum of the inverses by $K_{i'}$) and then E appears g_i times. Therefore,

$$c_{ij1} = \begin{cases} 0 & j \neq i' \\ g_i & j = i'. \end{cases} \tag{A.7-3.8}$$

If we sum eqn (A.7-3.7) over all ν, from $\nu = 1$ to $\nu = r$ (r = the number of irreducible representations) we get

$$g_i g_j \sum_{\nu=1}^{r} \chi^\nu(C_i)\chi^\nu(C_j) = \sum_{p=1}^{k} c_{ijp} g_p \sum_{\nu=1}^{r}\chi^\nu(E)\chi^\nu(C_p). \tag{A.7-3.9}$$

Since

$$\begin{aligned} \chi^{\text{reg}}(R) &= \sum_{\nu=1}^{r} a_\nu \chi^\nu(R) \qquad (\Gamma^{\text{reg}} = \text{regular representation}) \\ &= \sum_{\nu=1}^{r} n_\nu \chi^\nu(R) \\ &= \sum_{\nu=1}^{r} \chi^\nu(E)\chi^\nu(R) \end{aligned}$$

and

$$\chi^{\text{reg}}(R) = \begin{cases} 0 & \text{if } R \neq E \\ g & \text{if } R = E \end{cases}$$

we have

$$\sum_{\nu=1}^{r} \chi^{\nu}(E)\chi^{\nu}(R) = \begin{cases} 0 & \text{if } R \neq E \\ g & \text{if } R = E \end{cases}$$

eqn (A.7-3.9) now becomes

$$\begin{aligned} g_i g_j \sum_{\nu=1}^{r} \chi^{\nu}(C_i)\chi^{\nu}(C_j) &= \sum_{p=1}^{k} c_{ijp} g_p g \delta_{p1} \\ &= c_{ij1} g_1 g \\ &= c_{ij1} g \end{aligned}$$

and using eqn (A.7-3.8),

$$g_i g_j \sum_{\nu=1}^{r} \chi^{\nu}(C_i)\chi^{\nu}(C_j) = \begin{cases} 0 & \text{if } j \neq i' \\ g_i g & \text{if } j = i' \end{cases}$$

and

$$\sum_{\nu=1}^{r} \chi^{\nu}(C_i)\chi^{\nu}(C_j) = (g/g_j)\, \delta_{ji'}$$

and for a unitary representation, since $\chi^{\nu}(R^{-1}) = \chi^{\nu}(R)^*$ (see the line below eqn (7-3.4)),

$$\sum_{\nu=1}^{r} \chi^{\nu}(C_i)\chi^{\nu}(C_j)^* = (g/g_j)\, \delta_{ij}.$$

This final equation shows that k r-dimensional orthogonal vectors can be formed by using the r characters $\chi^{\nu}(C_i)$, $\nu = 1, 2, \ldots r$, as components and hence $k \leqslant r$. But we have already shown (eqn (7-3.7)) that $r \leqslant k$, so therefore $k = r$ as well as

$$\sum_{\nu=1}^{k} \chi^{\nu}(C_i)\chi^{\nu}(C_j)^* = (g/g_j)\delta_{ij}. \qquad \text{(A.7-3.10)}$$

PROBLEMS

7.1. Given the characters χ of a reducible representation Γ of the indicated point group $\mathscr{G}$ for the various classes of $\mathscr{G}$ in the order in which these classes appear in the character table, find the number of times each irreducible representation occurs in Γ.

(a) $\mathscr{C}_{2v}$ $\chi = 4, -2, 0, -2,$
(b) $\mathscr{C}_{3h}$ $\chi = 4, 1, 1, 2, -1, -1,$
(c) $\mathscr{D}_{4d}$ $\chi = 6, 0, -2, 0, -2, 0, 0,$
(d) $\mathscr{O}_h$ $\chi = 15, 0, -1, 1, 1, -3, 0, 5, -1, 3.$

7.2. Consider the four functions of Problem 5.2 which form a basis for a reducible representation Γ of $\mathscr{D}_4$. Using projection operators find the orthonormal basis functions which reduce Γ. Assume $(f_i, f_j) = \delta_{ij}$.

7.3. Show that the characters of $\mathscr{C}_{4v}$ obey the orthogonality rules of eqns (7-3.5) and (A.7-3.10).

7.4. How many times does each irreducible representation of the $\mathscr{C}_{2v}$ point group occur in the nine-dimensional representation found in Problem 5.3?

7.5. Consider the group whose group table is

	E	*A*	*B*	*C*
E	*E*	*A*	*B*	*C*
A	*A*	*C*	*E*	*B*
B	*B*	*E*	*C*	*A*
C	*C*	*B*	*A*	*E*

write out the matrices and characters for the regular representation of this group.

7.6. Determine the irreducible representations to which the following real orbitals belong for the indicated point group:

(*a*) p_1, p_2, p_3 in $\mathscr{D}_4$ and $\mathscr{D}_{2h}$,
(*b*) d_1, d_2, d_3, d_4, d_5 in $\mathscr{O}_h$,
(*c*) d_1, d_2, d_3, d_4, d_5 in $\mathscr{D}_{3h}$,
(*d*) d_1, d_2, d_3, d_4, d_5 in $\mathscr{T}_d$.

8. Representations and quantum mechanics

8-1. Introduction

In this chapter we introduce the Schrödinger equation; this equation is fundamental to all applications of quantum mechanics to chemical problems. For molecules of chemical interest it is an equation which is exceedingly difficult to solve and any possible simplifications due to the symmetry of the system concerned are very welcome. We are able to introduce symmetry, and thereby the results of the previous chapters, by proving one single but immensely valuable fact: the transformation operators O_R commute with the Hamiltonian operator, $\mathscr{H}$. It is by this subtle thread that we can then deduce some of the properties of the solutions of the Schrödinger equation without even solving it.

Further, we will find in this chapter that wavefunctions (nuclear or electronic) must be functions which form bases for the irreducible representations of the point group to which the molecule belongs. With this knowledge we are able to determine which integrals over molecular wavefunctions are necessarily zero and this in turn (next chapter) leads to well known spectroscopic selection rules.

8-2. The invariance of Hamiltonian operators under O_R

Aside from relativistic and quantum electrodynamic effects, a single molecule in free space is completely described by the Schrödinger equation

$$\mathscr{H}\Psi = E\Psi, \tag{8-2.1}$$

where $\mathscr{H}$, the Hamiltonian, is an operator defined by certain quantum mechanical rules and which can be determined solely from a knowledge of the number of electrons and nuclei and the charges and masses of the nuclei. Ψ, the total wavefunction, is a function of the coordinates of the electrons and nuclei; it defines through further quantum mechanical rules all the properties of the molecule. E is a constant and is the total energy of the molecule. Eqn (8-2.1) is similar to eqn (4-4.1) and E and Ψ are called the eigenvalues and eigenfunctions of the Schrödinger equation. The equation is solved by finding functions Ψ such that when they are acted upon by $\mathscr{H}$ the functions are simply regenerated multiplied by a constant. For molecules containing more than three particles, solution of Schrödinger equation is no easy matter (because of mathematical difficulties) and one has to resort to approximate methods. The reader might note that if one is interested in systems which are changing with time, then eqn (8-2.1) takes a slightly different form: the time-dependent Schrödinger equation and the eigenfunctions of this equation are functions of time.

One way of simplifying eqn (8-2.1) is to use the Born–Oppenheimer approximation; we will not go into the details of this approximation but baldly state the results which come from its application; these are:

$$\Psi = \psi_{\text{el}}\psi_{\text{nuc}}$$

where explicitly ψ_{el} is a function of the coordinates of the electrons alone and ψ_{nuc} is a function of the coordinates of the nuclei alone. If we consider a molecule with n electrons and N nuclei, we can let X_{el} symbolize the collection of $3n$ electronic coordinates $x_1^{(1)}, x_2^{(1)}, x_3^{(1)}, x_1^{(2)}, x_2^{(2)}, x_3^{(2)}, \ldots x_1^{(n)}, x_2^{(n)}, x_3^{(n)}$ and X_{nuc} symbolize the collection of $3N$ nuclear coordinates (or displacements of the nuclei from certain equilibrium positions) $\xi_1^{(1)}, \xi_2^{(1)}, \xi_3^{(1)}, \xi_1^{(2)}, \xi_2^{(2)}, \xi_3^{(2)} \ldots \xi_1^{(N)}, \xi_2^{(N)}, \xi_3^{(N)}$; so that

$$\psi_{\text{el}} = \psi_{\text{el}}(X_{\text{el}}, X_{\text{nuc}})$$

and

$$\psi_{\text{nuc}} = \psi_{\text{nuc}}(X_{\text{nuc}})$$

and we refer to X_{el} as the electronic configuration and to X_{nuc} as the nuclear configuration.

In the Born–Oppenheimer approximation ψ_{el} and ψ_{nuc} are determined by two separate equations, an electronic equation

$$H_{\text{el}}\psi_{\text{el}} = E_{\text{el}}\psi_{\text{el}} \tag{8-2.2}$$

and a nuclear equation

$$H_{\text{nuc}}\psi_{\text{nuc}} = E\psi_{\text{nuc}} \tag{8-2.3}$$

where

$$H_{\text{el}} = T_{\text{el}} + V_{\text{el}} \tag{8-2.4}$$

T_{el} = electronic kinetic-energy operator

$$= \frac{-h^2}{8\pi^2 m}\sum_{i=1}^{n}\nabla_i^2 \tag{8-2.5}$$

V_{el} = potential-energy operator (contains electron–electron and electron–nuclear terms)

m = mass of the electron

∇_i^2 = Laplacian operator for the ith electron

$$= \frac{\partial^2}{\partial x_1^{(i)2}} + \frac{\partial^2}{\partial x_2^{(i)2}} + \frac{\partial^2}{\partial x_3^{(i)2}} \tag{8-2.6}$$

and

$$H_{\text{nuc}} = T_{\text{nuc}} + V_{\text{nuc}} \tag{8-2.7}$$

$$T_{\text{nuc}} = \text{nuclear kinetic-energy operator}$$

$$= \frac{-h^2}{8\pi^2} \sum_{i=1}^{N} \frac{1}{M_i} \nabla_i^2 \tag{8-2.8}$$

$$V_{\text{nuc}} = E_{\text{el}}(X_{\text{nuc}}) + \text{nuclear repulsion terms}$$

$$M_i = \text{mass of the } i\text{th nucleus}$$

$$\nabla_i^2 = \text{Laplacian operator for the } i\text{th nucleus}$$

$$= \frac{\partial^2}{\partial \xi_1^{(i)2}} + \frac{\partial^2}{\partial \xi_2^{(i)2}} + \frac{\partial^2}{\partial \xi_3^{(i)2}}. \tag{8-2.9}$$

Essentially Chapter 9 is concerned with the solution of the nuclear equation (eqn (8-2.3)), which involves the subject of molecular vibrations, and Chapter 10 deals with examples of the solution of the electronic equation (eqn (8-2.2)). The reader will have observed that the eigenvalues of the electronic equation E_{el} which occur in V_{nuc} are normally required before the nuclear equation can be solved, the latter equation providing the final total molecular energy E.

From our point of view the most significant thing about the Hamiltonian operators H_{el} and H_{nuc} is that they both commute with the operators O_R, we say that H_{el} and H_{nuc} are invariant under all symmetry transformation operators of the point group of the molecular framework

$$O_R H_{\text{el}} f(X_{\text{el}}) = H_{\text{el}} O_R f(X_{\text{el}}) \tag{8-2.10}$$

and

$$O_R H_{\text{nuc}} f(X_{\text{nuc}}) = H_{\text{nuc}} O_R f(X_{\text{nuc}}), \tag{8-2.11}$$

where $f(X_{\text{el}})$ and $f(X_{\text{nuc}})$ are functions of the electronic and nuclear coordinates respectively. We can prove the truth of eqns (8-2.10) and (8-2.11) by showing that

$$O_R T_{\text{el}} f(X_{\text{el}}) = T_{\text{el}} O_R f(X_{\text{el}}) \tag{8-2.12}$$

$$O_R T_{\text{nuc}} f(X_{\text{nuc}}) = T_{\text{nuc}} O_R f(X_{\text{nuc}}) \tag{8-2.13}$$

$$O_R V_{\text{el}} f(X_{\text{el}}) = V_{\text{el}} O_R f(X_{\text{el}}) \tag{8-2.14}$$

$$O_R V_{\text{nuc}} f(X_{\text{nuc}}) = V_{\text{nuc}} O_R f(X_{\text{nuc}}) \tag{8-2.15}$$

and this is done in Appendix A.8-1.

Now consider the situation when there are degenerate solutions to the equation

$$H\psi = E\psi \tag{8-2.16}$$

(this equation stands for either eqn (8-2.2) or (8-2.3) and what follows is good for both electronic and nuclear cases). We have

$$H\psi_1^\nu = E_\nu \psi_1^\nu$$
$$H\psi_2^\nu = E_\nu \psi_2^\nu$$
$$\ldots \quad \ldots \quad \text{etc.}$$

and since H is a linear operator (see § 2-2) any linear combination of ψ_1^ν, ψ_2^ν, etc. will also be a solution with the same eigenvalue E_ν, i.e.

$$H(a\psi_1^\nu + b\psi_2^\nu + \ldots) = E_\nu(a\psi_1^\nu + b\psi_2^\nu + \ldots).$$

Therefore, all the solutions with $E = E_\nu$ form a function space associated with the energy E_ν (see § 5-5) and if n_ν of them $\psi_1^\nu, \psi_2^\nu, \ldots, \psi_{n_\nu}^\nu$ are linearly independent, the space will be n_ν-dimensional.

If we can show that the functions $O_R\psi_j^\nu$ (all R and $j = 1, 2, \ldots, n_\nu$) also belong to the function space for energy E_ν, then we can write

$$O_R\psi_j^\nu = \sum_{i=1}^{n_\nu} D_{ij}^\nu(R)\psi_i^\nu, \qquad j = 1, 2, \ldots, n_\nu \tag{8-2.17}$$

and the ψ_j^ν will form a basis for a representation (with matrices $D^\nu(R)$) of the point group whose operations are R. It is easy to prove this in the following way:

$$\begin{aligned} H(O_R\psi_j^\nu) &= O_R(H\psi_j^\nu) \qquad (H \text{ and } O_R \text{ commute}) \\ &= O_R E_\nu \psi_j^\nu \\ &= E_\nu(O_R\psi_j^\nu) \end{aligned}$$

and therefore $O_R\psi_j^\nu$ is a solution of eqn (8-2.16) with an eigenvalue E_ν and it is consequently a member of the function space; it can therefore be written in the form of eqn (8-2.17). We say that the linearly independent degenerate wavefunctions for energy level E_ν form a basis for the representation Γ^ν.

We *assume* that all the degenerate wavefunctions associated with E_ν can be obtained by all the O_R acting on a given wavefunction; this is known as *normal* degeneracy. Any degenerate wavefunctions which cannot be obtained in this way we consider to be *accidentally* degenerate, i.e. accidental degeneracy has no obvious origin in the symmetry of the system. Barring such an accidental degeneracy, the representations produced by eqn (8-2.17) will be irreducible since no smaller matrices could express the most general transformation. We can think of accidental degeneracy as the numerical coincidence of a number of different energy levels, each with its own irreducible representation.

Or, put another way, if the matrices in eqn (8-2.17) are not irreducible then the degeneracy is deemed to be accidental.†

Since we know all the irreducible representations of a point group, we can tell a lot about the possible solutions to eqn (8-2.16), without actually solving it. For example, for ammonia which belongs to the $\mathscr{C}_{3v}$ point group, we know from Table 7-7.1 that its electronic and nuclear wavefunctions must fall into the following three categories:

those which are non-degenerate and totally symmetric (unchanged by all the transformation operators O_R), they will belong to the Γ^{A_1} representation;

those which are non-degenerate, unchanged by O_{C_3} but which change sign under O_{σ_v}, they will belong to Γ^{A_2};

pairs which are doubly degenerate and under both O_{C_3} and O_{σ_v} each wavefunction of a pair produces a linear combination of itself and its partner, they will belong to Γ^E.

8-3. Direct product representations within a group

It is always possible to form a new, and in general reducible, representation Γ of a given point group from any two given representations Γ^μ and Γ^ν of the group. This is done by forming a new function space for which the basis functions are *all* possible products of the basis functions of Γ^μ and Γ^ν. Let the basis functions of Γ^μ and Γ^ν be

$$f_1^\mu, f_2^\mu, \dots f_{n_\mu}^\mu$$

and

$$f_1^\nu, f_2^\nu, \dots f_{n_\nu}^\nu$$

respectively, where n_μ and n_ν are the dimensions of Γ^μ and Γ^ν. Then the basis functions for Γ will be $f_i^\mu f_j^\nu$ $(i = 1, 2, \dots, n_\mu; j = 1, 2, \dots n_\nu)$. We will put these functions in what is called *dictionary order*

$$\begin{aligned} g_1 &= f_1^\mu f_1^\nu, \quad g_2 = f_1^\mu f_2^\nu, \dots \; g_{n_\nu} = f_1^\mu f_{n_\nu}^\nu \\ g_{n_\nu+1} &= f_2^\mu f_1^\nu, \; g_{n_\nu+2} = f_2^\mu f_2^\nu, \dots \; g_{2n_\nu} = f_2^\mu f_{n_\nu}^\nu \\ &\dots\dots\dots\dots\dots\dots \; g_{n_\mu n_\nu} = f_{n_\mu}^\mu f_{n_\nu}^\nu, \end{aligned}$$

† Examples of accidental degeneracy are rare except in problems in which the Hamiltonian involves a continuously variable parameter, such as the strength of an electric or magnetic field. In such cases accidental degeneracy can occur for certain specific values of the parameter in question at which a pair of energy levels cross. Such degeneracy is exceptional, unpredictable, and easily recognised when it does occur. The reader might note that some authors classify the degeneracy of hydrogenic wavefunctions of the same n value as accidental. However, this in fact is *not* accidental degeneracy since Fock (*Zeitschrift fur Physik*, **98**, 145 (1936)) has shown that the degeneracy can be considered to arise from a four-dimensional rotational symmetry of the Hamiltonian in momentum space. A complete discussion of group theory and the hydrogen atom has been given by Bander and Itzykson (*Reviews of modern Physics*, **38**, 330 (1966)).

so that with this convention our new basis functions are

$$g_1, g_2, \dots g_{n_\mu n_\nu}.$$

The new functions will form a basis for a $n_\mu n_\nu$-dimensional representation which we will symbolize by

$$\Gamma = \Gamma^\mu \otimes \Gamma^\nu. \tag{8-3.1}$$

$\Gamma^\mu \otimes \Gamma^\nu$ is called the *direct product* (or *Kronecker product*) of the representations Γ^μ and Γ^ν. The sign $\otimes$ does not mean multiplication, it is simply a signal that the direct product of two representations has been formed in the manner given above.

That the g functions do indeed form a basis for a representation of the point group can be verified by considering the effect of the transformation operators O_R

$$O_R(f_i^\mu f_j^\nu) = (O_R f_i^\mu)(O_R f_j^\nu) = \sum_{p=1}^{n_\mu} \sum_{q=1}^{n_\nu} D_{pi}^\mu(R) D_{qj}^\nu(R) f_p^\mu f_q^\nu \tag{8-3.2}$$

or, if we define

$$D_{sr}^{\mu\otimes\nu}(R) = D_{pi}^\mu(R) D_{qj}^\nu(R), \tag{8-3.3}$$

$$O_R g_r = \sum_{s=1}^{n_\mu n_\nu} D_{sr}^{\mu\otimes\nu}(R) g_s \tag{8-3.4}$$

where, by using dictionary order, the sum over p and q becomes a sum over s. The $n_\mu n_\nu$ product-functions g_r are thus transformed into linear combinations of one another by the transformation operators O_R of the group; they therefore form a basis for the Γ $(= \Gamma^\mu \otimes \Gamma^\nu)$ $n_\mu n_\nu$-dimensional representation of the point group.

The matrix $D^{\mu\otimes\nu}(R)$ that corresponds to the symmetry operation R is a square matrix of order $n_\mu n_\nu$ and its elements are the $(n_\mu n_\nu)^2$ possible products of each of the n_μ^2 elements of $D^\mu(R)$ with each of the n_ν^2 elements of $D^\nu(R)$. The matrix $D^{\mu\otimes\nu}(R)$ is called the direct product of $D^\mu(R)$ and $D^\nu(R)$:

$$D^{\mu\otimes\nu}(R) = D^\mu(R) \otimes D^\nu(R). \tag{8-3.5}$$

The direct product of two matrices is quite different from the ordinary matrix product. First, let us consider how the indices of the various matrix elements are related. By comparing eqns (8-3.2) and (8-3.4) we have

$$f_i^\mu f_j^\nu = g_r$$

and

$$f_p^\mu f_q^\nu = g_s.$$

Hence the subscript r in $D_{sr}^{\mu\otimes\nu}(R)$ is determined by the subscripts i and j, while s is determined by p and q. The dictionary-order convention then gives the following table:

i	j	r
1	1	1
1	2	2
.	.	.
1	n_ν	n_ν
2	1	$n_\nu+1$
2	2	$n_\nu+2$
.	.	.
2	n_ν	$2n_\nu$
.	.	.
n_μ	n_ν	$n_\mu n_\nu$

The same table holds true with i, j, and r replaced by p, q, and s respectively.

As an example, consider the direct product of a 2×2 matrix A and a 3×3 matrix B, then eqn (8-3.3) gives:

$$\left\|\begin{matrix} A_{11} & A_{12} \\ A_{21} & A_{22} \end{matrix}\right\| \otimes \left\|\begin{matrix} B_{11} & B_{12} & B_{13} \\ B_{21} & B_{22} & B_{23} \\ B_{31} & B_{32} & B_{33} \end{matrix}\right\|$$

$$= \left\|\begin{matrix} A_{11}B_{11} & A_{11}B_{12} & A_{11}B_{13} & A_{12}B_{11} & A_{12}B_{12} & A_{12}B_{13} \\ A_{11}B_{21} & A_{11}B_{22} & A_{11}B_{23} & A_{12}B_{21} & A_{12}B_{22} & A_{12}B_{23} \\ A_{11}B_{31} & A_{11}B_{32} & A_{11}B_{33} & A_{12}B_{31} & A_{12}B_{32} & A_{12}B_{33} \\ A_{21}B_{11} & A_{21}B_{12} & A_{21}B_{13} & A_{22}B_{11} & A_{22}B_{12} & A_{22}B_{13} \\ A_{21}B_{21} & A_{21}B_{22} & A_{21}B_{23} & A_{22}B_{21} & A_{22}B_{22} & A_{22}B_{23} \\ A_{21}B_{31} & A_{21}B_{32} & A_{21}B_{33} & A_{22}B_{31} & A_{22}B_{32} & A_{22}B_{33} \end{matrix}\right\| \tag{8-3.5}$$

which can be partitioned into four sub-matrices

$$A\otimes B = \left\|\begin{array}{c|c} A_{11}B & A_{12}B \\ \hline A_{21}B & A_{22}B \end{array}\right\|$$

and is very different from ordinary matrix multiplication.

The characters of the direct product representation will be

$$\begin{aligned} \chi^{\mu\otimes\nu}(\boldsymbol{R}) &= \sum_{r=1}^{n_\mu n_\nu} D_{rr}^{\mu\otimes\nu}(\boldsymbol{R}) = \sum_{i=1}^{n_\mu}\sum_{j=1}^{n_\nu} D_{ii}^{\mu}(\boldsymbol{R}) D_{jj}^{\nu}(\boldsymbol{R}) \\ &= \chi^{\mu}(\boldsymbol{R})\chi^{\nu}(\boldsymbol{R}) \end{aligned} \tag{8-3.6}$$

(the easiest way of confirming the subscripting used here is to look at eqn (8-3.5)).

We can, of course, take the direct product of more than two representations, for example

$$\Gamma = \Gamma^{\sigma} \otimes \Gamma^{\mu} \otimes \Gamma^{\nu} \tag{8-3.7}$$

will, by extension, also be a representation of the point group. We simply apply the direct product rule twice. The characters for this representation will be

$$\chi^{\sigma\otimes\mu\otimes\nu}(R) = \chi^{\sigma}(R)\chi^{\mu}(R)\chi^{\nu}(R).$$

In general, the direct product representations are reducible and using the formulae of § 7-4 we have, if Γ^i are irreducible representations

$$\Gamma^{\mu} \otimes \Gamma^{\nu} = \sum_{i=1}^{k} a_i \Gamma^i, \tag{8-3.9}$$

(k = number of classes = number of irreducible representations)

$$\chi^{\mu\otimes\nu}(R) = \sum_{i=1}^{k} a_i \chi^i(R), \tag{8-3.10}$$

$$a_i = g^{-1} \sum_{R} \chi^{\mu\otimes\nu}(R)\chi^i(R)^* = g^{-1} \sum_{R} \chi^{\mu}(R)\chi^{\nu}(R)\chi^i(R)^*. \tag{8-3.11}$$

This technique of decomposing a direct product representation will be of great use in the next section.

The reader is cautioned that the term *direct product* has a second meaning in group theory. If the group $\mathscr{G}$ has the elements $g_1, g_2, \ldots g_n$ and the group $\mathscr{H}$ has the elements $h_1, h_2, \ldots, h_m$ and if $g_i h_k = h_k g_i$ for all i and k, then the group whose elements are

$$g_i h_k \qquad \begin{array}{l} i = 1, 2, \ldots n \\ k = 1, 2, \ldots m \end{array}$$

is said to be the direct product of group $\mathscr{G}$ and $\mathscr{H}$, e.g. $\mathscr{D}_{3d} = \mathscr{D}_3 \otimes \mathscr{C}_i$, $\mathscr{C}_{3h} = \mathscr{C}_3 \otimes \mathscr{C}_s$.† We have already come across this situation in § 7-8.

8-4. Vanishing integrals

In this section we derive certain rules which will determine whether or not an integral over given electronic or nuclear wavefunctions vanishes; from such rules we can deduce spectroscopic selection rules.

Consider the integral

$$\int \psi^{\sigma}(X)^* F^{\lambda}(X) \psi^{\rho}(X) \, \mathrm{d}\tau$$

† The reader may check that the operations of $\mathscr{C}_i$ (**E** and ***i***) commute with all the operations of $\mathscr{D}_3$ and that the operations of $\mathscr{C}_s$ (**E** and σ_h) commute with all the operations of $\mathscr{C}_3$.

where $\psi^\sigma(X)$ and $\psi^\rho(X)$ are electronic or nuclear wavefunctions which belong to energy levels E_σ and E_ρ respectively and form a basis for the irreducible representations Γ^σ and Γ^ρ. X is the usual electronic or nuclear configuration and $F^\lambda(X)$ is a given function of X belonging to a function space which we will assume generates the Γ^λ irreducible representation. The integration is carried out over all of the electronic or nuclear coordinates. It is apparent that the integrand,

$$\psi^\sigma(X)^* F^\lambda(X)\psi^\rho(X),$$

is one of the basis functions for the direct product representation $\Gamma^{\sigma *}\otimes\Gamma^\lambda\otimes\Gamma^\rho$. (Note that $\psi^\sigma(X)^*$ belongs to $\Gamma^{\sigma *}$.)† If we carry out the reduction of this representation

$$\Gamma^{\sigma *}\otimes\Gamma^\lambda\otimes\Gamma^\rho = a_\alpha\Gamma^\alpha + a_\beta\Gamma^\beta + \ldots \tag{8-4.1}$$

then the original basis functions, $\psi^\sigma(X)^* F^\lambda(X)\psi^\rho(X)$, can be expressed in terms of the basis functions which generate the irreducible representations Γ^α, Γ^β etc. i.e.

$$\psi^\sigma(X)^* F^\lambda(X)\psi^\rho(X) = c_\alpha f^\alpha + c_\beta f^\beta + \ldots \tag{8-4.2}$$

If we apply to the integrand the projection operator P^μ (see § 7-6) where

$$P^\mu = \sum_R \chi^\mu(R)^* O_R$$

and it happens that

$$P^\mu\{\psi^\sigma(X)^* F^\lambda(X)\psi^\rho(X)\} = 0$$

then Γ^μ does not appear in eqn (8-4.1) and f^μ does not appear in eqn (8-4.2). If $\Gamma^{\sigma *}\otimes\Gamma^\lambda\otimes\Gamma^\rho$ does not contain Γ^1, the totally symmetric representation (for which $\chi^1(R) = 1$ for all R and $P^1 = \sum_R O_R$), we will have

$$0 = P^1\{\psi^\sigma(X)^* F^\lambda(X)\psi^\rho(X)\} = \sum_R O_R\{\psi^\sigma(X)^* F^\lambda(X)\psi^\rho(X)\}. \tag{8-4.3}$$

But the O_R are such (see § 5-7) that for any function $Q(X)$

$$\int Q(X)\,d\tau = \int O_R Q(X')\,d\tau' = \int O_R Q(X)\,d\tau,$$

therefore, summing over all R,

$$g\int Q(X)\,d\tau = \int \sum_R O_R Q(X)\,d\tau$$

where g is the order of the group, and consequently

$$\int \psi^\sigma(X)^* F^\lambda(X)\psi^\rho(X)\,d\tau = g^{-1}\int \sum_R O_R\{\psi^\sigma(X)^* F^\lambda(X)\psi^\rho(X)\}\,d\tau.$$

† $\Gamma^{\sigma *}$ is the irreducible representation whose matrices $D^\sigma(R)^*$ are the conjugate complex of the matrices of Γ^σ. If $D^\sigma(R)$ are real then $\Gamma^{\sigma *} = \Gamma^\sigma$.

Hence from eqn (8-4.3) we have the important result that the integral $\int \psi^\sigma(X)^* F^\lambda(X)\psi^\rho(X)\,d\tau$ will be zero if Γ^1 does not appear in $\Gamma^{\sigma *}\otimes\Gamma^\lambda\otimes\Gamma^\rho$ (this is a sufficient but not necessary condition).

This same condition may be expressed in an alternative way. If $\chi^\mu(C_i)$ is the character for any element in the ith class in the Γ^μ irreducible representation, then

$$\chi^{\sigma*\otimes\mu}(C_i) = \chi^\sigma(C_i)^*\chi^\mu(C_i)$$

(note that $\chi^{\sigma *}(C_i)$ = conjugate complex of $\chi^\sigma(C_i) = \chi^\sigma(C_i)^*$) and the number of times the totally symmetric irreducible representation Γ^1 occurs in the reduction of $\Gamma^{\sigma *}\otimes\Gamma^\mu$ is

$$a_1 = g^{-1}\sum_{i=1}^{k} g_i\chi^{\sigma*\otimes\mu}(C_i)\chi^1(C_i)^* = g^{-1}\sum_{i=1}^{k} g_i\chi^\sigma(C_i)^*\chi^\mu(C_i)$$

and recalling eqn (7-3.5) and the fact that Γ^σ and Γ^μ are irreducible representations, we have

$$a_1 = \delta_{\sigma\mu}. \tag{8-4.4}$$

Hence Γ^1 appears once in $\Gamma^{\sigma *}\otimes\Gamma^\mu$ if $\mu = \sigma$ and not at all if $\mu \neq \sigma$. Now consider the direct product representation $\Gamma^{\sigma *}\otimes\Gamma^\lambda\otimes\Gamma^\rho$. If in the reduction of $\Gamma^\lambda\otimes\Gamma^\rho$ the representation Γ^σ does not occur, then by eqn (8-4.4), $a_1 = 0$ and $\Gamma^{\sigma *}\otimes\Gamma^\lambda\otimes\Gamma^\rho$ does not contain Γ^1 and

$$\int \psi^\sigma(X)^* F^\lambda(X)\psi^\rho(X)\,d\tau$$

is zero. So that reduction of $\Gamma^\lambda\otimes\Gamma^\rho$ and checking whether it contains Γ^σ or not is all that is required to see if the integral vanishes. Also, if $F^\lambda(X)$ is replaced by an operator $\boldsymbol{H}$ which belongs to the totally symmetric irreducible representation $\Gamma^1[\chi^1(\boldsymbol{R}) = 1$, all $\boldsymbol{R}]$,† then

$$\int \psi^\sigma(X)^*\boldsymbol{H}\psi^\rho(X)\,d\tau = 0 \tag{8-4.5}$$

unless $\Gamma^\sigma = \Gamma^\rho$, i.e. unless the two wavefunctions belong to the same irreducible representation.

Appendix

A.8-1. Proof of eqns (8-2.12) to (8-2.15)

The proof of these equations follows that given by Schonland.

To prove eqn (8-2.12), consider first a single point with coordinates x_1, x_2, and x_3. Under the operation $\boldsymbol{R}$ this point moves to x_1', x_2', x_3' where, by eqn (5-2.17),

$$x_i' = \sum_{j=1}^{3} D_{ij}(\boldsymbol{R})x_j, \qquad i = 1, 2, 3 \tag{A.8-1.1}$$

†See page 218.

and if the coordinate system has mutually perpendicular axes, the matrix $D(\boldsymbol{R})$ will be orthogonal, so that

$$\sum_{k=1}^{3} D_{ik}(\boldsymbol{R})D_{jk}(\boldsymbol{R}) = \delta_{ij} \qquad \begin{matrix} i = 1, 2, 3 \\ j = 1, 2, 3. \end{matrix} \tag{A.8-1.2}$$

Now, by definition, if f is some function of the coordinates $f(x_1, x_2, x_3)$ then

$$\boldsymbol{O}_{\boldsymbol{R}} f(x_1', x_2', x_3') = f(x_1, x_2, x_3)$$

and if we form a new function $\nabla^2 f$, where ∇^2 is the Laplacian operator, then:

$$\begin{aligned} \boldsymbol{O}_{\boldsymbol{R}} \nabla^2 f(x_1', x_2', x_3') &= \nabla^2 f(x_1, x_2, x_3) \\ &= \nabla^2 \boldsymbol{O}_{\boldsymbol{R}} f(x_1', x_2', x_3'). \end{aligned} \tag{A.8-1.3}$$

The right-hand side of this equation has the form $\nabla^2 g'$, where

$$g' = \boldsymbol{O}_{\boldsymbol{R}} f(x_1', x_2', x_3')$$

and ∇^2 refers to differentiation with respect to x_1, x_2, x_3.

Now

$$\frac{\partial g'}{\partial x_k} = \frac{\partial g'}{\partial x_1'}\frac{\partial x_1'}{\partial x_k} + \frac{\partial g'}{\partial x_2'}\frac{\partial x_2'}{\partial x_k} + \frac{\partial g'}{\partial x_3'}\frac{\partial x_3'}{\partial x_k}$$

and since, by eqn (A.8-1.1),

$$\frac{\partial x_i'}{\partial x_k} = D_{ik}(\boldsymbol{R})$$

we have

$$\frac{\partial g'}{\partial x_k} = \sum_{i=1}^{3} D_{ik}(\boldsymbol{R}) \frac{\partial g'}{\partial x_i'}.$$

Differentiating once more with respect to x_k gives

$$\frac{\partial^2 g'}{\partial x_k^2} = \sum_{j=1}^{3}\sum_{i=1}^{3} D_{jk}(\boldsymbol{R})D_{ik}(\boldsymbol{R}) \frac{\partial}{\partial x_j'}\left(\frac{\partial g'}{\partial x_i'}\right)$$

and summing this equation over the three values of k gives

$$\begin{aligned} \nabla^2 g' &= \sum_{k=1}^{3} \frac{\partial^2 g'}{\partial x_k^2} \\ &= \sum_{k=1}^{3}\sum_{j=1}^{3}\sum_{i=1}^{3} D_{jk}(\boldsymbol{R})D_{ik}(\boldsymbol{R}) \frac{\partial}{\partial x_j'}\left(\frac{\partial g'}{\partial x_i'}\right) \\ &= \sum_{j=1}^{3}\sum_{i=1}^{3} \delta_{ij} \frac{\partial}{\partial x_j'}\left(\frac{\partial g'}{\partial x_i'}\right) \qquad \text{(see eqn (A.8-1.2))} \\ &= \sum_{i=1}^{3} \frac{\partial}{\partial x_i'}\left(\frac{\partial g'}{\partial x_i'}\right) \\ &= \nabla'^2 g'. \end{aligned}$$

If we put back $g' = \boldsymbol{O}_{\boldsymbol{R}} f(x_1', x_2', x_3')$ in this equation we get

$$\nabla^2 \boldsymbol{O}_{\boldsymbol{R}} f(x_1', x_2', x_3') = \nabla'^2 \boldsymbol{O}_{\boldsymbol{R}} f(x_1', x_2', x_3').$$

Likewise $\nabla^2 f(x_1', x_2', x_3') = \nabla'^2 f(x_1', x_2', x_3')$ and eqn (A.8-1.3) becomes

$$[O_R(\nabla'^2 f)](x_1', x_2', x_3') = \nabla'^2[(O_R f)(x_1', x_2', x_3')]$$

and since x_1', x_2', x_3' occur throughout this equation, we conclude that

$$O_R \nabla^2 f = \nabla^2 O_R f. \qquad \text{(A.8-1.4)}$$

Taking an equation like eqn (A.8-1.4) for each electron and multiplying by $-h^2/8\pi^2 m$ and adding we obtain eqn (8-2.12).

To prove eqn (8-2.13), let us suppose that $\boldsymbol{R}$, when it is applied to the nuclear framework, changes any general nuclear configuration from X_{nuc} to X'_{nuc}, then if the base vectors are transferred as in § 5-4(2) (see also Fig. 5-4.3), we have, in terms of coordinates rather than base vectors,

$$\xi_i^{(q)'} = \sum_{j=1}^{3} D_{ij}(\boldsymbol{R})\xi_j^{(p)}, \qquad i = 1, 2, 3 \qquad \text{(A.8-1.5)}$$

where a displacement from the equilibrium position of nucleus q has been transferred to where p was before the operation was carried out [in § 5-4(2) we combined the N equations (one for each nucleus) like eqn (A.8-1.5) for the base vectors together to obtain a $3N$-dimensional matrix]. A slight change in the derivation of eqn (A.8-1.4) then leads to

$$O_R \nabla_p^2 f(X_{\text{nuc}}) = \nabla_q^2 O_R f(X_{\text{nuc}}).$$

Because of the nature of $\boldsymbol{R}$, p and q must be physically identical and therefore have the same mass, so that

$$O_R \frac{1}{M_p} \nabla_p^2 f(X_{\text{nuc}}) = \frac{1}{M_q} \nabla_q^2 O_R f(X_{\text{nuc}})$$

and eqn (8-2.13) follows by addition.

Let us now consider eqn (8-2.15). V_{nuc} is solely a function of the relative positions of the nuclei, i.e. $V_{\text{nuc}} = V_{\text{nuc}}(X_{\text{nuc}})$. Any symmetry operation must leave these *relative* positions, and hence V_{nuc}, unaltered, i.e. if under $\boldsymbol{R}$ any general nuclear configuration X_{nuc} becomes X'_{nuc} then

$$V_{\text{nuc}}(X_{\text{nuc}}) = V_{\text{nuc}}(X'_{\text{nuc}}). \qquad \text{(A.8-1.6)}$$

From the definition of O_R we have

$$O_R V_{\text{nuc}}(X'_{\text{nuc}}) = V_{\text{nuc}}(X_{\text{nuc}})$$
$$= V_{\text{nuc}}(X'_{\text{nuc}})$$

or

$$O_R V_{\text{nuc}} = V_{\text{nuc}}$$

and

$$O_R V_{\text{nuc}} f(X_{\text{nuc}}) = O_R V_{\text{nuc}} O_R f(X_{\text{nuc}})$$
$$= V_{\text{nuc}} O_R f(X_{\text{nuc}})$$

which is eqn (8-2.15).

Last we must prove eqn (8-2.14). V_{el} is a function of the relative positions of the electrons and nuclei, that is $V_{\text{el}} = V_{\text{el}}(X_{\text{el}}, X_{\text{nuc}})$ where X_{nuc} in this case is the specific nuclear configuration used to define the molecule's symmetry. If a symmetry operation R is first applied to the whole molecule, all particles (electrons and nuclei), then the relative positions of the particles are unchanged and so is V_{el}

$$V_{el}(X_{el}, X_{nuc}) = V_{el}(X'_{el}, X'_{nuc}).$$

If we now apply R^{-1} to the nuclei *alone* then, since this only interchanges like nuclei and by definition leaves the nuclear framework physically unchanged, V_{el} still remains the same

$$V_{el}(X_{el}, X_{nuc}) = V_{el}(X'_{el}, X'_{nuc})$$
$$= V_{el}(X'_{el}, X_{nuc}).$$

So that for the fixed nuclear configuration which defines the molecule's symmetry, the change of electronic configuration caused by R, $X_{el} \rightarrow X'_{el}$, leaves V_{el} unchanged. The rest of the proof of eqn (8-2.14) is the same as the proof of eqn (8-2.15).

PROBLEMS

8.1. To what irreducible representations can the following direct product representations be reduced for the specified point group?

(*a*) $\Gamma^{A_1}\otimes\Gamma^{A_1}$, $\Gamma^{A_1}\otimes\Gamma^{A_2}$, $\Gamma^{A_2}\otimes\Gamma^{E}$, $\Gamma^{E}\otimes\Gamma^{E}$ for $\mathscr{C}_{3v}$

(*b*) $\Gamma^{E'}\otimes\Gamma^{E'}$, $\Gamma^{A''_1}\otimes\Gamma^{A''_2}$, $\Gamma^{A''_2}\otimes\Gamma^{E''}$ for $\mathscr{D}_{3h}$

(*c*) $\Gamma^{E_1}\otimes\Gamma^{E_1}$, $\Gamma^{E_1}\otimes\Gamma^{E_2}$, $\Gamma^{E_2}\otimes\Gamma^{E_2}$ for $\mathscr{C}_{5v}$.

8.2. To what irreducible representation must ψ^{σ} belong if the integral

$$\int \psi^{\sigma}(X)^* F^{\lambda}(X)\psi^{\rho}(X)\, d\tau$$

is to be non-zero in the following cases?

(*a*) $\mathscr{C}_{4v}$ $\Gamma^{\lambda} = \Gamma^{E}$; $\Gamma^{\rho} = \Gamma^{A_1}, \Gamma^{A_2}, \Gamma^{B_1}, \Gamma^{B_2}$

(*b*) $\mathscr{D}_{6h}$ $\Gamma^{\lambda} = \Gamma^{E_{1u}}$; $\Gamma^{\rho} = \Gamma^{E_{2u}}$

(*c*) $\mathscr{T}_{d}$ $\Gamma^{\lambda} = \Gamma^{T_2}$; $\Gamma^{\rho} = \Gamma^{A_2}, \Gamma^{E}, \Gamma^{T_1}, \Gamma^{T_2}$.

9. Molecular vibrations

9-1. Introduction

In this chapter we apply the results of the previous chapters to the problem of molecular vibrations. Before doing so, however, it is necessary to have some knowledge of the quantum-mechanical equations which govern the way in which a molecule vibrates. We find that the solution of these equations is greatly simplified by changing the coordinates of the nuclei from Cartesian coordinates to a new type, defined in a special way, called the *normal coordinates*. This change is no more mysterious than changing, say, from Cartesian coordinates to polar coordinates when solving the Schrödinger equation for the hydrogen atom; the basic principle is the same, namely the mathematics is made easier. So we start this chapter with a discussion of normal coordinates.

We then discover an extremely important fact; each normal coordinate belongs to one of the irreducible representations of the point group of the molecule concerned and is a part of a basis which can be used to produce that representation. Because of their relationship with the normal coordinates, the vibrational wavefunctions associated with the fundamental vibrational energy levels also behave in the same way. We are therefore able to classify both the normal coordinates and fundamental vibrational wavefunctions according to their symmetry species and to predict from the character tables the degeneracies and symmetry types which can, in principle, exist.

Furthermore, knowledge of the irreducible representations to which the vibrational wavefunctions belong coupled with the vanishing integral rule tells us a good deal about the infra-red and Raman spectra of the molecule under consideration.

9-2. Normal coordinates

If we consider a molecule with N nuclei, then the displacements of the nuclei from their equilibrium positions in Cartesian coordinates can be written as

$$\xi_1^{(1)}, \xi_2^{(1)}, \xi_3^{(1)}, \xi_1^{(2)}, \ldots \xi_3^{(N)}$$

and the corresponding velocities as

$$\dot{\xi}_1^{(1)}, \dot{\xi}_2^{(1)}, \dot{\xi}_3^{(1)}, \dot{\xi}_1^{(2)}, \ldots \dot{\xi}_3^{(N)}$$

where

$$\dot{\xi}_j^{(i)} = \mathrm{d}\xi_j^{(i)}/\mathrm{d}t.$$

Or we can use the so-called mass-weighted displacement coordinates

$$q_1^{(1)}, q_2^{(1)}, q_3^{(1)}, \ldots q_3^{(N)} \tag{9-2.1}$$

with velocities:

$$\dot{q}_1^{(1)}, \dot{q}_2^{(1)}, \dot{q}_3^{(1)}, \ldots \dot{q}_3^{(N)} \tag{9-2.2}$$

where

$$q_j^{(i)} = M_i^{\frac{1}{2}} \xi_j^{(i)}$$

and M_i is the mass of the ith nucleus. In actual fact it will be more convenient to let the subscript on the q's and $\dot{q}$'s run over all the coordinates and velocities, i.e. from 1 to $3N$, so that we have:

$$q_1, q_2, q_3, \ldots q_{3N}$$

and

$$\dot{q}_1, \dot{q}_2, \dot{q}_3, \ldots \dot{q}_{3N}$$

in place of eqns (9-2.1) and (9-2.2).

In classical terms, if we use the mass-weighted Cartesian displacement coordinates, the kinetic energy of the moving nuclei is†

$$T = \tfrac{1}{2} \sum_{i=1}^{3N} \sum_{j=1}^{3N} \delta_{ij} \dot{q}_i \dot{q}_j \tag{9-2.3}$$

(these terms are of the familiar $\frac{1}{2}mv^2$ type) and the potential energy, relative to its value when the nuclei are in their equilibrium positions, is V, which can be expanded in a Taylor series as:

$$V = \sum_{i=1}^{3N} \left(\frac{\partial V}{\partial q_i}\right)_0 q_i + \tfrac{1}{2} \sum_{i=1}^{3N} \sum_{j=1}^{3N} \left(\frac{\partial^2 V}{\partial q_i \partial q_j}\right)_0 q_i q_j + \ldots \tag{9-2.4}$$

where the subscript 0 denotes that the derivative is evaluated when the nuclei are in their equilibrium positions. Since, by definition, V is minimal for the equilibrium configuration, we know that

$$\left(\frac{\partial V}{\partial q_i}\right)_0 = 0 \qquad i = 1, 2, \ldots 3N \tag{9-2.5}$$

and if we replace the second derivatives (which are called the *harmonic force constants* and are intrinsic properties of the molecule under consideration) by

$$B_{ij} = \left(\frac{\partial^2 V}{\partial q_i \partial q_j}\right)_0 \qquad \begin{array}{l} i = 1, 2, \ldots 3N \\ j = 1, 2, \ldots 3N \end{array} \tag{9-2.6}$$

and stop the expansion after the quadratic terms (the *harmonic oscillator approximation*), we have

$$V = \tfrac{1}{2} \sum_{i=1}^{3N} \sum_{j=1}^{3N} B_{ij} q_i q_j \ddagger . \tag{9-2.7}$$

† T is the classical analogue of the quantum mechanical operator T_{nuc} defined in eqn (8-2.7).

‡ If the potential energy of the nuclei in their equilibrium positions is W_{eq}, then $V + W_{\text{eq}} = V_{\text{nuc}}$, where V_{nuc} is defined in eqn (8-2.7).

The classical equation of motion for the moving nuclei is

$$\frac{\mathrm{d}}{\mathrm{d}t}\left(\frac{\partial T}{\partial \dot{q}_i}\right)+\frac{\partial V}{\partial q_i}=0 \qquad i=1,2,\ldots 3N \tag{9-2.8}$$

and using eqns (9-2.3) and (9-2.7) this becomes

$$\sum_{j=1}^{3N}\delta_{ij}\frac{\mathrm{d}^2q_j}{\mathrm{d}t^2}+\sum_{j=1}^{3N}B_{ij}q_j=0 \qquad i=1,2\ldots 3N. \tag{9-2.9}$$

Now let us choose a set of $3N$ coefficients $C_1, C_2,\ldots$ and C_{3N} such that when each of the eqns (9-2.9) is multiplied by the appropriate C_i and the $3N$ equations are added, we obtain

$$\frac{\mathrm{d}^2Q}{\mathrm{d}t^2}+\lambda Q=0, \tag{9-2.10}$$

where

$$Q=\sum_{j=1}^{3N}h_jq_j \tag{9-2.11}$$

(i.e. Q is a linear combination of the mass-weighted displacement coordinates) and λ is a constant. There will be, in fact, $3N$ ways of making the choice of the $3N$ coefficients. We can see the reason for this by looking at the equalities which must exist between eqns (9-2.9) and (9-2.10), that is we must have

$$\sum_{i=1}^{3N}C_i\sum_{j=1}^{3N}\delta_{ij}\frac{\mathrm{d}^2q_j}{\mathrm{d}t^2}=\frac{\mathrm{d}^2}{\mathrm{d}t^2}\left(\sum_{j=1}^{3N}h_jq_j\right)$$

or

$$\sum_{i=1}^{3N}C_i\delta_{ij}=h_j \qquad j=1,2,\ldots 3N \tag{9-2.12}$$

and

$$\sum_{i=1}^{3N}C_i\sum_{j=1}^{3N}B_{ij}q_j=\lambda\sum_{j=1}^{3N}h_jq_j$$

or

$$\sum_{i=1}^{3N}C_iB_{ij}=\lambda h_j \qquad j=1,2,\ldots 3N. \tag{9-2.13}$$

From eqn (9-2.12) we get

$$C_j=h_j$$

and hence

$$Q=\sum_{j=1}^{3N}C_jq_j$$

and by combining eqns (9-2.12) and (9-2.13) we have

$$\sum_{i=1}^{3N}(B_{ij}-\lambda\delta_{ij})C_i=0 \qquad j=1,2,\ldots 3N. \tag{9-2.14}$$

For this set of $3N$ simultaneous equations to have non-trivial solutions for the C_i, the following equation must hold true (see Appendix A.4-3(a)):

$$\det(B-\lambda E) = 0 \qquad (9\text{-}2.15)$$

where B is the matrix formed from the elements B_{ij} and E is the unit matrix. There will be $3N$ roots (values of λ) of eqn (9-2.15) which, when found, can be used in turn to solve eqns (9-2.14) for the C_i (one additional equation, a normalization equation, $\sum_{i=1}^{3N} C_i^2 = 1$ is required to determine *all* of the $3N$ C's). Since there are $3N$ λ values, there are $3N$ sets of C_i which will produce eqn (9-2.10). For convenience, we will add a subscript to λ and Q to distinguish the different solutions and an additional subscript to the C's to show with which λ value they are associated, i.e.

$$\lambda_1 : C_{11},\quad C_{21},\ldots\quad C_{3N\,1} :\quad Q_1 = \sum_{i=1}^{3N} C_{i1}q_i$$

$$\lambda_2 : C_{12},\quad C_{22},\ldots\quad C_{3N\,2} :\quad Q_2 = \sum_{i=1}^{3N} C_{i2}q_i$$

$$\begin{matrix} \cdot & \cdot & \cdot & \cdot & \cdot & \cdot \\ \cdot & \cdot & \cdot & \cdot & \cdot & \cdot \\ \cdot & \cdot & \cdot & \cdot & \cdot & \cdot \end{matrix}$$

$$\lambda_{3N} : C_{1\,3N}, C_{2\,3N},\ldots C_{3N\,3N} : Q_{3N} = \sum_{i=1}^{3N} C_{i\,3N}q_i.$$

The $Q_1, Q_2,\ldots Q_{3N}$ are called *normal coordinates* and what we have done is to transform the coordinates q_i to another set Q_i such that eqn (9-2.10) is true. We can form the matrix C by using the coefficients for each λ value as *columns:*

$$C = \begin{Vmatrix} C_{11} & C_{12} & \ldots & C_{1\,3N} \\ C_{21} & C_{22} & \ldots & C_{2\,3N} \\ \cdot & \cdot & & \cdot \\ \cdot & \cdot & & \cdot \\ \cdot & \cdot & & \cdot \\ C_{3N\,1} & C_{3N\,2} & \ldots & C_{3N\,3N} \end{Vmatrix}$$

and since B is symmetric, this matrix will be orthogonal (see Appendix A.4-3(c)).

As well as satisfying

$$\frac{d^2Q_i}{dt^2} + \lambda_i Q_i = 0 \qquad i = 1, 2,\ldots 3N \qquad (9\text{-}2.16)$$

the normal coordinates also satisfy:

$$T = \tfrac{1}{2}\sum_{i=1}^{3N} \dot{Q}_i^2 \tag{9-2.17}$$

and

$$V = \tfrac{1}{2}\sum_{i=1}^{3N} \lambda_i Q_i^2 \tag{9-2.18}$$

(these equations are proved in Appendix A.9-1).

The solutions of the equations of motion (eqn (9-2.16)) are easily found to be:

$$Q_i = A_i \cos(\lambda_i^{\frac{1}{2}} t + \varepsilon_i) \qquad i = 1, 2, \ldots 3N \tag{9-2.19}$$

where A_i and ε_i are constants and t is the time. Since

$$Q_i = \sum_{j=1}^{3N} C_{ji} q_j \qquad i = 1, 2, \ldots 3N \tag{9-2.20}$$

and C is orthogonal, we have

$$\begin{aligned} q_j &= \sum_{i=1}^{3N} C_{ji} Q_i \qquad j = 1, 2, \ldots 3N \\ &= \sum_{i=1}^{3N} C_{ji} A_i \cos(\lambda_i^{\frac{1}{2}} t + \varepsilon_i) \end{aligned}$$

and if all the normal coordinates are zero except one, say Q_k (that is $A_i = 0$ except for $i = k$), then

$$q_j = C_{jk} A_k \cos(\lambda_k^{\frac{1}{2}} t + \varepsilon_k) \tag{9-2.21}$$

and each mass weighted Cartesian displacement coordinate is varying sinusoidally with time with a frequency of ν_k where $2\pi\nu_k = \lambda_k^{\frac{1}{2}}$. Such a motion is called a *normal mode* and there are clearly $3N$ such modes, each one associated with one of the $3N$ normal coordinates.

Some nuclear displacements will be such that the bond lengths and angles in the molecule are unchanged from their equilibrium values and V is consequently unchanged; such will be the case for translation and rotation of the molecule as a whole. For a non-linear molecule, a rigid movement of the molecule as a whole may be expressed as a combination of translations along, and of rotations about, three chosen axes. To describe any general translation–rotation movement we will necessarily require six coordinates and therefore at least six normal coordinates; hence V is zero for at least six of the Q_i having non-zero values. Since V is measured relative to the minimum potential energy (the value for the equilibrium nuclear framework) $V \geqslant 0$ and since

$$V = \tfrac{1}{2}\sum_{i=1}^{3N} \lambda_i Q_i^2,$$

we have $\lambda_i \geqslant 0$ (Q_i^2 is always positive). Now the only way in which $\frac{1}{2}\sum_{i=1}^{3N} \lambda_i Q_i^2$ can be zero for at least six of the Q_i's having non-zero values, if $\lambda_i \geqslant 0$, is for there to be at least six λ_i which are zero. In fact, there must be exactly six λ_i which are zero because if there were more one would be able to carry out a normal mode which is not necessarily a combination of translations and rotations and still have $V = 0$, this cannot be done because if the bond lengths or the angles change then $V > 0$.

We therefore associate $Q_1, Q_2, \ldots Q_{3N-6}$ and $\lambda_1, \lambda_2, \ldots \lambda_{3N-6}$ (all positive) with vibrations and $Q_{3N-5}, Q_{3N-4}, \ldots Q_{3N}$ and

$$\lambda_{3N-5} = \lambda_{3N-4} = \ldots \lambda_{3N} = 0$$

with translations and rotations and write

$$V = \tfrac{1}{2} \sum_{i=1}^{3N-6} \lambda_i Q_i^2.$$

For a linear molecule there are only two independent rotations and an argument similar to the one above leads to:

$$V = \tfrac{1}{2} \sum_{i=1}^{3N-5} \lambda_i Q_i^2.$$

The point of changing from Cartesian displacement coordinates to normal coordinates is that it brings about a great simplification of the vibrational equation. Furthermore, we will see that the normal coordinates provide a basis for a representation of the point group to which molecule belongs.

9-3. The vibrational equation

The nuclear equation (eqn (8-2.3)) written out in full is

$$\frac{-h^2}{8\pi^2} \sum_{i=1}^{N} \frac{1}{M_i} \nabla_i^2 \psi_{\text{nuc}}(X_{\text{nuc}}) + V_{\text{nuc}} \psi_{\text{nuc}}(X_{\text{nuc}}) = E \psi_{\text{nuc}}(X_{\text{nuc}}) \quad (9\text{-}3.1)$$

where the symbols have been defined in § 8-2. It is possible to approximate ψ_{nuc} as the product of three functions, one a function of the coordinates describing translation ψ^{tr}, another a function of the coordinates describing rotation ψ^{rot} and the third a function of the normal coordinates describing vibration ψ^{vib}, that is:

$$\psi_{\text{nuc}} = \psi^{\text{tr}} \psi^{\text{rot}} \psi^{\text{vib}}.$$

Eqn (9-3.1) can then be separated into three eigenvalue equations for the three types of motion. The three eigenvalues will be W^{tr} (translational energy), W^{rot} (rotational energy) and W^{vib} (vibrational energy) and the total molecular energy E is given by

$$E = W^{\text{tr}} + W^{\text{rot}} + W^{\text{vib}} + W^{\text{eq}}$$

where W^{eq} is the energy of the molecule when the nuclei are kept fixed in their equilibrium positions (V in eqn (9-2.4) is relative to W^{eq}).

Of the three eigenvalue equations, the one of interest to us is the vibrational equation. It has a particularly simple form when normal coordinates are employed because the classical kinetic and potential energies then have no cross terms (see eqns (9-2.17) and (9-2.18)) and this fact leads to a simple form for their quantum mechanical analogues (the kinetic energy and potential energy operators). The vibrational equation is thus

$$\sum_{i=1}^{3N-6} \frac{\partial^2}{\partial Q_i^2}\psi^{\text{vib}}(Q_1,\ldots Q_{3N-6}) + \frac{8\pi^2}{h^2}\left(W^{\text{vib}} - \tfrac{1}{2}\sum_{i=1}^{3N-6}\lambda_i Q_i^2\right)\psi^{\text{vib}}(Q_1,\ldots Q_{3N-6}) = 0 \tag{9-3.2}$$

and if we replace W^{vib} by a sum of terms

$$W^{\text{vib}} = \sum_{i=1}^{3N-6} W_i$$

and $\psi^{\text{vib}}(Q_1,\ldots Q_{3N-6})$ by a product of functions, each of which is a function of a *single* vibrational normal coordinate

$$\psi^{\text{vib}}(Q_1,\ldots Q_{3N-6}) = \prod_{i=1}^{3N-6}\psi_i(Q_i)$$

and divide by $\psi^{\text{vib}}(Q_1,\ldots Q_{3N-6})$, then we have

$$\sum_{i=1}^{3N-6}\left\{\frac{1}{\psi_i(Q_i)}\frac{d^2\psi_i(Q_i)}{dQ_i^2} + \frac{8\pi^2}{h^2}(W_i - \tfrac{1}{2}\lambda_i Q_i^2)\right\} = 0.$$

Each term of this sum is a function of just one normal coordinate and is independent of all the other terms. Since the sum is zero, each term must be zero

$$\frac{1}{\psi_i(Q_i)}\frac{d^2\psi_i(Q_i)}{dQ_i^2} + \frac{8\pi^2}{h^2}(W_i - \tfrac{1}{2}\lambda_i Q_i^2) = 0$$

or

$$\frac{d^2\psi_i(Q_i)}{dQ_i^2} + \frac{8\pi^2}{h^2}(W_i - \tfrac{1}{2}\lambda_i Q_i^2)\psi_i(Q_i) = 0$$

$$i = 1, 2,\ldots 3N-6. \tag{9-3.3}$$

Eqn (9-3.3) is the same as the well known one-dimensional harmonic oscillator equation and has as its solutions

$$\psi_i(Q_i) = \mathcal{N}_i \exp(-\tfrac{1}{2}\alpha_i Q_i^2) H_{n_i}(\alpha_i^{\frac{1}{2}} Q_i) \tag{9-3.4}$$

and

$$W_i = (n_i + \tfrac{1}{2})h\nu_i$$

where:

$\mathcal{N}_i$ is a normalizing constant and is chosen such that

$\int_{-\infty}^{+\infty} \psi_i^2(Q_i)\, dQ_i = 1,$

$\alpha_i = \frac{2\pi}{h}\lambda_i^{\frac{1}{2}} = \frac{4\pi^2\nu_i}{h},$

$\nu_i = \frac{\lambda_i^{\frac{1}{2}}}{2\pi}$ (fundamental frequency),

$n_i = 0, 1, 2, \ldots$ (vibrational quantum number),

$H_{n_i}(x)$ = a Hermite orthogonal polynomial, e.g. $H_0(x) = 1$, $H_1(x) = 2x$, $H_2(x) = 4x^2-2$, $H_3(x) = 8x^3-12x$, etc.

The fundamental frequencies ν_i $(i = 1, 2, \ldots 3N-6)$ are related to λ_i and since λ_i are the roots of $\det(B-\lambda E) = 0$, ν_i are related to the matrix B and to the molecular force constants B_{ij}. Hence the vibrational energy levels for a non-linear polyatomic molecule in the harmonic oscillator approximation are given by

$$W^{\text{vib}} = \sum_{i=1}^{3N-6} (n_i+\tfrac{1}{2})h\nu_i \tag{9-3.6}$$

and the corresponding vibrational wavefunctions are given by

$$\psi^{\text{vib}} = \psi_{n_1 n_2 \cdots n_{3N-6}} = \mathcal{N} \exp\left(-\tfrac{1}{2}\sum_{i=1}^{3N-6} \alpha_i Q_i^2\right) \prod_{i=1}^{3N-6} H_{n_i}(\alpha_i^{\frac{1}{2}} Q_i) \tag{9-3.7}$$

and both W^{vib} and ψ^{vib} are characterized by the values of the $3N-6$ vibrational quantum numbers $n_1, n_2, \ldots n_{3N-6}$.

We see from eqn (9-3.7) that the lowest energy state, the *ground state*, occurs when $n_1 = n_2 = \ldots n_{3N-6} = 0$ and the energy is then

$$W_0^{\text{vib}} = \sum_{i=1}^{3N-6} \tfrac{1}{2}h\nu_i. \tag{9-3.8}$$

W_0^{vib} is called the zero-point energy. The ground state wavefunction is

$$\psi_0^{\text{vib}} = \mathcal{N} \exp\left(-\tfrac{1}{2}\sum_{i=1}^{3N-6} \alpha_i Q_i^2\right) \tag{9-3.9}$$

where $\mathcal{N}$ is a normalization constant.

The vibrational energy levels where all the quantum numbers are zero except for one which is unity are called the *fundamental levels.* There will, in principle, be $3N-6$ such levels and the energy of the pth one $(n_1 = 0, n_2 = 0, \ldots n_p = 1, \ldots n_{3N-6} = 0)$ will be

$$W_p^{\text{vib}} = \sum_{i\neq p}^{3N-6} \tfrac{1}{2}h\nu_i + (1+\tfrac{1}{2})h\nu_p = W_0^{\text{vib}} + h\nu_p. \tag{9-3.10}$$

Therefore the frequency of the radiation absorbed or emitted in a transition between the ground state and a fundamental level is $(W_p^{\text{vib}} - W_0^{\text{vib}})/h = \nu_p$, the pth fundamental frequency. The corresponding wavefunction will be

$$\psi_p^{\text{vib}} = \psi_{00\cdots1\cdots0} \propto \psi_0^{\text{vib}} H_1(\alpha_p^{\frac{1}{2}} Q_p) \propto \psi_0^{\text{vib}} Q_p. \tag{9-3.11}$$

The infra-red and Raman spectra of molecules are dominated by transitions between the ground state and the fundamental levels but, in practice, the number of fundamental frequencies observed does not reach $3N-6$ since (a) some of the λ_i are identical (leading to degenerate fundamental levels) and (b) selection rules forbid certain transitions. Both (a) and (b) are determined by the symmetry of the molecule.

9-4. The Γ^0 (or Γ^{3N}) representation

The mass weighted displacements of the nuclei of a molecule from their equilibrium positions q_i can be used to generate a representation of the point group to which the molecule belongs. If under the symmetry operation $\boldsymbol{R}$, the mass weighted nuclear displacements

$$q_1, q_2, \dots q_{3N}$$

become

$$q_1', q_2', \dots q_{3N}'$$

then we can write

$$q_i' = \sum_{j=1}^{3N} D_{ij}^0(\boldsymbol{R}) q_j, \qquad i = 1, 2, \dots 3N \tag{9-4.1}$$

and the matrices $D^0(\boldsymbol{R})$ will form a representation Γ^0 of the point group. This is just a generalization of what we did for a single point in § 5-2. In some books the notation $D^{3N}(\boldsymbol{R})$ and Γ^{3N} is used for this representation.

Alternatively, one can set up mass weighted base vectors with their origins at the equilibrium nuclear positions, and transform these vectors with the symmetry operation $\boldsymbol{R}$, then

$$\boldsymbol{q}_1, \boldsymbol{q}_2, \dots \boldsymbol{q}_{3N}$$

(if we have three base vectors for each of the N nuclei) will become

$$\boldsymbol{q}_1', \boldsymbol{q}_2', \dots \boldsymbol{q}_{3N}'$$

and

$$\boldsymbol{R}\boldsymbol{q}_i = \boldsymbol{q}_i' = \sum_{j=1}^{3N} D_{ji}^0(\boldsymbol{R}) \boldsymbol{q}_j, \qquad i = 1, 2, \dots 3N \tag{9-4.2}$$

(see § 5-4) and we will obtain a representation Γ^0 identical to the one before. The matrices $D^0(\boldsymbol{R})$ will be unitary and Γ^0 will be a unitary representation (as the matrix elements are real, the $D^0(\boldsymbol{R})$ will, in actual fact, be orthogonal).

We can change the basis of the above representation by switching to normal coordinates (or normal coordinate vectors) and obtain a representation Γ^n which is equivalent to Γ^0 (see § 6-2). The change of basis is defined by (see eqn (9-2.20))

$$Q_i = \sum_{j=1}^{3N} C_{ji} q_j, \qquad i = 1, 2, \ldots 3N$$

or

$$\mathbf{Q}_i = \sum_{j=1}^{3N} C_{ji} \mathbf{q}_j, \qquad i = 1, 2, \ldots 3N$$

and the matrices $D^n(\boldsymbol{R})$ of the representation Γ^n, which is equivalent to Γ^0, are found from

$$Q_k' = \sum_{r=1}^{3N} D_{kr}^n(\boldsymbol{R}) Q_r \qquad (9\text{-}4.3)$$

or

$$\boldsymbol{R}\mathbf{Q}_k = \mathbf{Q}_k' = \sum_{r=1}^{3N} D_{rk}^n(\boldsymbol{R}) \mathbf{Q}_r. \qquad (9\text{-}4.4)$$

Since the representations are equivalent, the $D^n(\boldsymbol{R})$ matrices can be found from the $D^0(\boldsymbol{R})$ matrices by a similarity transformation and in Appendix A.9-2 we show that the matrix which does the transforming is the matrix C (formed from the coefficients C_{ij}) so that

$$D^n(\boldsymbol{R}) = C^{-1} D^0(\boldsymbol{R}) C \qquad (\text{all } \boldsymbol{R}). \qquad (9\text{-}4.5)$$

Since both C and $D^0(\boldsymbol{R})$ are orthogonal so is $D^n(\boldsymbol{R})$ (see Appendix A.4-4(g)).

As an example, let us consider three sets of base vectors associated with three identical nuclei which, in their equilibrium positions, are at the corners of an equilateral triangle (such a system belongs to the $\mathscr{D}_{3h}$ point group); cf. Fig. 5-4.3. For the symmetry operation $\boldsymbol{C}_3$ we have from eqn (5-4.3)

$$D^0(\boldsymbol{C}_3) = \begin{Vmatrix} 0 & 0 & 0 & 0 & 0 & 0 & -1/2 & \sqrt{3}/2 & 0 \\ 0 & 0 & 0 & 0 & 0 & 0 & -\sqrt{3}/2 & -1/2 & 0 \\ 0 & 0 & 0 & 0 & 0 & 0 & 0 & 0 & 1 \\ -1/2 & \sqrt{3}/2 & 0 & 0 & 0 & 0 & 0 & 0 & 0 \\ -\sqrt{3}/2 & -1/2 & 0 & 0 & 0 & 0 & 0 & 0 & 0 \\ 0 & 0 & 1 & 0 & 0 & 0 & 0 & 0 & 0 \\ 0 & 0 & 0 & -1/2 & \sqrt{3}/2 & 0 & 0 & 0 & 0 \\ 0 & 0 & 0 & -\sqrt{3}/2 & -1/2 & 0 & 0 & 0 & 0 \\ 0 & 0 & 0 & 0 & 0 & 1 & 0 & 0 & 0 \end{Vmatrix}$$

If we make some reasonable assumptions about the force constants B_{ij} for three nuclei in such a framework and solve the equations:

$$\det(B-\lambda E) = 0,$$

$$\sum_{i=1}^{3N}(B_{ij}-\lambda_k\delta_{ij})C_{ik} = 0, \qquad \begin{array}{l} j = 1, 2, \ldots 3N \\ k = 1, 2, \ldots 3N \end{array}$$

$$\sum_{i=1}^{3N} C_{ik}^2 = 1, \qquad k = 1, 2, \ldots 3N$$

we obtain:

$$Q_1 = (-2q_1+q_4+\sqrt{3}q_5+q_7-\sqrt{3}q_8)/2\sqrt{3} \qquad (\lambda_1 = f_1)$$
$$Q_2 = (2q_2-\sqrt{3}q_4-q_5+\sqrt{3}q_7-q_8)/2\sqrt{3} \qquad (\lambda_2 = f_2)$$
$$Q_3 = (-2q_1+q_4-\sqrt{3}q_5+q_7+\sqrt{3}q_8)/2\sqrt{3} \qquad (\lambda_3 = f_2)$$
$$Q_4 = (2q_2+2q_5+2q_8)/2\sqrt{3} \qquad (\lambda_4 = 0)$$
$$Q_5 = (-2q_1-2q_4-2q_7)/2\sqrt{3} \qquad (\lambda_5 = 0)$$
$$Q_6 = (2q_2+\sqrt{3}q_4-q_5-\sqrt{3}q_7-q_8)/2\sqrt{3} \qquad (\lambda_6 = 0)$$
$$Q_7 = (q_3+q_6+q_9)/\sqrt{3} \qquad (\lambda_7 = 0)$$
$$Q_8 = (2q_3-q_6-q_9)/\sqrt{6} \qquad (\lambda_8 = 0)$$
$$Q_9 = (q_6-q_9)/\sqrt{2} \qquad (\lambda_9 = 0)$$

where the numerical values of f_1 and f_2 depend on the particular choice of force constants, and where λ_4 to λ_9 correspond to translations and rotations. Hence

$$C = \frac{1}{2\sqrt{3}}\left\Vert\begin{array}{ccccccccc}
-2 & 0 & -2 & 0 & -2 & 0 & 0 & 0 & 0 \\
0 & 2 & 0 & 2 & 0 & 2 & 0 & 0 & 0 \\
0 & 0 & 0 & 0 & 0 & 0 & 2 & 2\sqrt{2} & 0 \\
1 & -\sqrt{3} & 1 & 0 & -2 & \sqrt{3} & 0 & 0 & 0 \\
\sqrt{3} & -1 & -\sqrt{3} & 2 & 0 & -1 & 0 & 0 & 0 \\
0 & 0 & 0 & 0 & 0 & 0 & 2 & -\sqrt{2} & \sqrt{6} \\
1 & \sqrt{3} & 1 & 0 & -2 & -\sqrt{3} & 0 & 0 & 0 \\
-\sqrt{3} & -1 & \sqrt{3} & 2 & 0 & -1 & 0 & 0 & 0 \\
0 & 0 & 0 & 0 & 0 & 0 & 2 & -\sqrt{2} & -\sqrt{6}
\end{array}\right\Vert$$

and since C is orthogonal (see Appendix A.4-3(c)) $C^{-1} = \tilde{C}$. We can now evaluate $D^n(C_3)$ from:

$$D^n(C_3) = C^{-1}D^0(C_3)C = \tilde{C}D^0(C_3)C$$

$$= \begin{Vmatrix} 1 & 0 & 0 & 0 & 0 & 0 & 0 & 0 & 0 \\ 0 & -1/2 & -\sqrt{3}/2 & 0 & 0 & 0 & 0 & 0 & 0 \\ 0 & \sqrt{3}/2 & -1/2 & 0 & 0 & 0 & 0 & 0 & 0 \\ 0 & 0 & 0 & -1/2 & \sqrt{3}/2 & 0 & 0 & 0 & 0 \\ 0 & 0 & 0 & -\sqrt{3}/2 & -1/2 & 0 & 0 & 0 & 0 \\ 0 & 0 & 0 & 0 & 0 & 1 & 0 & 0 & 0 \\ 0 & 0 & 0 & 0 & 0 & 0 & 1 & 0 & 0 \\ 0 & 0 & 0 & 0 & 0 & 0 & 0 & -1/2 & -\sqrt{3}/2 \\ 0 & 0 & 0 & 0 & 0 & 0 & 0 & \sqrt{3}/2 & -1/2 \end{Vmatrix} \tag{9-4.6}$$

It is apparent that $D^n(C_3)$ is in block form and since the same block form appears for all the other symmetry operations of the point group, the Γ^0 representation has been reduced by the change to the normal coordinate basis. That such a reduction will *always* occur, is a point taken up in the next section. Needless to say Γ^0 and Γ^n have identical characters i.e. $\chi^0(\boldsymbol{R}) = \chi^n(\boldsymbol{R})$, for all $\boldsymbol{R}$.

The reader might note that once the matrix C has been determined it is possible, with the aid of eqn (9-2.21) to give a 'picture' of a normal mode. This is done by depicting the displacement of each nucleus, at some instant in the course of the normal vibration, by an arrow whose length is proportional to the displacement. In Fig. 9-4.1 pictures are given for the first three normal modes of the previous example.

9-5. The reduction of Γ^0

The previous example was not an exception: it is generally true that Γ^0 can be reduced by a change of coordinates (or base vectors) from

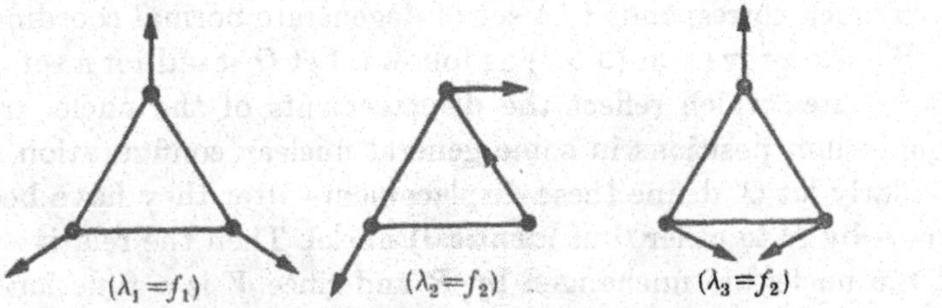

FIG. 9-4.1. The first three normal modes for the system described in § 9-4.

q to Q (or $\boldsymbol{q}$ to $\boldsymbol{Q}$). As has been mentioned before, the same λ value may occur more than once in the solution of eqn (9-2.15) (in fact, we know for sure that $\lambda = 0$ will occur six times), so let us consider that there are M *distinct* λ values $\lambda_1, \lambda_2, \ldots$ and λ_M and group together those normal coordinates which are associated with the same λ value by subscripting them in the following way:

$$\begin{array}{llll} Q_{1(1)}, & Q_{1(2)}, \ldots & Q_{1(n_1)}: & \lambda_1 \\ Q_{2(1)}, & Q_{2(2)}, \ldots & Q_{2(n_2)}: & \lambda_2 \\ \cdot & \cdot & \cdot & \cdot \\ \cdot & \cdot & \cdot & \cdot \\ \cdot & \cdot & \cdot & \cdot \\ Q_{\nu(1)}, & Q_{\nu(2)}, \ldots & Q_{\nu(n_\nu)}: & \lambda_\nu \\ \cdot & \cdot & \cdot & \cdot \\ \cdot & \cdot & \cdot & \cdot \\ \cdot & \cdot & \cdot & \cdot \\ Q_{M(1)}, & Q_{M(2)}, \ldots & Q_{M(n_M)}: & \lambda_M \end{array}$$

where n_ν is the number of normal coordinates associated with λ_ν or the number of times λ_ν occurs in the solution of eqn (9-2.15).

We can then replace eqn (9-2.18) by

$$V = \tfrac{1}{2} \sum_{\rho=1}^{M} \sum_{m=1}^{n_\rho} \lambda_\rho Q^2_{\rho(m)} \tag{9-5.1}$$

and eqn (9-4.3) by

$$Q'_{\mu(i)} = \sum_{\nu=1}^{M} \sum_{j=1}^{n_\nu} D^n_{\mu(i)\nu(j)}(\boldsymbol{R}) Q_{\nu(j)} \tag{9-5.2}$$

and we can prove that if $\mu \neq \nu$,

$$D^n_{\mu(i)\nu(j)}(\boldsymbol{R}) = 0 \qquad \text{(for all } \boldsymbol{R}\text{)}. \tag{9-5.3}$$

The implication of this last equation is that the $D^n(\boldsymbol{R})$ are in block form with each $n_\nu \times n_\nu$ block corresponding to n_ν identical λ values, i.e. each block corresponds to a set of degenerate normal coordinates.

We can prove eqn (9-5.3) as follows. Let $\boldsymbol{Q}$ stand for a set of normal coordinates which reflect the displacements of the nuclei from their equilibrium positions in some general nuclear configuration X_{nuc} and similarly let $\boldsymbol{Q}'$ define these displacements after they have been transferred by $\boldsymbol{R}$ to other (but identical) nuclei. Then the relative positions of the nuclei are unchanged by $\boldsymbol{R}$ and since V is a function solely of these relative positions (see the footnote to eqn (9-2.7)), we must have

$V(Q) = V(Q')$, (see also eqn (A.8-1.6)). Now

$$V(Q') = \tfrac{1}{2}\sum_{\rho=1}^{M}\sum_{m=1}^{n_\rho}\lambda_\rho Q'^2_{\rho(m)}$$

$$= \tfrac{1}{2}\sum_{\rho=1}^{M}\sum_{m=1}^{n_\rho}\lambda_\rho\left\{\sum_{\sigma=1}^{M}\sum_{k=1}^{n_\sigma}D^n_{\rho(m)\sigma(k)}(\boldsymbol{R})Q_{\sigma(k)}\right\}\left\{\sum_{\nu=1}^{M}\sum_{j=1}^{n_\nu}D^n_{\rho(m)\nu(j)}(\boldsymbol{R})Q_{\nu(j)}\right\}$$

$$= \tfrac{1}{2}\sum_{\rho=1}^{M}\sum_{m=1}^{n_\rho}\sum_{\sigma=1}^{M}\sum_{k=1}^{n_\sigma}\sum_{\nu=1}^{M}\sum_{j=1}^{n_\nu}\lambda_\rho D^n_{\rho(m)\sigma(k)}(\boldsymbol{R})D^n_{\rho(m)\nu(j)}(\boldsymbol{R})Q_{\sigma(k)}Q_{\nu(j)}$$

and

$$V(Q) = \tfrac{1}{2}\sum_{\nu=1}^{M}\sum_{j=1}^{n_\nu}\lambda_\nu Q^2_{\nu(j)}$$

hence, comparing terms,

$$\sum_{\rho=1}^{M}\sum_{m=1}^{n_\rho}\lambda_\rho D^n_{\rho(m)\sigma(k)}(\boldsymbol{R})D^n_{\rho(m)\nu(j)}(\boldsymbol{R}) = \lambda_\nu\delta_{\nu\sigma}\delta_{jk}. \qquad (9\text{-}5.4)$$

Since $D^n(\boldsymbol{R})$ is orthogonal

$$\sum_{\sigma=1}^{M}\sum_{k=1}^{n_\sigma}D^n_{\rho(m)\sigma(k)}(\boldsymbol{R})D^n_{\mu(i)\sigma(k)}(\boldsymbol{R}) = \sum_{\sigma=1}^{M}\sum_{k=1}^{n_\sigma}D^n_{\rho(m)\sigma(k)}(\boldsymbol{R})\{D^n(\boldsymbol{R})\}^{-1}_{\sigma(k)\mu(i)}$$
$$= \delta_{\rho\mu}\delta_{mi},$$

and multiplying eqn (9-5.4) by $D^n_{\mu(i)\sigma(k)}(\boldsymbol{R})$ and summing over σ and k, we obtain:

$$\sum_{\rho=1}^{M}\sum_{m=1}^{n_\rho}\lambda_\rho D^n_{\rho(m)\nu(j)}(\boldsymbol{R})\delta_{\rho\mu}\delta_{mi} = \sum_{\sigma=1}^{M}\sum_{k=1}^{n_\sigma}\lambda_\nu D^n_{\mu(i)\sigma(k)}(\boldsymbol{R})\delta_{\nu\sigma}\delta_{jk}$$

or

$$\lambda_\mu D^n_{\mu(i)\nu(j)}(\boldsymbol{R}) = \lambda_\nu D^n_{\mu(i)\nu(j)}(\boldsymbol{R}).$$

So that if $\mu \neq \nu$ and consequently $\lambda_\mu \neq \lambda_\nu$, then

$$D^n_{\mu(i)\nu(j)}(\boldsymbol{R}) = 0$$

and we have shown the truth of eqn (9-5.3).

Thus

$$C^{-1}D^0(\boldsymbol{R})C = D^n(\boldsymbol{R}) = \begin{Vmatrix} D^1(\boldsymbol{R}) & [0] & [0] & \cdot \\ [0] & D^2(\boldsymbol{R}) & [0] & \cdot \\ [0] & [0] & D^3(\boldsymbol{R}) & \cdot \\ \cdot & \cdot & \cdot & \cdot \\ \cdot & \cdot & \cdot & \cdot \\ \cdot & \cdot & \cdot & \cdot \end{Vmatrix}$$

with the same block form for all $\boldsymbol{R}$, where $D^\nu(\boldsymbol{R})$ is a $n_\nu \times n_\nu$ matrix associated with λ_ν which represents $\boldsymbol{R}$ in the representation Γ^ν. [0] is a rectangular array with all the elements zero. It turns out (see the next section) that the $D^\nu(\boldsymbol{R})$ can be assumed to be irreducible representations for the point group concerned.

Our previous example of three nuclei at the corners of an equilateral triangle (their equilibrium positions) belongs to the $\mathscr{D}_{3h}$ point group and eqn (9-4.6) can now be written as:

$$D^n(C_3) = \begin{Vmatrix} D^{A_1'}(C_3) & [0] & [0] & [0] & [0] & [0] \\ [0] & D^{E'}(C_3) & [0] & [0] & [0] & [0] \\ [0] & [0] & D^{E'}(C_3) & [0] & [0] & [0] \\ [0] & [0] & [0] & D^{A_2'}(C_3) & [0] & [0] \\ [0] & [0] & [0] & [0] & D^{A_2''}(C_3) & [0] \\ [0] & [0] & [0] & [0] & [0] & D^{E''}(C_3) \end{Vmatrix}.$$

There are similar equations for the other operations of the point group.† Therefore, Q_1 forms a basis for the $\Gamma^{A_1'}$ representation and is a *vibrational* normal coordinate, Q_2 and Q_3 together form a basis for the $\Gamma^{E'}$ representation and they are also *vibrational* normal coordinates, the rest correspond to $\lambda = 0$ and are *translational* or *rotational* normal coordinates, Q_4 and Q_5 belong to $\Gamma^{E'}$, Q_6 to $\Gamma^{A_2'}$, Q_7 to $\Gamma^{A_2''}$ and Q_8 and Q_9 to $\Gamma^{E''}$.

9-6. The classification of normal coordinates

We are now in the position of being able to determine the irreducible representations to which the different normal coordinates belong. From this knowledge it will be possible to find out which of the fundamental frequencies are infra-red or Raman active. The reduction of Γ^0 (which is equivalent to Γ^n) gives

$$\Gamma^0(\equiv \Gamma^n) = \Gamma^1 \oplus \Gamma^2 \dots \oplus \Gamma^{M-1} \oplus \Gamma^{t,r} = \Gamma^v \oplus \Gamma^{t,r}$$

where $\Gamma^{t,r}$ corresponds to the translational and rotational normal coordinates ($\lambda = 0$) and Γ^v to the vibrational normal coordinates. Of course, in the above sequence some of the Γ's may be the same even though the normal coordinates correspond to different λ's; there is no reason why more than one set of normal coordinates should not belong to the same representation. As was mentioned before, we will assume that Γ^1, Γ^2,... Γ^{M-1} are irreducible representations, that is we will ignore the possible occurrence of accidental degeneracies where the λ's are equal not because of symmetry but because a fortuitous set of

† To assign to the individual blocks the correct symmetry species, one simply finds the character of each block for each R and compares the result with the $\mathscr{D}_{3h}$ character table.

force constants B_{ij} exists. On the other hand $\Gamma^{t,r}$ will be reducible. From the characters $\chi^0(R)$, $\chi^{t,r}(R)$ we can find $\chi^v(R)$ by using

$$\chi^v(R) = \chi^0(R) - \chi^{t,r}(R)$$

and from the standard decomposition formula (eqn (7-4.2)) we can then find the number of distinct fundamental levels and their degeneracies. The irreducible representation Γ^v which is generated by $Q_{v(i)}$ ($i = 1, 2, \ldots n_v$) is called the *symmetry species* of the coordinate(s) $Q_{v(i)}$ ($i = 1, 2, \ldots n_v$) and it determines the vibrational selection rules.

There are simple formulae for finding the character of Γ^0, which can be written as

$$\chi^0(R) = \sum_{i=1}^{3N} D^0_{ii}(R).$$

In the first place, only the nuclei which are unmoved by R can contribute to the diagonal of $D^0(R)$ and, referring to § 5-2, for these there is a contribution per nucleus of:

R	E	$C(\theta)$	σ	$S(\theta)$	i
$\chi(R)$	3	$1+2\cos\theta$	1	$-1+2\cos\theta$	-3

Since $E = C(2\pi)$, $\sigma = S(2\pi)$ and $i = S(\pi)$, we can abbreviate things even further and put

$$\chi^0(R) = n_R\chi(R) \tag{9-6.1}$$

where

$$\chi(R) = 1+2\cos\theta \quad \text{for} \quad R = C(\theta)$$
$$\chi(R) = -1+2\cos\theta \quad \text{for} \quad R = S(\theta)$$
$$n_R = \text{number of unmoved nuclei.}$$

It is also quite straight forward to determine the character of $\Gamma^{t,r}$. For a non-linear molecule there are three translational normal coordinates and three rotational normal coordinates and we write $\Gamma^{t,r} = \Gamma^t \oplus \Gamma^r$. The translational motion corresponds to a displacement of the molecule in some arbitrary direction and it can be depicted by a single vector showing the displacement of the centre of mass. Let this vector be

$$x\mathbf{e}_1 + y\mathbf{e}_2 + z\mathbf{e}_3$$

(x, y, and z are Cartesian coordinates). We have shown before that a symmetry operation will transform this vector to

$$x'\mathbf{e}_1 + y'\mathbf{e}_2 + z'\mathbf{e}_3$$

where x', y', and z' are certain linear combinations of x, y, and z. Thus the three translational coordinates will generate exactly the same representation that is generated by position vectors in physical space and we conclude that the set of coordinates x, y, and z forms a basis for a three-dimensional representation of the point group; we call this representation Γ^t and $\chi^t(\boldsymbol{R})$ will be identical to the $\chi(\boldsymbol{R})$ given below eqn (9-6.1). The irreducible representation(s) to which x, y, and z belong are usually given in the character tables.† But note that 'x; y' in a character table has a different meaning from '(x, y)'; the absence of parentheses indicates that x forms a basis for a one-dimensional irreducible representation and y forms a basis for the same irreducible representation as x, while the presence of parentheses indicates that x and y together form a basis for a two-dimensional irreducible representation.

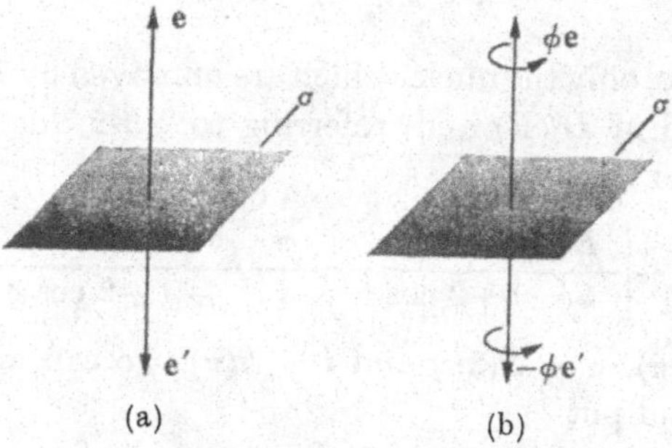

FIG. 9-6.1. Effect of a reflection on (a) a position vector; (b) an axial vector.

We must also find the symmetry species of the rotation and we follow the treatment given by Levine. An arbitrary rigid rotation of the molecule can be resolved into rotations by various angles about the three translational axes x, y, and z. We can represent the rotation about the x axis by a single vector $\boldsymbol{R}_x$ pointing along the x axis. The length of $\boldsymbol{R}_x$ is proportional to the angle of rotation and its direction is given by the direction the thumb points when the fingers of the right hand curl in the direction of rotation. The effect of any symmetry operation on the nuclear displacement vectors of a rotation defines its effect on $\boldsymbol{R}_x$. Thus, the C_2 rotation about the z axis sends the displacement vectors for rotation about the x axis into their negatives, thereby reversing the direction of this rotation; hence $\boldsymbol{R}_x$, whose direction is defined relative to the direction of rotation is reversed in direction. Reflection in the yz plane $\boldsymbol{\sigma}_{yz}$ has no effect on the displacement vectors for rotation about the x axis; hence $\boldsymbol{R}_x$ is unchanged by this symmetry operation. Thus the effect of $\boldsymbol{\sigma}_{yz}$ on $\boldsymbol{R}_x$ differs from its effect on a vector pointing along the x axis and representing translational motion; the latter vector is changed to its negative by $\boldsymbol{\sigma}_{yz}$ (see Fig. 9-6.1). Vectors

† Proper orientation of $\mathbf{e}_1$, $\mathbf{e}_2$ and $\mathbf{e}_3$ will ensure that Γ^t is in reduced form.

like R_x are called axial vectors or pseudo-vectors. The symmetry operations transform each of three axial vectors representing molecular rotation about the x, y, and z axes into linear combinations of one another, thereby generating the representation Γ^{r}, which corresponds to the molecular rotational modes.

The characters of Γ^{r} are related to those of Γ^{t}. Consider first the effect on, say R_x, of a $C(\theta)$ rotation about some axis, not necessarily the x axis. This rotation will move the rotation displacement vectors in such a manner as to transform R_x into a vector R'_x, where R'_x is the vector obtained by applying $C(\theta)$ directly to R_x. Thus, for proper rotations, the matrices describing how R_x, R_y, and R_z transform are exactly those matrices that describe how ordinary position vectors along the x, y, and z axes transform; the matrix in the representation Γ^{r} corresponding to any $C(\theta)$ is therefore the same as the matrix in Γ^{t} that corresponds to $C(\theta)$ and the characters for proper rotations are the same for Γ^{r} as for Γ^{t}.

Now consider the effect on R_x of a $S(\theta)$ operation. We have

$$S(\theta) = \sigma C(\theta).$$

The $C(\theta)$ part has the same effect on R_x as on a translational vector along the x axis. However the effect of a reflection on R_x is opposite to its effect on a translational vector along the x axis. Thus the matrices describing the effect of $S(\theta)$ on the rotation vectors are the negatives of the matrices describing the effect of $S(\theta)$ on the translational vectors. The character of any $S(\theta)$ for Γ^{r} is the negative of its character in Γ^{t}.

Summarizing, we have:

$$\begin{aligned}
\chi^{\mathrm{r}}\{C(\theta)\} &= \chi^{\mathrm{t}}\{C(\theta)\} \\
\chi^{\mathrm{r}}\{S(\theta)\} &= -\chi^{\mathrm{t}}\{S(\theta)\} \\
\chi^{\mathrm{t,r}}(R) &= 2(1+2\cos\theta) \quad \text{if} \quad R = C(\theta) \\
\chi^{\mathrm{t,r}}(R) &= 0 \quad \text{if} \quad R = S(\theta).
\end{aligned}$$

If we return to our equilateral triangle example and apply the above rules, we obtain the numbers in Table 9-6.1. Using this table in conjunction with the $\mathscr{D}_{3h}$ character table (Table 9-7.2) and the decomposition rule:

$$a_\mu = g^{-1} \sum_R \chi^0(R)\chi^\mu(R)^* = g^{-1} \sum_{i=1}^{k} g_i \chi^0(C_i)\chi^\mu(C_i)^*$$

we obtain

$$\Gamma^{\mathrm{t}} = \Gamma^{E'} \oplus \Gamma^{A_2''}$$

$$\Gamma^{\mathrm{r}} = \Gamma^{A_2'} \oplus \Gamma^{E''}$$

and

$$\Gamma^{\mathrm{v}} = \Gamma^{A_1'} \oplus \Gamma^{E'}.$$

TABLE 9-6.1

Characters for the Γ^0 representation of the $\mathscr{D}_{3h}$ point group

$\mathscr{D}_{3h}$	E	$2C_3$	$3C_2$	σ_h	$2S_3$	$3\sigma_v$
$\chi^0(C_i)$	9	0	-1	3	0	1
$\chi^t(C_i)$	3	0	-1	1	-2	1
$\chi^r(C_i)$	3	0	-1	-1	2	-1
$\chi^v(C_i)$	3	0	1	3	0	1

9-7. Further examples of normal coordinate classification

In this section three more examples of normal coordinate classification are given. We consider (1) H_2O, (2) CO_3^{2-} and (3) CH_4.

(1) The character table for the point group of H_2O $\mathscr{C}_{2v}$ is given in Table 9-7.1 and below it we show the characters for the Γ^0 representation (found from eqn (9-6.1)). From these characters we obtain

$$\Gamma^0 = 3\Gamma^{A_1} \oplus \Gamma^{A_2} \oplus 2\Gamma^{B_1} \oplus 3\Gamma^{B_2}.$$

As is usual practice, the representations for Γ^t and Γ^r are also given in the character table and we directly obtain

$$\Gamma^t = \Gamma^{A_1} \oplus \Gamma^{B_1} \oplus \Gamma^{B_2}$$

and

$$\Gamma^r = \Gamma^{A_2} \oplus \Gamma^{B_1} \oplus \Gamma^{B_2}.$$

Hence, by subtracting these from Γ^0 we have

$$\Gamma^v = 2\Gamma^{A_1} \oplus \Gamma^{B_2}.$$

So that there are three distinct (having different λ's or ν's) non-degenerate vibrational normal modes with symmetry species Γ^{A_1} (two) and Γ^{B_2} (one).

TABLE 9-7.1

Character table for $\mathscr{C}_{2v}$†

$\mathscr{C}_{2v}$		E	C_2	σ_v	σ_v'
A_1	z	1	1	1	1
A_2	R_z	1	1	-1	-1
B_1	x; R_y	1	-1	1	-1
B_2	y; R_x	1	-1	-1	1
$\chi^0(C_i)$–H_2O		9	-1	1	3

† $\sigma_v = \sigma_{xz}$ (perpendicular to the molecular plane); $\sigma_v' = \sigma_{yz}$ (molecular plane).

TABLE 9-7.2

Character table for $\mathscr{D}_{3h}$

$\mathscr{D}_{3h}$		E	$2C_3$	$3C_2$	σ_h	$2S_3$	$3\sigma_v$
A_1'		1	1	1	1	1	1
A_2'	R_z	1	1	−1	1	1	−1
E'	(x, y)	2	−1	0	2	−1	0
A_1''		1	1	1	−1	−1	−1
A_2''	z	1	1	−1	−1	−1	1
E''	(R_x, R_y)	2	−1	0	−2	1	0
$\chi^0(C_i)$–CO_3^{2-}		12	0	−2	4	−2	2

(2) The character table for the point group of $CO_3^{2-}(\mathscr{D}_{3h})$ is given in Table 9-7.2 and below it we show the characters for the Γ^0 representation. From these characters we obtain

$$\Gamma^0 = \Gamma^{A_1'} \oplus \Gamma^{A_2'} \oplus 3\Gamma^{E'} \oplus 2\Gamma^{A_2''} \oplus \Gamma^{E''}$$

and, by inspection of the table,

$$\Gamma^t = \Gamma^{E'} \oplus \Gamma^{A_2''}$$

and

$$\Gamma^r = \Gamma^{A_2'} \oplus \Gamma^{E''}.$$

Hence

$$\Gamma^v = \Gamma^{A_1'} \oplus 2\Gamma^{E'} \oplus \Gamma^{A_2''}$$

and there are two distinct non-degenerate vibrational modes of species $\Gamma^{A_1'}$ and $\Gamma^{A_2''}$ and two distinct doubly degenerate modes with the same symmetry species $\Gamma^{E'}$.

(3) The character table for the point group of CH_4 ($\mathscr{T}_d$) is given in Table 9-7.3 and below it we show the characters for the Γ^0 representation. From these characters we obtain

$$\Gamma^0 = \Gamma^{A_1} \oplus \Gamma^{E} \oplus \Gamma^{T_1} \oplus 3\Gamma^{T_2}$$

and, by inspection of the table,

$$\Gamma^t = \Gamma^{T_2}$$

TABLE 9-7.3

Character table for $\mathscr{T}_d$

$\mathscr{T}_d$	E	$8C_3$	$3C_2$	$6S_4$	$6\sigma_d$
A_1	1	1	1	1	1
A_2	1	1	1	−1	−1
E	2	−1	2	0	0
$T_1(R_x, R_y, R_z)$	3	0	−1	1	−1
$T_2(x, y, z)$	3	0	−1	−1	1
$\chi^0(C_i)$–CH_4	15	0	−1	−1	3

and

$$\Gamma^{r} = \Gamma^{T_1}.$$

Hence

$$\Gamma^{v} = \Gamma^{A_1} \oplus \Gamma^{E} \oplus 2\Gamma^{T_2}$$

and the nine vibrational normal coordinates (or modes) can be classified as follows: one gives rise to Γ^{A_1}, two are doubly degenerate and give rise to Γ^{E}, three are triply degenerate and give rise to Γ^{T_2}, and the last three give rise to the same representation Γ^{T_2} but with a different λ (or ν) value from the previous one.

9-8. Normal coordinates for linear molecules

Linear molecules belong to the $\mathscr{C}_{\infty v}$ point group if they are unsymmetrical and to the $\mathscr{D}_{\infty h}$ point group if they are symmetrical. Once again they are special cases and we will only state the results. There are two kinds of normal coordinate for linear molecules: longitudinal, in which the nuclei have undergone longitudinal displacements along the molecular axis and transverse, where the nuclei have been displaced perpendicularly to the molecular axis.

For an unsymmetrical linear molecule there are $N-1$ non-degenerate vibrational normal coordinates belonging to the symmetry species Σ^+ (see Table 7-8.1) and they are of the longitudinal type; and $N-2$ pairs of vibrational normal coordinates (doubly degenerate), each pair belonging to the Π symmetry species and they are of the transverse type.

For symmetrical linear molecules, if there is an even number of nuclei, there are $N/2$ vibrational normal coordinates belonging to Σ_g^+, $N/2-1$ belonging to Σ_u^+, (these are of the longitudinal type) and there are $N/2-1$ pairs belonging to Π_g and $N/2-1$ pairs belonging to Π_u (these are of the transverse type). If there is an odd number of nuclei, there are $(N-1)/2$ vibrational normal coordinates belonging to Σ_g^+ and $(N-1)/2$ belonging to Σ_u^+ (these are of the longitudinal type) and $(N-3)/2$ pairs belonging to Π_g and $(N-1)/2$ pairs belonging to Π_u (these are of the transverse type).

9-9. Classification of the vibrational levels

Classifying the vibrational energy levels means finding out to which irreducible representation of the molecular point group the vibrational wavefunction(s) associated with a given level belong.

First let us consider the vibrational ground state. The corresponding wavefunction is (see eqn (9-3.9))

$$\psi_0^{\text{vib}} = N \exp\left(-\tfrac{1}{2} \sum_{i=1}^{3N-6} \alpha_i Q_i^2\right)$$

which, written in the style of § 9-5, becomes

$$\psi_0^{\text{vib}} = N \exp(-\tfrac{1}{2} \sum_{\rho=1}^{M-1} \alpha_\rho \sum_{m=1}^{n_\rho} Q^2_{\rho(m)}) \tag{9-9.1}$$

where

$$\alpha_\rho = \frac{2\pi}{h} \lambda_\rho^{\frac{1}{2}} = \frac{4\pi^2}{h} \nu_\rho.$$

Under the symmetry operation $\boldsymbol{R}$ the normal coordinate $Q_{\rho(k)}$ is transformed to $Q'_{\rho(k)}$ which is a linear combination of $Q_{\rho(1)}, \ldots Q_{\rho(n_\rho)}$ and since these coordinates form a basis for the ρth irreducible representation, we have

$$Q'_{\rho(k)} = \sum_{m=1}^{n_\rho} D^\rho_{km}(\boldsymbol{R}) Q_{\rho(m)} \tag{9-9.2}$$

which is the 'reduced version' of eqn (9-4.3) and where $D^\rho(\boldsymbol{R})$ is the matrix representing $\boldsymbol{R}$ in the ρth irreducible representation of the point group to which the molecule belongs. Since the $D^n(\boldsymbol{R})$ are orthogonal, the individual blocks, e.g. $D^\rho(\boldsymbol{R})$, must also be orthogonal and hence

$$Q_{\rho(m)} = \sum_{k=1}^{n_\rho} D^\rho_{km}(\boldsymbol{R}) Q'_{\rho(k)}. \tag{9-9.3}$$

If we substitute eqn (9-9.3) in eqn (9-9.1), we find

$$\psi_0^{\text{vib}}(Q) = N \exp\left(-\tfrac{1}{2} \sum_{\rho=1}^{M-1} \alpha_\rho \sum_{m=1}^{n_\rho} \sum_{j=1}^{n_\rho} \sum_{k=1}^{n_\rho} D^\rho_{jm}(\boldsymbol{R}) D^\rho_{km}(\boldsymbol{R}) Q'_{\rho(j)} Q'_{\rho(k)}\right) \tag{9-9.4}$$

where the Q in $\psi_0^{\text{vib}}(Q)$ indicates that the wavefunction take the values of the normal coordinates before the symmetry operation $\boldsymbol{R}$ is applied. Since $D^\rho(\boldsymbol{R})$ is orthogonal

$$\sum_{m=1}^{n_\rho} D^\rho_{jm}(\boldsymbol{R}) D^\rho_{km}(\boldsymbol{R}) = \delta_{jk}$$

and we can replace eqn (9-9.4) by

$$\begin{aligned}\psi_0^{\text{vib}}(Q) &= N \exp\left(-\tfrac{1}{2} \sum_{\rho=1}^{M-1} \alpha_\rho \sum_{j=1}^{n_\rho} \sum_{k=1}^{n_\rho} \delta_{jk} Q'_{\rho(j)} Q'_{\rho(k)}\right) \\ &= N \exp\left(-\tfrac{1}{2} \sum_{\rho=1}^{M-1} \alpha_\rho \sum_{k=1}^{n_\rho} Q'^2_{\rho(k)}\right) = \psi_0^{\text{vib}}(Q'). \end{aligned} \tag{9-9.5}$$

If we define the symmetry transformation operator $\boldsymbol{O_R}$ in the usual way by

$$\boldsymbol{O_R} \psi_0^{\text{vib}}(Q') = \psi_0^{\text{vib}}(Q),$$

we have from eqn (9-9.5)

$$\boldsymbol{O_R} \psi_0^{\text{vib}}(Q') = \psi_0^{\text{vib}}(Q')$$

or

$$\boldsymbol{O_R} \psi_0^{\text{vib}} = \psi_0^{\text{vib}} \qquad (\text{all } \boldsymbol{R}).$$

Consequently ψ_0^{vib} generates the identical representation Γ^1 of the point group and we say that the ground state vibrational wavefunction is totally symmetric.

The fundamental vibrational energy levels lie at an energy $h\nu_\rho$ above the ground state. Each level has associated with it a certain fundamental frequency ν_ρ and therefore a certain λ value λ_ρ and each λ_ρ has associated with it n_ρ normal coordinates $Q_{\rho(m)}$ $(m = 1, 2, \ldots n_\rho)$. Coupled with each of these normal coordinates there is a vibrational wavefunction ψ_m^ρ which is proportional to $\psi_0^{\text{vib}} Q_{\rho(m)}$ (see eqn (9-3.11)). Thus the fundamental levels have an n_ρ-fold degeneracy; that is there are n_ρ wavefunctions having the same energy.

If we write these degenerate wavefunctions as

$$\psi_m^\rho(Q) = N\psi_0^{\text{vib}}(Q)Q_{\rho(m)} \qquad m = 1, 2, \ldots n_\rho$$

where N is a normalization constant, and substitute eqns (9-9.3) and (9-9.5) we get

$$\psi_m^\rho(Q) = N\psi_0^{\text{vib}}(Q') \sum_{k=1}^{n_\rho} D_{km}^\rho(R) Q'_{\rho(k)} = \sum_{k=1}^{n_\rho} D_{km}^\rho(R)\psi_k^\rho(Q').$$

Hence

$$O_R\psi_m^\rho(Q') = \psi_m^\rho(Q) = \sum_{k=1}^{n_\rho} D_{km}^\rho(R)\psi_k^\rho(Q')$$

and

$$O_R\psi_m^\rho = \sum_{k=1}^{n_\rho} D_{km}^\rho(R)\psi_k^\rho.$$

Therefore the wavefunctions ψ_m^ρ $(m = 1, 2, \ldots n_\rho)$ form a basis for the irreducible representation Γ^ρ, the same representation to which the normal coordinates $Q_{\rho(1)}, \ldots Q_{\rho(n_\rho)}$ associated with the fundamental frequency ν_ρ belong.

This result is tremendously useful, it not only leads to selection rules for vibrational spectroscopy but also, as was the case with electronic wavefunctions (see § 8-2), allows us to predict from inspection of the character table the degeneracies and symmetries which are allowed for the fundamental vibrational wavefunctions of any particular molecule.

9-10. Infra-red spectra

If incident radiation with a frequency equal to one of the fundamental frequencies falls on a molecule, it may make a transition from the ground state to the appropriate fundamental level. These normal frequencies usually occur in the infra-red spectral region. The probability of such a transition occurring, however, depends on the relationship between the molecule's electric dipole moment (as a function of the nuclear coordinates) and the wavefunctions of the ground state and of the fundamental level.

Consider a set of mutually perpendicular axes (specified by the vectors $\mathbf{e}_1$, $\mathbf{e}_2$, and $\mathbf{e}_3$) with their origins at that point in the molecule which is unmoved by all the symmetry operations of the point group and let us refer the position of any nucleus to a set of axes (parallel to $\mathbf{e}_1$, $\mathbf{e}_2$, and $\mathbf{e}_3$) whose origin is at the equilibrium position of the nucleus. Then the electric dipole moment of the molecule when the nuclei and electrons are in positions described by some configuration X, is the vector

$$\mu_1(X)\mathbf{e}_1+\mu_2(X)\mathbf{e}_2+\mu_3(X)\mathbf{e}_3$$

where

$$\mu_k(X) = \sum_{p=1}^{N} x_k^p e^p$$

and x_k^p is the coordinate of the pth nucleus in the direction of $\mathbf{e}_k$ and e^p is the effective charge on the pth nucleus.

Quantum mechanics tells us that the probability of an electric dipole induced transition occurring between the states described by ψ_n and ψ_m is proportional to

$$\sum_{k=1}^{3}\left(\int \psi_n^* \mu_k \psi_m \, \mathrm{d}\tau\right)^2$$

where the integration is over the whole range of the nuclear coordinates. Consequently a transition from the vibrational ground state to the ρth fundamental level is forbidden (has zero probability) if

$$\int \psi_0^{\text{vib}*} \mu_k \psi_m^\rho \, \mathrm{d}\tau = 0$$

for all m values (1, 2,... n_ρ) and all k values (1, 2, and 3). The vectors $\mathbf{e}_1$, $\mathbf{e}_2$, and $\mathbf{e}_3$ generate the same representation Γ^{t} as the translational normal coordinates and therefore

$$\boldsymbol{R}\mathbf{e}_i = \sum_{j=1}^{3} D_{ji}^{\mathrm{t}}(\boldsymbol{R})\mathbf{e}_j$$

and it can easily be shown that

$$\boldsymbol{O}_{\boldsymbol{R}}\mu_k = \sum_{i=1}^{3} D_{ik}^{\mathrm{t}}(\boldsymbol{R})\mu_i \qquad k = 1, 2, 3.$$

Therefore the $\mu_k(X)$ form a basis for Γ^{t}. There will be a particular choice of axes $\mathbf{e}_1$, $\mathbf{e}_2$, and $\mathbf{e}_3$ which will reduce Γ^{t} to the irreducible representations which it contains and this choice is the same as the choice of translational axes which reduces Γ^{t}. Therefore μ_1 belongs to the same irreducible representation as x, μ_2 belongs to the same one as y and μ_3 belongs to the same one as z, and these can be found by consulting the appropriate character table.

Since ψ_0^{vib} belongs to Γ^1 and ψ_m^ρ belongs to Γ^ρ, the conclusions of § 8-4 dictate that

$$\int \psi_0^{\text{vib}*}\mu_k\psi_m^\rho \, \mathrm{d}\tau = 0$$

unless Γ^ρ coincides with the irreducible representation to which μ_k belongs. We can therefore obtain the selection rule which governs the transitions from the ground state to the fundamental levels: ν_ρ is only infra-red active if its symmetry species Γ^ρ is the same as one of the irreducible representations contained in Γ^{t}.

An alternative but not so general selection rule (it is restricted to the harmonic oscillator approximation) is that $\int\psi_0^{\text{vib}*}\mu_k\psi_m^\rho \, \mathrm{d}\tau$ is zero if $\partial\mu_k/\partial Q_m^\rho$ (evaluated for the equilibrium nuclear configuration) is zero, i.e. if there is no linear dependence of the dipole moment on the normal coordinate Q_m^ρ.

This selection rule may be found by making a Taylor series expansion of $\mu_k(X)$ in the normal coordinates $Q_1, Q_2, \ldots Q_{3N-6}$ (we revert to our initial notation):

$$\mu_k = \mu_k^0 + \sum_{l=1}^{3N-6}\left(\frac{\partial\mu_k}{\partial Q_l}\right)^0 Q_l + \ldots, \qquad k = 1, 2, 3$$

where the superscripts indicate evaluation at the equilibrium configuration. If the vibrational wavefunctions are written as $\prod_{i=1}^{3N-6}\psi_i(n_i, Q_i)$ where n_i is the ith vibrational quantum number, then the ground state wavefunction is

$$\psi_0^{\text{vib}} = \prod_{i=1}^{3N-6}\psi_i(0, Q_i)$$

and the wavefunctions for excited states in which only one vibrational quantum number is non-zero are

$$\psi_m = \left\{\prod_{i\neq m}^{3N-6}\psi_i(0, Q_i)\right\}\psi_m(n_m, Q_m).$$

Since

$$\int \psi_i(j, Q_i)^*\psi_i(k, Q_i) \, \mathrm{d}Q_i = \delta_{jk}$$

(the harmonic oscillator functions are orthonormal) we have

$$\int \psi_0^{\text{vib}*}\mu_k\psi_m \, \mathrm{d}\tau =$$

$$= \mu_k^0\left\{\prod_{i\neq m}^{3N-6}\int \psi_i(0, Q_i)^*\psi_i(0, Q_i) \, \mathrm{d}Q_i\right\}\int \psi_m(0, Q_m)^*\psi_m(n_m, Q_m) \, \mathrm{d}Q_m +$$

$$+\sum_{l\neq m}^{3N-6}\left(\frac{\partial\mu_k}{\partial Q_l}\right)^0\left\{\prod_{i\neq m,l}^{3N-6}\int \psi_i(0, Q_i)^*\psi_i(0, Q_i) \, \mathrm{d}Q_i\right\}\times$$

$$\times \int \psi_l(0, Q_l)^* Q_l \psi_l(0, Q_l)\, dQ_l \int \psi_m(0, Q_m)^* \psi_m(n_m, Q_m)\, dQ_m +$$

$$+ \left(\frac{\partial \mu_k}{\partial Q_m}\right)^0 \left\{ \prod_{i \neq m}^{3N-6} \int \psi_i(0, Q_i)^* \psi_i(0, Q_i)\, dQ_i \right\} \int \psi_m(0, Q_m)^* Q_m \psi_m(n_m, Q_m)\, dQ_m +$$

$$+\text{higher terms,} \qquad (9\text{-}10.1)$$

and if $n_m \neq 0$ the first two terms and the higher terms are zero and

$$\int \psi_0^{\text{vib}*} \mu_k \psi_m\, d\tau = \left(\frac{\partial \mu_k}{\partial Q_m}\right)^0 \int \psi_m(0, Q_m)^* Q_m \psi_m(n_m, Q_m)\, dQ_m$$

$$k = 1, 2, 3$$

and hence, if $\left(\frac{\partial \mu_k}{\partial Q_m}\right)^0 = 0$, the probability of the transition from the ground state to the excited state occurring is zero, i.e. for the transition to take place the dipole moment must change linearly with the normal coordinate Q_m.

Since $\psi_m(0, Q_m)Q_m$ is proportional to $\psi_m(1, Q_m)$ and the functions $\psi_m(n_m, Q_m)$ are orthogonal, it is clear that

$$\int \psi_m(0, Q_m)^* Q_m \psi_m(n_m, Q_m)\, dQ_m = 0$$

unless $n_m = 1$. Hence only transitions to the fundamental levels are allowed from the ground state. This is strictly true only within the approximations we have been making and, in reality, transitions to overtone levels ($n_m \neq 1$) do take place (see § 9-13).

Furthermore, transitions from the ground state to excited states for which *more* than one vibrational quantum number is non-zero are forbidden, e.g. if $n_1 \neq 0$ and $n_2 \neq 0$ in the excited state, then either or both of the integrals

$$\int \psi_1(0, Q_1)^* \psi_1(n_1, Q_1)\, dQ_1 \text{ and } \int \psi_2(0, Q_2)^* \psi_2(n_2, Q_2)\, dQ_2$$

will appear in each term of eqn (9-10.1) and hence the probability of the transition is zero. Again this is only strictly true within the approximations we have been making and transitions to so-called combination levels do in actual practice occur (see § 9-13).

9-11. Raman spectra

When an incident beam of radiation of frequency ν falls on a molecule, some radiation is scattered and in this scattered radiation we get, as well as ν, frequencies $\nu \pm \nu_\rho$ where ν_ρ is a fundamental frequency. This is called the Raman effect and when a fundamental frequency appears in the Raman spectrum it is said to be Raman active.

An incident radiation field with an electric vector $\boldsymbol{E}$ induces a dipole moment M in a molecule. The components of M are:

$$M_k = \sum_{j=1}^{3} \alpha_{kj} E_j \qquad k = 1, 2, 3$$

where

$$\boldsymbol{E} = E_1\mathbf{e}_1 + E_2\mathbf{e}_2 + E_3\mathbf{e}_3$$

and

$$\alpha_{kj} = \alpha_{jk} \qquad \begin{matrix} k = 1, 2, 3 \\ j = 1, 2, 3 \end{matrix}$$

and α_{11}, α_{22}, α_{33}, α_{12}, α_{13}, and α_{23} define the polarizability of the molecule. The latter are transformed by the symmetry operations $\boldsymbol{R}$ in the same way as x_1^2, x_2^2, x_3^2, x_1x_2, x_1x_3, and x_2x_3 where x_1, x_2, and x_3 are the co-ordinates of a point in physical space (see Appendix A.9-3). The six polarizability functions generate a reducible representation which we will call Γ^{α} and the character tables give the irreducible representations to which x_1^2 (or x^2), x_2^2 (or y^2), x_3^2 (or z^2), x_1x_2 (or xy), x_1x_3 (or xz), x_2x_3 (or yz), or the necessary combinations, belong and therefore give the decomposition of Γ^{α}. For example, for CH_4 ($\mathscr{T}_d$) $x^2+y^2+z^2$ belongs to Γ^{A_1}, $2z^2-x^2-y^2$ and x^2-y^2 to Γ^E and xy, xz, and yz to Γ^{T_2}, hence

$$\Gamma^{\alpha} = \Gamma^{A_1} \oplus \Gamma^{E} \oplus \Gamma^{T_2}.$$

Quantum mechanics tells us that the probability that Raman scattering involves the fundamental frequency ν_ρ depends on the integrals

$$\int \psi_0^{\text{vib}*} \alpha_{ij} \psi_m^{\rho}\, d\tau \qquad \begin{matrix} i = 1, 2, 3 \\ j = 1, 2, 3 \\ m = 1, 2, \ldots n_\rho \end{matrix}$$

and therefore ν_ρ is only Raman active if Γ^{ρ} coincides with one of the irreducible representations contained in Γ^{α} (remember that ψ_0^{vib} belongs to Γ^1). This rule is equivalent to saying that ν_ρ is only Raman active if the polarizability changes during the ρth normal vibration.

In a molecule with a centre of inversion, the irreducible representations in Γ^t are of u-type and those in Γ^{α} are of g-type and since Γ^{ρ} cannot coincide with both a u- *and* a g-type irreducible representation, no fundamental frequency for this type of molecule can be both infra-red and Raman active.

9-12. The infra-red and Raman spectra of CH_4 and CH_3D

In this section we will determine the differences in the infra-red and Raman spectra of methane CH_4 and monodeuteromethane CH_3D by finding the number of active fundamentals and their symmetries for the two molecules.

CH_4 belongs to the $\mathscr{T}_d$ point group (the symmetry elements are shown in Fig. 3-6.2) and the reduction of Γ^0 was carried out in § 9-7(3) with the result that

$$\Gamma^0 = \Gamma^{A_1} \oplus \Gamma^E \oplus \Gamma^{T_1} \oplus 3\Gamma^{T_2}.$$

Furthermore, inspection of the character table shows that

$$\Gamma^t = \Gamma^{T_2}$$

and

$$\Gamma^r = \Gamma^{T_1}$$

and hence

$$\Gamma^v = \Gamma^{A_1} \oplus \Gamma^E \oplus 2\Gamma^{T_2}.$$

Also from the character table in Appendix I, we have

$$\Gamma^\alpha = \Gamma^{A_1} \oplus \Gamma^E \oplus \Gamma^{T_2}.$$

So that the non-degenerate fundamental level which belongs to Γ^{A_1} will only be Raman active (Γ^{A_1} is contained in Γ^α but not in Γ^t), the doubly-degenerate fundamental level which belongs to Γ^E will also only be Raman active (Γ^E is contained in Γ^α but not in Γ^t) and the two triply-degenerate fundamental levels which belong to Γ^{T_2} will be both infra-red and Raman active (Γ^{T_2} is contained in both Γ^t and Γ^α).

CH_3D belongs to the $\mathscr{C}_{3v}$ point group and the C–D axis is the C_3 axis. The characters for the Γ^0 representation may be found by using eqn (9-6.1) and they are

$\mathscr{C}_{3v}$	E	$2C_3$	$3\sigma_v$
$\chi^0(C_i)$	15	0	3

These characters together with those of the irreducible representations of the point group may be fed into eqn (7-4.2) and the reduction of Γ^0 carried out, when this is done we obtain

$$\Gamma^0 = 4\Gamma^{A_1} \oplus \Gamma^{A_2} \oplus 5\Gamma^E$$

and since by inspection of the character table

$$\Gamma^t = \Gamma^{A_1} \oplus \Gamma^E$$

and

$$\Gamma^r = \Gamma^{A_2} \oplus \Gamma^E$$

we have

$$\Gamma^v = 3\Gamma^{A_1} \oplus 3\Gamma^E.$$

Furthermore,

$$\Gamma^\alpha = 2\Gamma^{A_1} \oplus 2\Gamma^E.$$

Therefore the three non-degenerate fundamental levels belonging to Γ^{A_1} are both infra-red and Raman active and the three doubly degenerate fundamental levels belonging to Γ^E are also both infra-red and Raman active.

So for CH_3D the number of fundamental frequencies which appear in the infra-red spectrum and the Raman spectrum are the same, whereas for CH_4 this is not so. This is sufficient information to distinguish between the two molecules.

9-13. Combination and overtone levels and Fermi resonance

If it is possible to excite two normal modes simultaneously then a transition can occur to what is called a combination level. Such a level will be characterized by a set of vibrational quantum numbers which are all zero except for two which are unity (see eqn (9-3.6)); it will lie at an energy of $h(\nu_\rho + \nu_\sigma)$ above the ground state where ν_ρ and ν_σ are the relevant fundamental frequencies. The corresponding vibrational wavefunctions will be of the form

$$N\psi_0^{\text{vib}}(Q)Q_{\rho(i)}Q_{\sigma(j)} \qquad \begin{array}{l} i = 1, 2, \ldots n_\rho \\ j = 1, 2, \ldots n_\sigma \end{array}$$

and as there are $n_\rho n_\sigma$ products of $Q_{\rho(i)}$ and $Q_{\sigma(j)}$ the combination level, in the harmonic oscillator approximation, will have a degeneracy equal to $n_\rho n_\sigma$. The $n_\rho n_\sigma$ wavefunctions for a given combination level, taken together, will form a basis for the direct product representation $\Gamma^\rho \otimes \Gamma^\sigma$. This representation will, in general, be reducible. There will therefore be combinations of the functions $N\psi_0^{\text{vib}}(Q)Q_{\rho(i)}Q_{\sigma(j)}$ which will form bases for the irreducible representations contained in $\Gamma^\rho \otimes \Gamma^\sigma$. In the harmonic oscillator approximation these combinations are all degenerate; however, if anharmonic terms such as

$$\tfrac{1}{6}\sum_i^{3N}\sum_j^{3N}\sum_k^{3N}\left(\frac{\partial^3 V}{\partial q_i\,\partial q_j\,\partial q_k}\right)_0 q_i q_j q_k$$

are included in eqn (9-2.7), then this degeneracy is lifted and in place of a single combination level there will be a group of levels with energies *approximately* $h(\nu_\rho + \nu_\sigma)$ greater than that of the ground state and there will be one such level for each irreducible representation contained in $\Gamma^\rho \otimes \Gamma^\sigma$ (see § 8-3). If any of the irreducible representations coincide with those found in Γ^t or Γ^α then a frequency approximately equal to $\nu^\rho + \nu^\sigma$ will occur in the infra-red or Raman spectrum. Since these transitions are forbidden in the harmonic oscillator approximation they will be weaker than the fundamental ones.

An overtone level is characterized by a set of vibrational quantum numbers which are all zero except one which has a value greater than unity, say m. If the quantum number which is non-zero corresponds to a fundamental frequency ν_ρ, then from eqn (9-3.6) we see that the overtone level will lie $m\nu_\rho$ above the ground state. It can be shown that if the ρth fundamental level is non-degenerate then for m even the overtone level belongs to the totally symmetric representation and for m odd it belongs to the same representation as the ρth fundamental level. If ν_ρ is degenerate, then the symmetry species of the overtone is difficult to determine and once again anharmonic effects destroy the degeneracy predicted by the harmonic oscillator approximation and new levels are created which belong to some definite symmetry species. Though the probability of a transition from the ground state to an overtone level is zero in the harmonic oscillator approximation, the probability can be quite high when anharmonic terms are taken into account.

If an active fundamental level happens to lie close to an overtone or combination level with the *same* symmetry species, then anharmonic terms in the potential V (see eqn (9-2.4)) will have the effect of mixing the two levels. Two new levels will be produced whose wavefunctions consist of approximately equal amounts of the wavefunctions belonging to the fundamental and the overtone or combination level. In such circumstances rather than having one strong (fundamental) and one weak (overtone or combination) transition there will be two strong transitions lying close together in energy. This phenomena is called *Fermi resonance.*

Appendices

A.9-1. Proof of eqns (9-2.17) and (9-2.18)

We have:

$$\sum_{i=1}^{3N} \dot{Q}_i^2 = \sum_{i=1}^{3N} \left(\sum_{j=1}^{3N} C_{ji}\dot{q}_j\right)\left(\sum_{k=1}^{3N} C_{ki}\dot{q}_k\right)$$

$$= \sum_{j=1}^{3N}\sum_{k=1}^{3N}\left(\sum_{i=1}^{3N} C_{ji}C_{ki}\right)\dot{q}_j\dot{q}_k.$$

But since C is orthogonal

$$\sum_{i=1}^{3N} C_{ji}C_{ki} = \delta_{jk}$$

and

$$\sum_{i=1}^{3N} \dot{Q}_i^2 = \sum_{j=1}^{3N}\sum_{k=1}^{3N} \delta_{jk}\dot{q}_j\dot{q}_k$$

$$= \sum_{j=1}^{3N} \dot{q}_j^2$$

and hence we obtain eqn (9-2.17).

From eqn (9-2.13) we can get

$$\sum_{q=1}^{3N} C_{qi}B_{qj} = \lambda_i C_{ji}$$

and hence

$$\begin{aligned}\sum_{i=1}^{3N} \lambda_i Q_i^2 &= \sum_{j=1}^{3N}\sum_{k=1}^{3N}\left(\sum_{i=1}^{3N} \lambda_i C_{ji}C_{ki}\right) q_j q_k \\ &= \sum_{j=1}^{3N}\sum_{k=1}^{3N}\left(\sum_{i=1}^{3N}\sum_{q=1}^{3N} C_{qi}B_{qj}C_{ki}\right) q_j q_k \\ &= \sum_{j=1}^{3N}\sum_{k=1}^{3N}\left(\sum_{q=1}^{3N} \delta_{qk}B_{qj}\right) q_j q_k \\ &= \sum_{j=1}^{3N}\sum_{k=1}^{3N} B_{kj}q_j q_k\end{aligned}$$

which proves eqn (9-2.18).

A.9-2. Proof that $D(R) = C^{-1}D^0(R)C$

We have, by definition,

$$Rq_i = \sum_{j=1}^{3N} D^0_{ji}(R)q_j \qquad i = 1, 2, \ldots, 3N \qquad \text{(A.9-2.1)}$$

and

$$Q_i = \sum_{j=1}^{3N} C_{ji}q_j$$

or

$$\begin{aligned}q_j &= \sum_{m=1}^{3N} (C^{-1})_{mj} Q_m \\ &= \sum_{m=1}^{3N} C_{jm}Q_m \qquad \text{(A.9-2.2)}\end{aligned}$$

(C is orthogonal). Substituting eqn (A.9-2.2) twice into eqn (A.9-2.1), we have

$$\sum_{k=1}^{3N} C_{ik}RQ_k = \sum_{j=1}^{3N} D^0_{ji}(R) \sum_{m=1}^{3N} C_{jm}Q_m$$

$$\sum_{k=1}^{3N} C_{ik} \sum_{r=1}^{3N} D^n_{rk}(R)Q_r = \sum_{j=1}^{3N} D^0_{ji}(R) \sum_{m=1}^{3N} C_{jm}Q_m$$

and by equating the coefficients of the Q's:

$$\sum_{k=1}^{3N} C_{ik}D^n_{rk}(R) = \sum_{j=1}^{3N} D^0_{ji}(R)C_{jr}$$

$$\sum_{k=1}^{3N} D^n_{rk}(R)(C^{-1})_{ki} = \sum_{j=1}^{3N} (C^{-1})_{rj}D^0_{ji}(R)$$

and hence

$$D^n(R)C^{-1} = C^{-1}D^0(R)$$

and

$$D^n(R) = C^{-1}D^0(R)C.$$

A.9-3. Symmetry properties of polarizability functions

Let $\boldsymbol{R}$ transform the nuclear–electronic configuration X to X' and the electric vector $\boldsymbol{E}$ to $\boldsymbol{E}'$, then

$$\boldsymbol{E} = \sum_{i=1}^{3} E_i \mathbf{e}_i$$

$$\boldsymbol{E}' = \sum_{i=1}^{3} E_i' \mathbf{e}_i$$

and

$$E_m' = \sum_{j=1}^{3} D_{mj}(\boldsymbol{R}) E_j \qquad m = 1, 2, 3.$$

Also

$$M_i(X') = \sum_{m=1}^{3} \alpha_{im}(X') E_m' = \sum_{m=1}^{3} \alpha_{im}(X') \sum_{j=1}^{3} D_{mj}(\boldsymbol{R}) E_j$$

and

$$M_k(X) = \sum_{i=1}^{3} D_{ik}(\boldsymbol{R}) M_i(X')$$

therefore,

$$M_k(X) = \sum_{i=1}^{3} D_{ik}(\boldsymbol{R}) \sum_{m=1}^{3} \alpha_{im}(X') \sum_{j=1}^{3} D_{mj}(\boldsymbol{R}) E_j.$$

But

$$M_k(X) = \sum_{j=1}^{3} \alpha_{kj}(X) E_j$$

so

$$\alpha_{kj}(X) = \sum_{i=1}^{3} \sum_{m=1}^{3} D_{ik}(\boldsymbol{R}) D_{mj}(\boldsymbol{R}) \alpha_{im}(X')$$

and since, by definition of $\boldsymbol{O_R}$,

$$\boldsymbol{O_R} \alpha_{kj}(X') = \alpha_{kj}(X)$$

we have

$$\boldsymbol{O_R} \alpha_{kj}(X') = \sum_{i=1}^{3} \sum_{m=1}^{3} D_{ik}(\boldsymbol{R}) D_{mj}(\boldsymbol{R}) \alpha_{im}(X').$$

Now consider the functions

$$f_{kj}(x_1, x_2, x_3) = x_k x_j \qquad \begin{matrix} k = 1, 2, 3 \\ j = 1, 2, 3 \end{matrix}$$

it is easily shown that

$$\boldsymbol{O_R} f_{kj}(x_1', x_2', x_3') = \sum_{i=1}^{3} \sum_{m=1}^{3} D_{ik}(\boldsymbol{R}) D_{mj}(\boldsymbol{R}) f_{im}(x_1', x_2', x_3').$$

Therefore the effect of $\boldsymbol{O_R}$ on a polarizability function α_{kj} is the same as its effect on $x_k x_j$.

PROBLEMS

9.1. For ethylene:

(a) determine the point group;

(*b*) determine the number and symmetry of the vibrational normal coordinates;
(*c*) determine the spectroscopic activity of each fundamental level.

9.2. Show on the basis of infra-red and Raman spectra that it is possible to distinguish between the two crown forms of octachlorocyclooctane, one in which the hydrogen atoms are all equatorial ($\mathscr{D}_{4d}$) and the other in which the hydrogen atoms are alternating between axial and equatorial positions ($\mathscr{C}_{4v}$).

9.3. Discuss how the *cis* and *trans* isomers of N_2F_2 can be distinguished by infra-red and Raman measurements.

9.4. What will be the infra-red and Raman activity of the four fundamental levels of CO_3^{2-}?

9.5. Determine χ^0 and carry out the reduction of Γ^0 for the following molecules:
(*a*) NH_3 ($\mathscr{C}_{3v}$),
(*b*) $XeOF_4$ ($\mathscr{C}_{4v}$),
(*c*) $PtCl_4^{2-}$ ($\mathscr{D}_{4h}$),
(*d*) *trans*-glyoxal ($\mathscr{C}_{2h}$).

10. Molecular orbital theory

10-1. Introduction

In this chapter we will consider how to apply a knowledge of symmetry and its ramifications to the determination of electronic wavefunctions. We will do so by looking at a particular kind of approximate electronic wavefunction for conjugated molecules. The Schrödinger equation for the electrons of a molecule, our starting point, is just as hard to solve as the Schrödinger equation for the nuclei. In the latter case we were able to find approximate solutions by replacing the true potential energy V with a sum of quadratic terms in the nuclear coordinates ($\frac{1}{2}\sum_i \sum_j B_{ij}q_iq_j$). In dealing with the electronic case we must, for example, for a molecule like benzene, make a whole series of fairly drastic approximations. First we consider the electronic wavefunction to be made up of molecular orbitals (approximation 1), then we restrict the form of the molecular orbitals (MOs) to linear combinations of atomic orbitals (approximation 2), then we separate out the part of the wavefunction concerned with σ-electrons and deal only with the π-electron part (approximation 3), finally we solve the appropriate equations by making assumptions about certain integrals over the π-electronic MOs (approximation 4). The final step brings us to the Hückel molecular orbital method, which is familiar to all chemists.

Symmetry enters the approximate solution of the electronic Schrödinger equation in two ways. In the first place, the exact MOs are eigenfunctions of an operator which commutes with all $\boldsymbol{O_R}$ of the point group concerned, they therefore generate irreducible representations of that point group (see Chapter 8) and can be classified accordingly. The same is true for the approximate MOs and consequently one constructs them from combinations of atomic orbitals (symmetry orbitals) which generate irreducible representations.

In the second place, the Hamiltonian operators which occur and commute with all $\boldsymbol{O_R}$ belong to the totally symmetric irreducible representation Γ^1 (see Appendix A.10-3) and integrals over them $\int \psi^{\nu *}\boldsymbol{H}\psi^{\mu}\,d\tau$ vanish unless $\Gamma^{\nu} = \Gamma^{\mu}$ (see eqn (8-4.5)). Thus, in carrying out an approximate solution of the electronic Schrödinger equation, changing to a set of basis functions which belong to the irreducible representations will allow us, by inspection, to put many of the integrals which occur equal to zero. There will also, because of this, be an immediate factorization of the equations. Two examples, benzene and the trivinylmethyl radical, will be considered in detail.

10-2. The Hartree–Fock approximation

The starting point for any molecular electronic problem is the electronic Schrödinger equation:

$$H\Psi(1, 2,\dots n) = E\Psi(1, 2,\dots n).$$

In this equation H, the Hamiltonian, is defined by precise quantum mechanical rules and can be written in atomic units (Appendix A.10-1) as

$$H = \sum_{\mu=1}^{n} h_\mu + \sum_{\mu=1}^{n}\sum_{\nu>\mu}^{n} 1/r_{\mu\nu} \qquad (10\text{-}2.1)$$

where

$$h_\mu = -\tfrac{1}{2}\nabla_\mu^2 - \sum_{\alpha=1}^{N} Z_\alpha/r_{\mu\alpha}. \qquad (10\text{-}2.2)$$

In eqns (10-2.1) and (10-2.2) n is the number of electrons in the molecule $r_{\mu\nu}$ is the distance between electron μ and electron ν, $\sum_{\mu=1}^{n}\sum_{\nu>\mu}^{n} 1/r_{\mu\nu}$ is the potential-energy operator due to interactions between the electrons, ∇_μ^2 is a Laplacian operator involving the coordinates of electron μ, $-\frac{1}{2}\nabla_\mu^2$ is the kinetic-energy operator (in atomic units) for electron μ, $r_{\mu\alpha}$ is the distance between electron μ and nucleus α, Z_α is the charge on nucleus α and $-\sum_{\alpha=1}^{N} Z_\alpha/r_{\mu\alpha}$ is the potential-energy operator arising from the interactions of electron μ with all the N nuclei.

$\Psi(1, 2,\dots n)$ is the electronic wavefunction and explicitly is a function of the coordinates of all n electrons; in this notation the coordinates of a given electron are symbolized by a single number. E is the total electronic energy of the molecule.

The Hartree–Fock (HF) method, or self consistent field (SCF) method as it is sometimes called, approximates $\Psi(1, 2,\dots n)$ by expressing it solely in terms of functions each of which contains the coordinates of just one electron; these functions are called *molecular orbitals* (MOs). This is an approximation because in reality the position of one electron is always correlated with the positions of the others, so that the function which describes a given electron cannot be independent of the functions describing the other electrons. It is for this reason that the error in the electronic energy in the HF approximation is called the *correlation energy*.

The actual way in which the MOs are put together to form

$$\Psi(1,2,\dots n)$$

is restricted by two fundamental laws. One is that an electronic wavefunction must not distinguish, by treating in a different manner, one electron from another; this is the law of indistinguishability of identical particles. The other law is that the electronic wavefunction must change its sign when two electrons are interchanged; this is the antisymmetry law and in fact it leads to the Pauli exclusion principle.

Both these laws are satisfied by expressing $\Psi(1, 2, \ldots n)$ in terms of the MOs with a Slater determinant:

$$\Psi(1, 2, \ldots n) = 1/\sqrt{n!}\begin{vmatrix} \Phi_1(1) & \Phi_2(1) & \ldots & \Phi_n(1) \\ \Phi_1(2) & \Phi_2(2) & \ldots & \Phi_n(2) \\ \cdot & \cdot & & \cdot \\ \cdot & \cdot & & \cdot \\ \cdot & \cdot & & \cdot \\ \Phi_1(n) & \Phi_2(n) & \ldots & \Phi_n(n) \end{vmatrix}^{\dagger} . \quad (10\text{-}2.3)$$

In this determinant $\Phi_i(j)$ symbolizes the ith MO as a well-defined function of the coordinates of electron j; we say that electron j is occupying the MO Φ_i. If the determinant is multiplied out there will be $n!$ terms and each term corresponds to one of the $n!$ permutations of the n electrons amongst the n MOs; since all permutations are included, every electron is treated equally and the indistinguishability law is satisfied.

If two electrons are interchanged then, for a Slater determinant, this is equivalent to interchanging two rows and if two rows of a determinant are exchanged it changes sign, hence the antisymmetry law is also satisfied, e.g. if we exchange electrons 1 and 2, we get

$$\Psi(2, 1, \ldots n) = (1/\sqrt{n!})\begin{vmatrix} \Phi_1(2) & \Phi_2(2) & \ldots & \Phi_n(2) \\ \Phi_1(1) & \Phi_2(1) & \ldots & \Phi_n(1) \\ \cdot & \cdot & & \cdot \\ \cdot & \cdot & & \cdot \\ \cdot & \cdot & & \cdot \\ \Phi_1(n) & \Phi_2(n) & \ldots & \Phi_n(n) \end{vmatrix}$$

$$= -1/\sqrt{n!}\begin{vmatrix} \Phi_1(1) & \Phi_2(1) & \ldots & \Phi_n(1) \\ \Phi_1(2) & \Phi_2(2) & \ldots & \Phi_n(2) \\ \cdot & \cdot & & \cdot \\ \cdot & \cdot & & \cdot \\ \cdot & \cdot & & \cdot \\ \Phi_1(n) & \Phi_2(n) & \ldots & \Phi_n(n) \end{vmatrix}$$

$$= -\Psi(1,2,\ldots n).$$

† This equation is often abbreviated as

$$\Psi(1, 2, \ldots n) = |\Phi_1(1)\Phi_2(2)\Phi_3(3) \ldots \Phi_n(n)|.$$

That $\Psi(1, 2, \ldots n)$ necessarily satisfies the Pauli exclusion principle is evident from the fact that if two MOs are absolutely identical (including their spin components), then so are two columns in the Slater determinant and a determinant with two identical columns is zero.

The $1/\sqrt{n!}$ factor preceding the determinant takes account of normalization, since if the MOs are orthonormal, this factor ensures that

$$\int \ldots \iint \Psi^*(1,2,\ldots n)\Psi(1, 2,\ldots n)\, d\tau_1\, d\tau_2 \ldots d\tau_n = 1.$$

The way in which the Hartree–Fock MOs are determined is by using the variational method, that is the form of each MO is varied until the integral $\int \ldots \iint \Psi^*(1,2,\ldots n)\boldsymbol{H}\Psi(1, 2,\ldots n)\, d\tau_1\, d\tau_2 \ldots d\tau_n$ is as low as possible:

$$\delta \left\{\int \ldots \iint \Psi^*(1, 2,\ldots n)\boldsymbol{H}\Psi(1, 2,\ldots n)\, d\tau_1\, d\tau_2 \ldots d\tau_n\right\} = 0. \tag{10-2.4}$$

We will call the MOs that satisfy eqn (10-2.4) with respect to *complete* variation in their form the exact MOs.

By introducing eqn (10-2.3) into eqn (10-2.4) it is possible to arrive at a simple set of *n one-electron* eigenvalue equations, called the Hartree–Fock equations:

$$\boldsymbol{H}^{\text{eff}}(i)\Phi_i(i) = \varepsilon_i\Phi_i(i) \qquad i = 1, 2,\ldots n \tag{10-2.5}$$

where $\boldsymbol{H}^{\text{eff}}(i)$ (explicitly defined in the next section) is an effective Hamiltonian operator related to $\boldsymbol{H}$ and involving the coordinates of electron i. The eigenvalues ε_i are constants called orbital energies. Once the MOs have been found, the total electronic energy is obtained from the equation:

$$E = \int \ldots \iint \Psi^*(1, 2,\ldots n)\boldsymbol{H}\Psi(1, 2,\ldots n)\, d\tau_1\, d\tau_2 \ldots d\tau_n.$$

Since it can be shown that $\boldsymbol{H}^{\text{eff}}(i)$, like the original Hamiltonian $\boldsymbol{H}$, commutes with the transformation operators $\boldsymbol{O}_R$ for all operations $\boldsymbol{R}$ of the point group to which the molecule belongs, the MOs associated with a given orbital energy will form a function space whose basis generates a definite irreducible representation of the point group. This is exactly parallel to the situation for the exact total electronic wavefunctions.

The exact MOs may therefore be classified according to the irreducible representation to which they belong and usually the symbol for the irreducible representation (in lower case type) is used to label the MO with which it is associated. For example, for the point group $\mathscr{T}_d$ of methane a MO might be a non-degenerate 'a_1' orbital or a triply degenerate 't_2' orbital etc.

10-3. The LCAO MO approximation

Rather than find the most perfect MOs which satisfy eqn (10-2.4) (or eqns (10-2.5)), it is common practice to replace them by particular mathematical functions of a restricted nature. These functions will generally contain certain parameters which can then be optimized in accordance with eqn (10-2.4). Since these MOs are not completely flexible, we will have introduced a further approximation, the severity of which is determined by the degree of inflexibility in the form of our chosen functions. Typical of this kind of approximation is the one which expresses the *space part* of the MOs as various linear combinations of atomic orbitals centred on the same or different nuclei in the molecule. We write the space part of each of the approximate MOs as

$$\Phi_j(i) = \sum_{s=1}^{m} C_{sj}\phi_s(i) \qquad \left(m \geq \frac{n}{2}\right) \tag{10-3.1}$$

where the ϕ_s are atomic orbitals and the C_{sj} are linear coefficients.

These Φ_j are called Linear Combination of Atomic Orbitals Molecular Orbitals (LCAO MOs) and if they are introduced into the Hartree–Fock equations (eqns (10-2.5)), a simple set of equations (the Hartree–Fock–Roothaan equations) is obtained which can be used to determine the optimum coefficients C_{sj}. For those systems where the space part of each MO is doubly occupied, i.e. there are two electrons in each Φ_j with spin α and spin β respectively so that the complete MOs including spin are different, the total wavefunction is

$$\Psi(1, 2,\ldots n) = |\, \Phi_1(1)\alpha(1)\, \Phi_1(2)\beta(2)\, \Phi_2(3)\alpha(3) \ldots \Phi_{n/2}(n)\beta(n)\, |$$

and we have the following equations:

$$\sum_{k=1}^{m} (H_{jk}^{\text{eff}} - \varepsilon_i S_{jk}) C_{ki} = 0 \qquad \begin{matrix} j = 1, 2,\ldots m \\ i = 1, 2,\ldots m \end{matrix} \tag{10-3.2}$$

where

$$\left.\begin{aligned} H_{jk}^{\text{eff}} &= \int \phi_j^*(1) H^{\text{eff}}(1) \phi_k(1)\, \mathrm{d}\tau_1 \\ S_{jk} &= \int \phi_j^*(1)\phi_k(1)\, \mathrm{d}\tau_1 \end{aligned}\right\} \tag{10-3.3}$$

$$\left.\begin{aligned} H^{\text{eff}}(\mu) &= h_\mu + \sum_{i=1}^{n/2} \{2J_i(\mu) - K_i(\mu)\} \\ h_\mu &= -\tfrac{1}{2}\nabla_\mu^2 - \sum_{\alpha=1}^{N} Z_\alpha / r_{\mu\alpha} \end{aligned}\right\} \tag{10-3.4}$$

$J_i(\mu)$ and $K_i(\mu)$ are the Coulomb and exchange operator respectively, and are defined by the equations:

$$J_i(1)\phi_k(1) = \left\{\int \Phi_i^*(2)\Phi_i(2)r_{12}^{-1}\,d\tau_2\right\}\phi_k(1)$$

and

$$K_i(1)\phi_k(1) = \left\{\int \Phi_i^*(2)\phi_k(2)r_{12}^{-1}\,d\tau_2\right\}\Phi_i(1).$$

Non-trivial solutions of eqns (10-3.2) can be obtained provided that the eigenvalues ε_i, the LCAO MO *orbital energies* and approximations to the ε_i of eqn (10-2.5), satisfy the equation

$$\det(H_{jk}^{\text{eff}} - \varepsilon_i S_{jk}) = 0. \qquad (10\text{-}3.5)$$

With this proviso, eqns (10-3.2) coupled with a normalization equation can be solved to produce m sets of coefficients (each set corresponding to a particular MO and orbital energy) from which we can choose $n/2$ which correspond to the lowest orbital energies and to those MOs which are occupied in the electronic ground state.

The total electronic energy is then

$$E = \int \ldots \iint \Psi^*(1, 2, \ldots n)H\Psi(1, 2, \ldots n)\,d\tau_1\,d\tau_2 \ldots d\tau_n$$

$$= 2\sum_{i=1}^{n/2}\varepsilon_i - \sum_{i=1}^{n/2}\sum_{j=1}^{n/2}(2J_{ij} - K_{ij}) \qquad (10\text{-}3.6)$$

where

$$J_{ij} = \iint \Phi_i^*(1)\Phi_j^*(2)r_{12}^{-1}\,\Phi_i(1)\Phi_j(2)\,d\tau_1\,d\tau_2$$

and

$$K_{ij} = \iint \Phi_i^*(1)\Phi_j^*(2)r_{12}^{-1}\Phi_i(2)\Phi_j(1)\,d\tau_1\,d\tau_2.$$

An alternative notation for the preceding equations is given in Appendix A.10-2.

The reader should note that eqns (10-3.2) have to be solved iteratively since the coefficients C_{sj} appear in the operators $J_i(\mu)$ and $K_i(\mu)$ and hence in $H^{\text{eff}}(\mu)$ and H_{jk}^{eff}. What is done, therefore, is to guess sets of coefficients and with them calculate H_{jk}^{eff}, then solve eqns (10-3.2) for a new set of coefficients. These new coefficients can then be used as input to H_{jk}^{eff} and the process repeated until the input and output coefficients are consistent.

In the above equations, integration over the spin parts of the MOs has been carried out and the Φ_j refer only to the space part of the MOs.

In the previous section we stated that the exact MOs belonging to a given orbital energy must form a basis for one of the irreducible representations of the point group to which the molecule belongs and the same is true for the approximate MOs. Furthermore, if one ensures *a priori* that the approximate MOs do behave in this way, the calculations are greatly simplified because the vanishing integral rule comes into play. The way in which one makes sure that the approximate MOs form bases for the irreducible representations is by first forming linear combinations of atomic orbitals which do. These combinations are called, appropriately, *symmetry orbitals* and the coefficients of the atomic orbitals of which they are composed are totally determined by symmetry arguments. We will write a symmetry orbital as

$$\phi'_s = \sum_r c_{rs}\phi_r \tag{10-3.7}$$

where ϕ_r is an atomic orbital. The MOs are then formed from the symmetry orbitals by

$$\Phi_j = \sum_s C'_{sj}\phi'_s \tag{10-3.8}$$

and the coefficients C'_{sj} and total electronic energy are determined in the same fashion as before but with C'_{sj} replacing C_{sj} and ϕ'_s replacing ϕ_s. The simplification which results by doing this will become clear in § 10-6 and § 10-7.

10-4. The π-electron approximation

We now consider conjugated systems and approximate things even further by focusing attention upon only the π-electrons of such systems. The valence electrons of conjugated systems fall into two classes: σ-electrons and π-electrons. The σ-electrons are assumed to be fairly strongly localized in individual bonds and described by orbitals of σ-type symmetry (using the notation of linear molecules); they normally do not participate in those chemical reactions which do not involve bond breaking and they are regarded as relatively unreactive. The π-electrons, on the other hand, are highly delocalized over the carbon framework and play an important role in all reactions; they are often referred to as mobile electrons. In organic chemistry many of the properties of conjugated molecules can empirically be ascribed to the π-electrons alone and this indicates that it is not unreasonable in quantum mechanics to treat the π-electrons in an explicit fashion and to simply regard the σ-electrons as providing some kind of background potential field for them. The quantum mechanical separation of the electrons of a molecule into σ- and π-electrons is known as σ–π separability.

We therefore start the quantum mechanical treatment of conjugated systems by expressing the total electronic wavefunction in terms of a wavefunction for the σ-electrons and a wavefunction for the π-electrons:

$$\Psi(1, 2, \ldots n) = A(\sigma, \pi)\Psi_\sigma(1, 2, \ldots, n_\sigma)\Psi_\pi(1, 2, \ldots, n_\pi)$$

where n_σ and n_π are the number of σ- and π-electrons respectively. $A(\sigma, \pi)$ is an antisymmetrizer, which is an operator which 'exchanges' electrons between Ψ_σ and Ψ_π. It works in the same way as the Slater determinant did in § 10-2. In fact we could have written

$$\Psi(1, 2, \ldots n) = A\{\Phi_1(1)\Phi_2(2) \ldots \Phi_n(n)\}$$

in place of eqn (10-2.3). The π-electron approximation is then defined as that approximation in which the electronic wavefunctions for some set of molecular states are separable with the same Ψ_σ for all of them.

The total Hamiltonian H is then separated into two parts:

$$H = H_\sigma^0 + H_\pi$$

where

$$H_\sigma^0 = -\tfrac{1}{2}\sum_{\mu=1}^{n_\sigma} \nabla_\mu^2 - \sum_{\mu=1}^{n_\sigma}\sum_{\alpha=1}^{N} Z_\alpha/r_{\mu\alpha} + \sum_{\mu=1}^{n_\sigma}\sum_{\nu>\mu}^{n_\sigma} 1/r_{\mu\nu}$$

$$H_\pi = H_\pi^0 + \sum_{\mu=1}^{n_\pi}\sum_{\nu=1}^{n_\sigma} 1/r_{\mu\nu}$$

and

$$H_\pi^0 = -\tfrac{1}{2}\sum_{\mu=1}^{n_\pi} \nabla_\mu^2 - \sum_{\mu=1}^{n_\pi}\sum_{\alpha=1}^{N} Z_\alpha/r_{\mu\alpha} + \sum_{\mu=1}^{n_\pi}\sum_{\nu>\mu}^{n_\pi} 1/r_{\mu\nu}.$$

H_σ^0 and H_π^0 refer respectively to σ and π-electrons exclusively and the Hamiltonian H_π can also be written as

$$H_\pi = \sum_{\mu=1}^{n_\pi} h_\mu^{\text{core}} + \sum_{\mu=1}^{n_\pi}\sum_{\nu>\mu}^{n_\pi} 1/r_{\mu\nu} \qquad (10\text{-}4.1)$$

where

$$h_\mu^{\text{core}} = -\tfrac{1}{2}\nabla_\mu^2 - \sum_{\alpha=1}^{N} Z_\alpha/r_{\mu\alpha} + \sum_{\nu=1}^{n_\sigma} 1/r_{\mu\nu}. \qquad (10\text{-}4.2)$$

These last two equations have the same form as eqns (10-2.1) and (10-2.2) except that the core Hamiltonian h_μ^{core} includes an additional term, $\sum_{\nu=1}^{n_\sigma} 1/r_{\mu\nu}$, not present in h_μ which comes from the interactions between the μth π-electron and the n_σ σ-electrons.

The total electronic energy E is given by the sum of two terms

$$E = E_\sigma + E_\pi$$

where

$$E_\sigma = \int \Psi_\sigma^* H_\sigma^0 \Psi_\sigma \, d\tau_\sigma$$

and

$$E_\pi = \int \Psi_\pi^* H_\pi \Psi_\pi \, d\tau_\pi. \qquad (10\text{-}4.3)$$

Within the framework of the π-electron approximation E_σ is assumed to be simply a constant and the expression for E_π is used to find the optimum π-electron LCAO MOs; that is, the Hartree–Fock–Roothaan equations (eqns (10-3.2) to (10-3.4)) are applied to the π-electrons by replacing $\boldsymbol{H}$ by $\boldsymbol{H}_\pi$ and Ψ by Ψ_π, where Ψ_π is written as

$$\Psi_\pi = A\{\Phi_1^\pi(1)\Phi_2^\pi(2)\ldots\Phi_{n_\pi}^\pi(n_\pi)\}$$

where

$$\Phi_j^\pi(k) = \sum_s C_{sj}\phi_s(k) \tag{10-4.4}$$

and

$$\phi_s(k) = \pi\text{-atomic orbital}$$

(or, for symmetry orbitals, $\Phi_j^\pi(k) = \sum_s C'_{sj}\phi'_s(k)$).

10-5. Hückel molecular orbital method

Our approximations so far (the orbital approximation, LCAO MO approximation, π-electron approximation) have led us to a π-electronic wavefunction composed of LCAO MOs which, in turn, are composed of π-electron atomic orbitals. We still, however, have to solve the Hartree–Fock–Roothaan equations in order to find the orbital energies and coefficients in the MOs and this requires the calculation of integrals like (cf. eqns (10-3.3)):

$$H_{jk}^{\mathrm{eff},\pi} = \int \phi_j^*(1)\boldsymbol{H}^{\mathrm{eff},\pi}(1)\phi_k(1)\,\mathrm{d}\tau_1$$

and

$$S_{jk} = \int \phi_j^*(1)\phi_k(1)\,\mathrm{d}\tau_1.$$

In these integrals the additional superscript π indicates that we are now within the framework of the π-electron approximation and that essentially $\boldsymbol{H}$ has been replaced by $\boldsymbol{H}_\pi$ (see eqn (10-4.1)) and consequently $\boldsymbol{H}^{\mathrm{eff}}(\mu)$ (see eqn (10-3.4)) by

$$\boldsymbol{H}^{\mathrm{eff},\pi}(\mu) = \boldsymbol{h}_\mu^{\mathrm{core}} + \sum_i \{2\boldsymbol{J}_i(\mu) - \boldsymbol{K}_i(\mu)\}.$$

These integrals are difficult to evaluate exactly and the Hückel molecular orbital method centres on approximations to them.

It is assumed that each carbon atom contributes one π-electron and one $2p_z$ atomic orbital to the system, so that

$$\Phi_j^\pi(k) = \sum_{s=1}^{n_c} C_{sj}\phi_s(k)$$

where n_c equals the number of carbon atoms and ϕ_s is a $2p_z$ orbital located at the s carbon atom. The theory then makes the following important approximations:

$$H_{rr}^{\mathrm{eff},\pi} = \alpha$$

$$H_{rs}^{\mathrm{eff},\pi} = \begin{cases} \beta & \text{(if } r \text{ and } s \text{ signify nearest neighbour carbon atoms)} \\ 0 & \text{(otherwise)} \end{cases}$$

$$S_{rs} = \delta_{rs}$$

α and β are called the Coulombic and resonance integrals respectively and they are strictly empirical quantities which are determined by comparing the results of the theory with experimental data.

With these approximations the equation which corresponds to eqn (10-3.5)

$$\det(H_{jk}^{\text{eff},\pi} - \varepsilon^{\pi} S_{jk}) = 0$$

is solved. The roots of this equation correspond to the π-electron orbital-energies ε_i^{π} and they will be functions of α and β. Finally, the equations

$$\sum_k (H_{jk}^{\text{eff},\pi} - \varepsilon_i^{\pi} S_{jk}) C_{ki} = 0 \qquad j = 1,2, \dots n_c$$

are solved for the coefficients C_{ki}. The total π-electron energy is then given by

$$E_{\pi} = 2\sum_{i=1}^{n_c/2} \varepsilon_i^{\pi} - \sum_{i=1}^{n_c/2}\sum_{j=1}^{n_c/2} (2J_{ij} - K_{ij}) = 2\sum_{i=1}^{n_c/2} \varepsilon_i^{\pi} - G. \qquad (10\text{-}5.1)$$

Since G is assumed to be constant for all electronic states of a given molecule, the important part of E_{π} is the sum of the π-electron orbital energies.

If symmetry orbitals are used in place of atomic orbitals, then $H_{jk}^{\text{eff},\pi}$ and S_{jk} will become integrals over these orbitals and they will have to be broken down to integrals over the atomic orbitals before the Hückel approximations are made.

10-6. Hückel molecular orbital method for benzene

We will consider the application of the Hückel molecular orbital method to the benzene molecule and we will first see what happens when we do *not* make use of symmetry. The benzene molecule has a framework of six carbon atoms at the corners of a hexagon and each carbon atom contributes one π-electron. The π-electron MOs will be constructed from six $2p_z$ atomic orbitals, each located at one of the carbon atoms, thus,

$$\Phi_j^{\pi} = \sum_{s=1}^{6} C_{sj}\phi_s,$$

$$\phi_s = (2p_z)_s.$$

If we use the following Hückel approximations

$$H_{rr}^{\text{eff},\pi} = \alpha$$

$$H_{rs}^{\text{eff},\pi} = \begin{cases} \beta & (r \text{ and } s \text{ nearest neighbours}) \\ 0 & (\text{otherwise}) \end{cases}$$

$$S_{rs} = \delta_{rs}$$

then the equation which determines the π-electron orbital energies

$$\det(H_{jk}^{\text{eff},\pi} - \varepsilon^{\pi} S_{jk}) = 0 \qquad (10\text{-}6.1)$$

becomes eqn (10-6.2). This equation can be simplified by dividing each

$$\begin{vmatrix} \alpha-\varepsilon^{\pi} & \beta & 0 & 0 & 0 & \beta \\ \beta & \alpha-\varepsilon^{\pi} & \beta & 0 & 0 & 0 \\ 0 & \beta & \alpha-\varepsilon^{\pi} & \beta & 0 & 0 \\ 0 & 0 & \beta & \alpha-\varepsilon^{\pi} & \beta & 0 \\ 0 & 0 & 0 & \beta & \alpha-\varepsilon^{\pi} & \beta \\ \beta & 0 & 0 & 0 & \beta & \alpha-\varepsilon^{\pi} \end{vmatrix} = 0 \qquad (10\text{-}6.2)$$

element by β and letting

$$x = (\alpha - \varepsilon^{\pi})/\beta$$

to give eqn (10-6.3) which can be solved and the six roots $x_1, x_2, \ldots, x_6$

$$\begin{vmatrix} x & 1 & 0 & 0 & 0 & 1 \\ 1 & x & 1 & 0 & 0 & 0 \\ 0 & 1 & x & 1 & 0 & 0 \\ 0 & 0 & 1 & x & 1 & 0 \\ 0 & 0 & 0 & 1 & x & 1 \\ 1 & 0 & 0 & 0 & 1 & x \end{vmatrix} = 0 \qquad (10\text{-}6.3)$$

(and hence the six π-electron orbital energies) determined. The solution of eqn (10-6.3) requires multiplying out the determinant, obtaining a sixth order polynomial equation in x and then finding the six roots. This can be quite time consuming.

Now let us see what happens if we apply symmetry rules to the problem. Essentially what we will do is to write

$$\Phi_j^{\pi} = \sum_{s=1}^{6} C'_{sj} \phi'_s$$

where the ϕ'_s (symmetry orbitals) are symmetry-adapted combinations of $2p_z$ atomic orbitals which generate irreducible representations of the point group to which the molecule belongs. This is the same thing as saying that we will change the basis functions used for Φ^{π} from ϕ_s to ϕ'_s. Though benzene belongs to the $\mathscr{D}_{6h}(=\mathscr{D}_6 \otimes \mathscr{C}_i)$ point group, we can, in fact, get all the information we require from the simpler point group $\mathscr{D}_6$, to which benzene also belongs.

The six $2p_z$ atomic orbitals ($\phi_1, \phi_2, \ldots \phi_6$) form a basis for a reducible representation Γ^{AO} of $\mathscr{D}_6$, since by applying the usual techniques (§ 5-7) we find that the transformation operators O_R transform ϕ_i either into itself or the negative of itself or into one of the other five atomic orbitals or the negative of one of the others:

$$O_R\phi_i = \pm\phi_i$$

or

$$O_R\phi_i = \pm\phi_k$$

so that

$$O_R\phi_i = \sum_{j=1}^{6} D_{ji}^{AO}(R)\phi_j.$$

The diagonal elements of the matrices $D^{AO}(R)$ will only be non-zero if an orbital is transformed into the positive or negative of itself, hence we obtain the following characters for Γ^{AO}

$\mathscr{D}_6$	E	$2C_6$	$2C_3$	C_2	$3C_2'$	$3C_2''$
$\chi^{AO}(C_i)$	6	0	0	0	-2	0

(the C_6–C_3–C_2 axis is perpendicular to the molecule and through its centre, the three C_2' and three C_2'' axes are in the molecular plane with the C_2' axes running through opposite carbon atoms and the C_2'' axes bisecting opposite bonds). Using these characters and the $\mathscr{D}_6$ character table (see Appendix I), the standard reduction formula (eqn (7-4.2)) leads to:

$$\Gamma^{AO} = \Gamma^{A_2} \oplus \Gamma^{B_2} \oplus \Gamma^{E_1} \oplus \Gamma^{E_2}$$

and therefore it must be possible to find combinations of $\phi_1, \phi_2, \ldots \phi_6$ which will serve as bases for the irreducible representations Γ^{A_2}, Γ^{B_2}, Γ^{E_1}, Γ^{E_2} of $\mathscr{D}_6$. [The same combinations will also necessarily generate irreducible representations of $\mathscr{D}_{6h}$ and since each ϕ_i changes sign under O_{σ_h}, the corresponding irreducible representations from $\mathscr{D}_{6h}$ must be such that $\chi(\sigma_h)$ is negative.† From the character table for $\mathscr{D}_{6h}$ it is clear that we must have

$$\Gamma^{AO} = \Gamma^{A_{2u}} \oplus \Gamma^{B_{2g}} \oplus \Gamma^{E_{1g}} \oplus \Gamma^{E_{2u}}.]$$

To find the particular combinations of ϕ_i which form a basis for Γ^{A_2}, Γ^{B_2}, Γ^{E_1} and Γ^{E_2} we make use of the projection operator technique and define the following operators (see eqn (7-6.6)):

$$P^{\mu} = \sum_R \chi^{\mu}(R)^* O_R$$

$$P^{A_2} = O_E + (O_{C_6} + O_{C_6^{-1}}) + (O_{C_3} + O_{C_3^{-1}}) + O_{C_2} - \sum O_{C_2'} - \sum O_{C_2''}.$$

†See footnote on page 216.

(the $\sum$ indicates, for example, the three operators associated with the three C_2' symmetry elements)

$$P^{B_2} = O_E-(O_{C_6}+O_{C_6^{-1}})+(O_{C_3}+O_{C_3^{-1}})-O_{C_2}-\sum O_{C_2'}+\sum O_{C_2''}$$
$$P^{E_1} = 2O_E+(O_{C_6}+O_{C_6^{-1}})-(O_{C_3}+O_{C_3^{-1}})-2O_{C_2}$$
$$P^{E_2} = 2O_E-(O_{C_6}+O_{C_6^{-1}})-(O_{C_3}+O_{C_3^{-1}})+2O_{C_2}.$$

Since there is only one basis function for the one-dimensional irreducible representations Γ^{A_2} and Γ^{B_2}, we need only apply P^{A_2} and P^{B_2} to one of the starting functions ϕ_i:

$$\begin{aligned} P^{A_2}\phi_1 &= \phi_1+(\phi_2+\phi_6)+(\phi_3+\phi_5)+\phi_4 \\ &\quad -(-\phi_1-\phi_3-\phi_5)-(-\phi_2-\phi_4-\phi_6) \\ &= 2(\phi_1+\phi_2+\phi_3+\phi_4+\phi_5+\phi_6) \\ P^{B_2}\phi_1 &= 2(\phi_1-\phi_2+\phi_3-\phi_4+\phi_5-\phi_6) \end{aligned}$$

and these two linear combinations will form a basis for Γ^{A_2} and Γ^{B_2}, respectively. For Γ^{E_1} and Γ^{E_2} one must apply P^{E_1} and P^{E_2} to at least two ϕ_i in order to produce two linearly-independent basis functions for each of these two-dimensional irreducible representations. Hence

$$\begin{aligned} P^{E_1}\phi_1 &= 2\phi_1+\phi_2-\phi_3-2\phi_4-\phi_5+\phi_6, \\ P^{E_1}\phi_2 &= \phi_1+2\phi_2+\phi_3-\phi_4-2\phi_5-\phi_6, \\ P^{E_2}\phi_1 &= 2\phi_1-\phi_2-\phi_3+2\phi_4-\phi_5-\phi_6, \end{aligned}$$

and

$$P^{E_2}\phi_2 = -\phi_1+2\phi_2-\phi_3-\phi_4+2\phi_5-\phi_6.$$

Since in Hückel molecular orbital theory it is assumed that

$$\int \phi_i^*\phi_j \, d\tau = \delta_{ij},$$

the symmetry orbitals are readily normalized. For example, for the first symmetry orbital the normalization constant N is given by the equation

$$\begin{aligned} 1 &= \int N^*2(\phi_1+\phi_2+\phi_3+\phi_4+\phi_5+\phi_6)^*N2(\phi_1+\phi_2+\phi_3+\phi_4+\phi_5+\phi_6) \, d\tau \\ &= 4N^2\sum_{i=1}^{6}\sum_{j=1}^{6}\int \phi_i^*\phi_j \, d\tau = 4N^2\sum_{i=1}^{6}\sum_{j=1}^{6}\delta_{ij} = 24N^2 \end{aligned}$$

and therefore $N = (24)^{-\frac{1}{2}}$ and the normalized symmetry orbital is

$$(\phi_1+\phi_2+\phi_3+\phi_4+\phi_5+\phi_6)/\sqrt{6}.$$

It is convenient if the symmetry orbitals belonging to a degenerate irreducible representation are made orthogonal to each other and this is achieved in the present case by taking combinations which are the sum and difference of the original combinations. This works since if F and G are two real normalized functions, then

$$\int (F+G)(F-G)\,\mathrm{d}\tau = \int F^2\,\mathrm{d}\tau + \int GF\,\mathrm{d}\tau - \int FG\,\mathrm{d}\tau - \int G^2\,\mathrm{d}\tau$$
$$= 1-1$$
$$= 0.$$

The orthonormal symmetry orbitals are therefore:

$$\phi_1' = (\phi_1+\phi_2+\phi_3+\phi_4+\phi_5+\phi_6)/\sqrt{6}, \qquad (\Gamma^{A_2})$$
$$\phi_2' = (\phi_1-\phi_2+\phi_3-\phi_4+\phi_5-\phi_6)/\sqrt{6}, \qquad (\Gamma^{B_2})$$
$$\phi_3' = (\phi_1-\phi_2-2\phi_3-\phi_4+\phi_5+2\phi_6)/2\sqrt{3}, \qquad (\Gamma^{E_1})$$
$$\phi_4' = (\phi_1+\phi_2-\phi_4-\phi_5)/2, \qquad (\Gamma^{E_1})$$
$$\phi_5' = (\phi_1+\phi_2-2\phi_3+\phi_4+\phi_5-2\phi_6)/2\sqrt{3}, \qquad (\Gamma^{E_2})$$
$$\phi_6' = (\phi_1-\phi_2+\phi_4-\phi_5)/2. \qquad (\Gamma^{E_2})$$

These six orthonormal functions are an equivalent orthonormal basis to that of ϕ_i (they describe the same function space) and if we use them in place of ϕ_i by writing

$$\Phi_j^{\pi} = \sum_{s=1}^{6} C'_{sj}\phi'_s$$

then eqn (10-6.1) becomes

$$\det(H'_{jk} - \varepsilon^{\pi} S'_{jk}) = 0$$

where H'_{jk} and S'_{jk} have the same form as $H_{jk}^{\mathrm{eff},\pi}$ and S_{jk} except that in the integrands each ϕ_i is now replaced by ϕ'_i. Thus:

$$H'_{11} = \int \phi_1'^{*} H^{\mathrm{eff},\pi} \phi_1'\,\mathrm{d}\tau$$
$$= \tfrac{1}{6}\int (\phi_1+\phi_2+\phi_3+\phi_4+\phi_5+\phi_6) H^{\mathrm{eff},\pi} (\phi_1+\phi_2+\phi_3+\phi_4+\phi_5+\phi_6)\,\mathrm{d}\tau$$
$$= \tfrac{1}{6}(\alpha+\beta+\beta+\beta+\alpha+\beta+\beta+\alpha+\beta+\beta+\alpha+\beta+\beta+\alpha+\beta+\beta+\beta+\alpha)$$
$$= \alpha+2\beta$$

$$H'_{22} = \alpha-2\beta$$
$$H'_{33} = \alpha+\beta$$
$$H'_{44} = \alpha+\beta$$
$$H'_{55} = \alpha-\beta$$
$$H'_{66} = \alpha-\beta$$

and, most importantly,

$$H'_{ij} = 0 \qquad \text{for } i \neq j$$

and

$$S'_{ij} = \delta_{ij}.$$

If we divide each element in the determinant by β and put

$$x = (\alpha - \varepsilon^{\pi})/\beta,$$

then we obtain eqn (10-6.4) which is in block form. Any determinant in

$$\begin{vmatrix} x+2 & 0 & 0 & 0 & 0 & 0 \\ 0 & x-2 & 0 & 0 & 0 & 0 \\ 0 & 0 & x+1 & 0 & 0 & 0 \\ 0 & 0 & 0 & x+1 & 0 & 0 \\ 0 & 0 & 0 & 0 & x-1 & 0 \\ 0 & 0 & 0 & 0 & 0 & x-1 \end{vmatrix} = 0 \qquad (10\text{-}6.4)$$

block form can be factorized, for example if

$$\begin{vmatrix} A_1 & [0] & [0] & . \\ [0] & A_2 & [0] & . \\ [0] & [0] & A_3 & . \\ . & . & . & . \end{vmatrix} = 0$$

then $|A_1| = 0$ or $|A_2| = 0$ or $|A_3| = 0$ etc. (see eqn (5-9.8)). So that from eqn (10-6.4) we obtain:

$$x+2 = 0$$

or

$$x-2 = 0$$

or

$$x+1 = 0 \qquad \text{(twice)}$$

or

$$x-1 = 0 \qquad \text{(twice).}$$

It is clear that eqn (10-6.4) is much easier to solve than eqn (10-6.3), though the results, of course, are identical. In general, by using symmetry orbitals ϕ'_s which are a basis for one of the irreducible representations of the point group, the matrix whose elements are

$$(H'_{jk} - \varepsilon^{\pi} S'_{jk})$$

will be in block form with each block corresponding to the symmetry orbitals which belong to a given irreducible representation. This occurs since the Hamiltonian $H^{\text{eff.}\pi}$ commutes with all O_R and this means that it belongs to the totally symmetric representation Γ^1 (see Appendix A.10-3). The vanishing integral rule (§ 8-4) then predicts that $\int \phi_j'^* H^{\text{eff.}\pi} \phi'_k \, d\tau$ is zero if ϕ'_j and ϕ'_k belong to different irreducible representations. Similar arguments also hold for S'_{jk}.

The π-electron orbital energies for benzene are therefore, in order of increasing energy: $\alpha+2\beta$, $\alpha+\beta$, $\alpha+\beta$, $\alpha-\beta$, $\alpha-\beta$, and $\alpha-2\beta$ (β is negative) and these energy levels are labelled by using the lower case notation of the irreducible representations of $\mathscr{D}_{6h}$ which are associated with them:

ϵ

$\alpha - 2\beta$	————	(b_{2g})
$\alpha - \beta$	——— ———	(e_{2u})
$\alpha + \beta$	—✗✗— —✗✗—	(e_{1g})
$\alpha + 2\beta$	—✗✗—	(a_{2u})

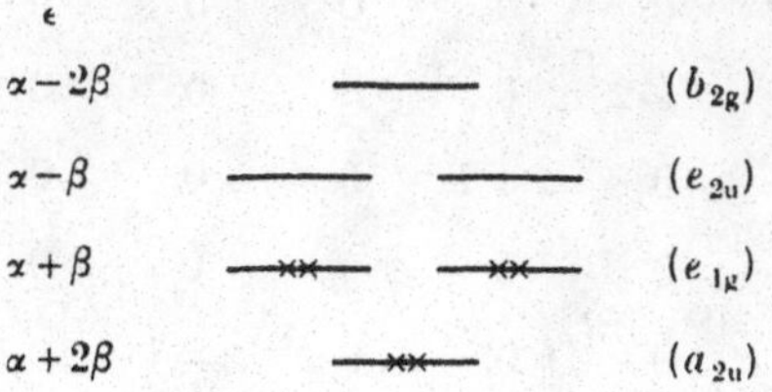

The crosses indicate that two electrons are fed into the non-degenerate a_{2u} level (one with spin α, the other with spin β) and that four electrons are fed into the doubly degenerate e_{1g} level; the whole making up the ground state π-electron configuration $(a_{2u})^2(e_{1g})^4$.

Ignoring G (see eqn (10-5.1)), the π-electron energy for the ground state will be $6\alpha + 8\beta$ which, when compared with the π-electron energy of three ethylene molecules ($6\alpha + 6\beta$), shows that the delocalization energy of benzene is 2β.

In this problem we are fortunate that the factorization of the determinant is complete and as a consequence the π-electron MOs are, in fact, identical with the symmetry orbitals:

$$\Phi_i^7 = \phi_i'$$

i.e. $C_{sj}' = \delta_{sj}$.

10-7. Hückel molecular orbital method for the trivinylmethyl radical

The trivinylmethyl radical $\cdot C(CH=CH_2)_3$ has seven carbon atoms and seven π-electrons and belongs to the $\mathscr{D}_{3h}$ point group. We will however use the lower symmetry point group $\mathscr{C}_3$ to which the molecule also belongs. The labeling of the carbon atoms is shown in Fig. 10-7.1. In Table 10-7.1 we show how the seven $2p_z$ atomic orbitals (ϕ_1, ϕ_2,... ϕ_7) transform under the operators O_R and from these results we obtain the characters of Γ^{AO}; they are given together with the $\mathscr{C}_3$ character table in Table 10-7.2. It will be noticed that the Γ^E representation has been

F ɪ ɢ. 10-7.1. The trivinylmethyl radical.

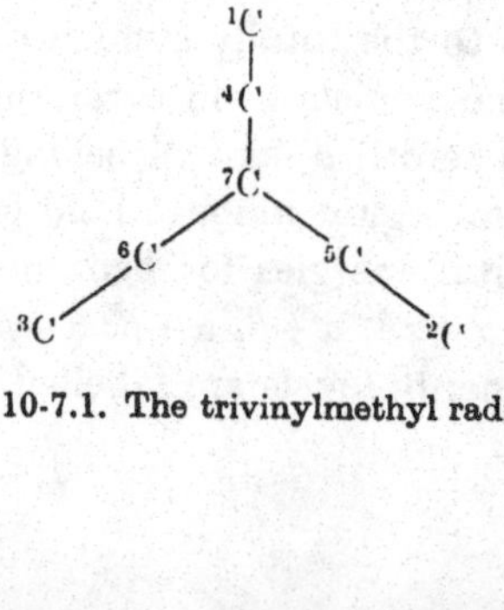

TABLE 10-7.1

Transformation of ϕ_i under O_R for the trivinylmethyl radical

	O_E	O_{C_3}	$O_{C_3^2}$
ϕ_1	ϕ_1	ϕ_2	ϕ_3
ϕ_2	ϕ_2	ϕ_3	ϕ_1
ϕ_3	ϕ_3	ϕ_1	ϕ_2
ϕ_4	ϕ_4	ϕ_5	ϕ_6
ϕ_5	ϕ_5	ϕ_6	ϕ_4
ϕ_6	ϕ_6	ϕ_4	ϕ_5
ϕ_7	ϕ_7	ϕ_7	ϕ_7

split into two one-dimensional representations (Γ^{E_1} and Γ^{E_2}), where the characters of one are merely the conjugate complexes of the characters of the other. Use of the reduction formula (eqn (7-4.2)) leads to

$$\Gamma^{AO} = 3\Gamma^{A}\oplus 2\Gamma^{E_1}\oplus 2\Gamma^{E_2}$$

so that the seven atomic orbitals can be combined into seven symmetry orbitals: three forming a basis for Γ^{A}, two for Γ^{E_1} and two for Γ^{E_2}.

These symmetry adapted combinations are found by using the projection operators:

$$P^{\mu} = \sum_{R} \chi^{\mu}(R)^{*} O_R$$

on the ϕ_i. The following linearly independent combinations are then obtained:

$$\begin{aligned}
P^{A}\phi_1 &= \phi_1+\phi_2+\phi_3,\\
P^{A}\phi_4 &= \phi_4+\phi_5+\phi_6,\\
P^{A}\phi_7 &= \phi_7+\phi_7+\phi_7,\\
P^{E_1}\phi_1 &= \phi_1+\varepsilon^{*}\phi_2+\varepsilon\phi_3,\\
P^{E_1}\phi_4 &= \phi_4+\varepsilon^{*}\phi_5+\varepsilon\phi_6,\\
P^{E_2}\phi_1 &= \phi_1+\varepsilon\phi_2+\varepsilon^{*}\phi_3,\\
P^{E_2}\phi_4 &= \phi_4+\varepsilon\phi_5+\varepsilon^{*}\phi_6,
\end{aligned}$$

TABLE 10-7.2

Character table $\mathscr{C}_3$ and χ^{AO} for the trivinylmethyl radical†

$\mathscr{C}_3$		E	C_3	C_3^2
	A	1	1	1
E	E_1	1	ε	ε^{*}
	E_2	1	ε^{*}	ε
$\chi^{AO}(C_i)$		7	1	1

† $\varepsilon = \exp(2\pi i/3) = \frac{1}{2}(\sqrt{3}i-1)$
$\varepsilon^{*} = \exp(-2\pi i/3) = -\frac{1}{2}(1+\sqrt{3}i)$
$\varepsilon+\varepsilon^{*} = -1$

where $\varepsilon = \exp(2\pi i/3)$ and all other combinations, e.g. $P^A\phi_2$, will either be one of these combinations or some combination of these combinations. When normalized, the symmetry orbitals are:

$$
\begin{aligned}
\phi_1' &= (\phi_1+\phi_2+\phi_3)/\sqrt{3}, && (\Gamma^A)\\
\phi_2' &= (\phi_4+\phi_5+\phi_6)/\sqrt{3}, && (\Gamma^A)\\
\phi_3' &= \phi_7, && (\Gamma^A)\\
\phi_4' &= (\phi_1+\varepsilon^*\phi_2+\varepsilon\phi_3)/\sqrt{3}, && (\Gamma^{E_1})\\
\phi_5' &= (\phi_4+\varepsilon^*\phi_5+\varepsilon\phi_6)/\sqrt{3}, && (\Gamma^{E_1})\\
\phi_6' &= (\phi_1+\varepsilon\phi_2+\varepsilon^*\phi_3)/\sqrt{3}, && (\Gamma^{E_2})\\
\phi_7' &= (\phi_4+\varepsilon\phi_5+\varepsilon^*\phi_6)/\sqrt{3}. && (\Gamma^{E_2})
\end{aligned}
$$

Since $\int\phi_i^*\phi_j\,d\tau = \delta_{ij}$ (Hückel approximation), it is easily confirmed that these ϕ_i' satisfy $\int\phi_i'^*\phi_i'\,d\tau = 1$. Because the symmetry orbitals belong to the irreducible representations of $\mathscr{C}_3$, the $\det(H_{jk}'-\varepsilon^{\pi}S_{jk}')$ will be in the following block form:

$$
\begin{vmatrix}
+ & + & + & & & & \\
+ & + & + & & & & \\
+ & + & + & & & & \\
 & & & + & + & & \\
 & & & + & + & & \\
 & & & & & + & + \\
 & & & & & + & +
\end{vmatrix}
$$

where the first block corresponds to the three symmetry orbitals of Γ^A type, the second block to the two symmetry orbitals of Γ^{E_1} type and the third block to the two symmetry orbitals of Γ^{E_2} type.

If we evaluate the elements of this determinant in accord with the Hückel approximations, divide each element by β, replace $(\alpha-\varepsilon^{\pi})/\beta$ by x and factorize the determinant into its three parts, we obtain† eqn

† Care has to be taken with the conjugate complexes in the Γ^{E_1} and Γ^{E_2} type symmetry orbitals, e.g.

$$
\begin{aligned}
H_{44}' &= \int\phi_4'^*H^{\text{eff},\pi}\phi_4'\,d\tau\\
&= \tfrac{1}{3}\int(\phi_1+\varepsilon\phi_2+\varepsilon^*\phi_3)H^{\text{eff},\pi}(\phi_1+\varepsilon^*\phi_2+\varepsilon\phi_3)\,d\tau\\
&= \alpha
\end{aligned}
$$

and,

$$
\begin{aligned}
H_{45}' &= \int\phi_4'^*H^{\text{eff},\pi}\phi_5'\,d\tau\\
&= \tfrac{1}{3}\int(\phi_1+\varepsilon\phi_2+\varepsilon^*\phi_3)H^{\text{eff},\pi}(\phi_4+\varepsilon^*\phi_5+\varepsilon\phi_6)\,d\tau\\
&= \beta
\end{aligned}
$$

(10-7.1) for the Γ^A symmetry orbitals, eqn (10-7.2) for the Γ^{E_1} symmetry orbitals, and eqn (10-7.3) for the Γ^{E_2} symmetry orbitals.

$$\begin{vmatrix} x & 1 & 0 \\ 1 & x & \sqrt{3} \\ 0 & \sqrt{3} & x \end{vmatrix} = 0 \qquad (10\text{-}7.1)$$

$$\begin{vmatrix} x & 1 \\ 1 & x \end{vmatrix} = 0 \qquad (10\text{-}7.2)$$

$$\begin{vmatrix} x & 1 \\ 1 & x \end{vmatrix} = 0 \qquad (10\text{-}7.3)$$

Clearly it is much easier to solve one 3×3 and two 2×2 determinantal equations than the 7×7 determinantal equation which occurs when no use is made of symmetry.

Multiplying out eqn (10-7.1), we get $x^3 - 4x = 0$ and the roots of this equation are $x_1 = 0$, $x_2 = 2$, and $x_3 = -2$ or $\varepsilon_1^\pi = \alpha$, $\varepsilon_2^\pi = \alpha - 2\beta$, and $\varepsilon_3^\pi = \alpha + 2\beta$. Substituting these roots, in turn, into

$$\sum_k (H'_{jk} - \varepsilon_i^\pi S'_{jk}) C'_{ki} = 0$$

and using the normalization equation

$$\sum_k C'^2_{ki} = 1,$$

we obtain for $\varepsilon_1^\pi = \alpha$,

$$C'_{11} = -\sqrt{3}/2, \qquad C'_{21} = 0, \qquad C'_{31} = 1/2$$

or

$$\Phi_1^\pi = -(\sqrt{3}/2)\phi'_1 + 0 \times \phi'_2 + \tfrac{1}{2}\phi'_3 = (\phi_1 + \phi_2 + \phi_3 - \phi_7)/2$$

for $\varepsilon_2^\pi = \alpha - 2\beta$,

$$C'_{12} = \sqrt{2}/4, \qquad C'_{22} = -1/\sqrt{2}, \qquad C'_{32} = \sqrt{6}/4$$

and

$$\Phi_2^\pi = (\phi_1 + \phi_2 + \phi_3 - 2\phi_4 - 2\phi_5 - 2\phi_6 + 3\phi_7)/2\sqrt{6}.$$

for $\varepsilon_3^\pi = \alpha + 2\beta$

$$C'_{13} = \sqrt{2}/4, \qquad C'_{23} = 1/\sqrt{2}, \qquad C'_{33} = \sqrt{6}/4,$$

and

$$\Phi_3^\pi = (\phi_1 + \phi_2 + \phi_3 + 2\phi_4 + 2\phi_5 + 2\phi_6 + 3\phi_7)/2\sqrt{6}.$$

Multiplying out eqn (10-7.2), we get $x^2 - 1 = 0$ or $x_4 = 1$ and $x_5 = -1$ or $\varepsilon_4^\pi = \alpha - \beta$ and $\varepsilon_5^\pi = \alpha + \beta$, thus

$$\Phi_4^\pi = (\phi_1 + \varepsilon^* \phi_2 + \varepsilon \phi_3 - \phi_4 - \varepsilon^* \phi_5 - \varepsilon \phi_6)/\sqrt{6}$$

and

$$\Phi_5^\pi = (\phi_1 + \varepsilon^* \phi_2 + \varepsilon \phi_3 + \phi_4 + \varepsilon^* \phi_5 + \varepsilon \phi_6)/\sqrt{6}.$$

From eqn (10-7.3) we get $\varepsilon_6^\pi = \alpha - \beta$ and $\varepsilon_7^\pi = \alpha + \beta$, and

$$\Phi_6^\pi = (\phi_1 + \varepsilon \phi_2 + \varepsilon^* \phi_3 - \phi_4 - \varepsilon \phi_5 - \varepsilon^* \phi_6)/\sqrt{6}$$

and

$$\Phi_7^\pi = (\phi_1+\varepsilon\phi_2+\varepsilon^*\phi_3+\phi_4+\varepsilon\phi_5+\varepsilon^*\phi_6)/\sqrt{6}.$$

The π-electron MOs Φ_4^π and Φ_6^π are degenerate, in that they correspond to the same π-electron orbital energy $(\alpha-\beta)$. As is the case with any degenerate wavefunctions, any combination of Φ_4^π and Φ_6^π will be equally valid. It is convenient to combine Φ_4^π and Φ_6^π in such a way as to obtain real MOs, this can be done by:

$$\Phi_4^\pi(\text{real}) = (\Phi_4^\pi+\Phi_6^\pi)/\sqrt{2} = (2\phi_1-\phi_2-\phi_3-2\phi_4+\phi_5+\phi_6)/\sqrt{12}$$
$$\Phi_6^\pi(\text{real}) = (\Phi_4^\pi-\Phi_6^\pi)/\sqrt{2}\mathrm{i} = (\phi_2-\phi_3-\phi_5+\phi_6)/2$$

(note that $\varepsilon+\varepsilon^* = -1$). Similarly for Φ_5^π and Φ_7^π (both corresponding to an orbital energy of $\alpha+\beta$), we can find the following two real MOs:

$$\Phi_5^\pi(\text{real}) = (\Phi_5^\pi+\Phi_7^\pi)/\sqrt{2} = (2\phi_1-\phi_2-\phi_3+2\phi_4-\phi_5-\phi_6)/\sqrt{12}$$
$$\Phi_7^\pi(\text{real}) = (\Phi_5^\pi-\Phi_7^\pi)/\sqrt{2}\mathrm{i} = (\phi_2-\phi_3+\phi_5-\phi_6)/2.$$

The π-electron orbital energy level diagram will be:

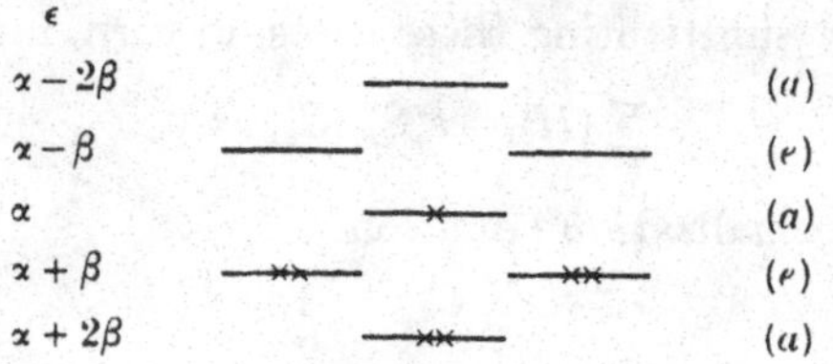

and if we distinguish the MOs of the same symmetry by preceding the irreducible representation notation by a number which ascends with increasing orbital energy, the π-electron configuration in the ground state will be $(1a)^2(1e)^4(2a)^1$ and the total π-electronic energy will be $E_\pi = 7\alpha+8\beta-G$.

If we consider the $\mathscr{D}_{3h}$ point group, we find that under the transformation operator O_{σ_h} all the molecular orbitals change sign:†

$$O_{\sigma_h}\Phi_i^\pi = -\Phi_i^\pi \qquad i = 1, 2,\ldots 7$$

and that under O_{C_2}, Φ_1^π, Φ_2^π, and Φ_3^π change sign:

$$O_{C_2}\Phi_i^\pi = -\Phi_i^\pi \qquad i = 1, 2, 3$$

and that under O_{σ_v}, Φ_1^π, Φ_2^π, and Φ_3^π are unchanged:

$$O_{\sigma_v}\Phi_i^\pi = \Phi_i^\pi \qquad i = 1, 2, 3.$$

This information is sufficient to classify the MOs with respect to the irreducible representations of $\mathscr{D}_{3h}$ and using the character table, we see that Φ_1^π, Φ_2^π, and Φ_3^π belong to $\Gamma^{A_2''}$ and Φ_4^π, Φ_5^π, Φ_6^π, and Φ_7^π belong to

† The $2p_z$ orbitals are perpendicular to the molecular plane σ_h and $O_{\sigma_h}\phi_i = -\phi_i$.

$\Gamma^{E''}$. The electronic configuration of the ground state will therefore be

$$(1a_2'')^2(1e'')^4(2a_2'')^1.$$

Appendices

A.10-1. Atomic units

The atomic unit of length is the radius of the first Bohr orbit in the hydrogen atom when the reduced mass of the electron is replaced by the rest mass m_e. Thus the atomic unit of length is

$$a_0 = \frac{h^2}{4\pi^2 m_e e^2} = 0{\cdot}052918 \text{ nm}.$$

The atomic unit of energy is

$$e^2/a_0 = 27{\cdot}210 \text{ eV}\dagger = 1 \text{ a.u. (of energy)}.$$

This is just twice the ionization potential of the hydrogen atom if the reduced mass of the electron is replaced by the rest mass. One atomic unit of energy is equivalent to twice the Rydberg constant for infinite mass.

When atomic units are used, one sets $e = h/2\pi = m_e = 1$ in quantum mechanical equations. For example $-h^2\nabla^2/8\pi^2 m_e$ becomes $-\frac{1}{2}\nabla^2$.

The advantage of atomic units is that if all the calculations are directly expressed in such units, the results do not vary with the subsequent revision of the numerical values of the fundamental constants.

A.10-2. An alternative notation for the LCAO MO method

Define the MOs by

$$\Phi_i = \sum_{\mu}^{m} C_{\mu i}\phi_\mu$$

where ϕ_μ are atomic orbitals, then the coefficients $C_{\mu i}$ are determined by

$$\sum_{\nu}^{m} (F_{\mu\nu} - \varepsilon_i S_{\mu\nu})C_{\nu i} = 0 \qquad \begin{array}{l} \mu = 1, 2, \ldots, m \\ i = 1, 2, \ldots, m \end{array}$$

where

$$F_{\mu\nu} = H_{\mu\nu} + \sum_{\lambda}^{m}\sum_{\sigma}^{m} P_{\lambda\sigma}\{(\mu\nu \mid \lambda\sigma) - \tfrac{1}{2}(\mu\sigma \mid \lambda\nu)\}$$

$$H_{\mu\nu} = \int \phi_\mu^*(1) h_1 \phi_\nu(1)\, d\tau_1$$

$$h_\mu = -\tfrac{1}{2}\nabla_\mu^2 - \sum_{\alpha=1}^{N} Z_\alpha / r_{\mu\alpha}$$

$$P_{\mu\nu} = 2\sum_{i=1}^{n/2} C_{\mu i}^* C_{\nu i} \qquad (n = \text{number of electrons})$$

$$(\mu\nu \mid \lambda\sigma) = \iint \phi_\mu^*(1)\phi_\nu(1) r_{12}^{-1} \phi_\lambda^*(2)\phi_\sigma(2)\, d\tau_1\, d\tau_2$$

$$S_{\mu\nu} = \int \phi_\mu^*(1)\phi_\nu(1)\, d\tau_1.$$

† 1 eV = $1{\cdot}60219 \times 10^{-19}$ J.

The total electronic energy is given by:

$$E = \sum_{\mu}^{m}\sum_{\nu}^{m} P_{\mu\nu}H_{\mu\nu} + \tfrac{1}{2}\sum_{\mu}^{m}\sum_{\nu}^{m}\sum_{\lambda}^{m}\sum_{\sigma}^{m} P_{\mu\nu}P_{\lambda\sigma}\{(\mu\nu \mid \lambda\sigma) - \tfrac{1}{2}(\mu\sigma \mid \lambda\nu)\}$$

$$= 2\sum_{i=1}^{\text{occ}} \varepsilon_i - \tfrac{1}{2}\sum_{\mu}^{m}\sum_{\nu}^{m}\sum_{\lambda}^{m}\sum_{\sigma}^{m} P_{\mu\nu}P_{\lambda\sigma}\{(\mu\nu \mid \lambda\sigma) - \tfrac{1}{2}(\mu\sigma \mid \lambda\nu)\}.$$

The reader may confirm that the content of these equations is the same as that of eqns (10-3.1) to (10-3.6).

A.10-3. Proof that the matrix elements of an operator H which commutes with all O_R of a group vanish between functions belonging to different irreducible representations

Let ψ_i^μ be a set of functions belonging to the irreducible representation Γ^μ and H an operator which commutes with all the transformation operators O_R, then

$$O_R\psi_i^\mu = \sum_j D_{ji}^\mu(R)\psi_j^\mu$$

and

$$O_R(H\psi_i^\mu) = HO_R\psi_i^\mu$$

$$= H\Big(\sum_j D_{ji}^\mu(R)\psi_j^\mu\Big)$$

$$= \sum_j D_{ji}^\mu(R)(H\psi_j^\mu).$$

Consequently, the functions $H\psi_i^\mu$ also form a basis for Γ^μ and hence by invoking the vanishing integral rule, the integral

$$\int \psi_k^{\nu *} H\psi_i^\mu \, d\tau = \int \psi_k^{\nu *}(H\psi_i)^\mu \, d\tau$$

vanishes unless $\Gamma^\nu = \Gamma^\mu$.

If we consider $H\psi_i^\mu$ in the direct product representation $\Gamma^H \otimes \Gamma^\mu$ then since $H\psi_i^\mu$ belong to Γ^μ, $\Gamma^H \otimes \Gamma^\mu = \Gamma^\mu$ and therefore $\Gamma^H = \Gamma^1$. Hence, any operator which commutes with all O_R of a point group can be said to belong to the totally symmetric irreducible representation Γ^1.

PROBLEM

10.1. For the following molecules, determine the point group and the symmetry of the MOs for the π-electrons, and, using Hückel theory, obtain the MOs and orbital energies:

(a) *trans*-1,3-butadiene,

(b) ethylene,

(c) cyclobutadiene,

(d) cyclopentadienyl radical C_5H_5,

(e) naphthalene,

(f) phenanthrene.

11. Hybrid orbitals

11-1. Introduction

IN this chapter we explore how symmetry considerations can be applied to one of the most pervasive concepts in all of chemistry: bonding between atoms by the sharing of pairs of electrons. Though the idea of an electron-pair bond was first introduced in 1916 by G. N. Lewis, it was only after the advent of quantum mechanics that it could be given a proper theoretical basis. This came about through the development of two theories: valence bond (VB) theory and localized MO theory; both of which describe the electron pair in terms of orbitals of the component atoms of the bond.

In VB theory the pair of bonding electrons in the bond A—B of some polyatomic molecule is described by the wavefunction

$$\psi_A(1)\psi_B(2)+\psi_B(1)\psi_A(2)$$

where ψ_A is an orbital centred on nucleus A and ψ_B is an orbital centred on nucleus B and the 1 and 2 indicate the two electrons (we ignore electron-spin considerations). In localized MO theory the electron pair is described by the wavefunction

$$\Psi(1)\Psi(2)$$

where Ψ is a localized MO extending over both nucleus A and nucleus B and which can be synthesized from an orbital centred on A(ψ_A) and an orbital centred on B (ψ_B), i.e.

$$\Psi(1) = c_1\psi_A(1)+c_2\psi_B(1), \qquad (11\text{-}1.1)$$

where c_1 and c_2 are numerical coefficients. Both of these bond descriptions are approximations and at first sight appear to be quite different, but, if we carry the approximations a stage further, the methods converge and become completely equivalent. For this reason we will only consider one of them and choose for our purposes the localized MO method.

When considering a polyatomic molecule, the general MO method (see Chapter 10) would describe the n electrons of the molecule by the wavefunction (see eqn (10-2.3))

$$\Psi(1, 2, \ldots n) = 1/\sqrt{n!}\begin{vmatrix} \Phi_1(1) & \Phi_2(1) & \ldots & \Phi_n(1) \\ \Phi_1(2) & \Phi_2(2) & \ldots & \Phi_n(2) \\ \cdot & \cdot & & \cdot \\ \cdot & \cdot & & \cdot \\ \cdot & \cdot & & \cdot \\ \Phi_1(n) & \Phi_2(n) & \ldots & \Phi_n(n) \end{vmatrix}$$

where the Φ_i are MOs which extend over the *entire* molecule, not just a single bond as in localized MO theory, and can be approximated by linear combinations of atomic orbitals centred on *all* the nuclei. Indeed, most quantum mechanical calculations done to-day use such wavefunctions. Clearly, the localized MO method, where the electrons in a polyatomic molecule are divided up into bond pairs, each described by MOs of the form of eqn (11-1.1), is an approximation to this more general treatment. The question arises therefore: why do we bother with it? The answer is two-fold. In the first place, chemical intuition and experience tells us that many properties of molecules are properties of the bonds and that these properties are often constant from one molecule to another, e.g. the existence of a characteristic infra-red absorption band near 3 μm due to a C—H valence stretching mode is used to detect the presence of C—H bonds in an unknown molecule. Such constancy would seem to imply localized distributions of charge which are transferable and which could be adequately described by localized MOs. In the second place, localized MOs are easier to imagine and handle and they preserve the conventional idea of a bond which is typified by the symbol A—B.

Symmetry plays an important role in localized MO theory since the orbitals used in the construction of the MOs ψ_A and ψ_B of eqn (11-1.1), must be symmetric about the bond axis (for the present we will limit our discussion to σ-bonding). The most natural, though not mandatory, building blocks to use for ψ_A and ψ_B are the atomic orbitals (AOs) of the component atoms (A and B). In some cases there is available a single AO on A and a single AO on B, both of which are symmetric about the bond axis and therefore meet our requirements. But more often, and particularly when A has to form several bonds, there are not the required number of atomic orbitals with the appropriate symmetry and it is necessary to synthesize ψ_A (or ψ_B) from several AOs of A (or B). For example methane CH_4 is a tetrahedral molecule with four equivalent C—H bonds pointing to the corners of a tetrahedron and each localized MO is made up of an orbital from the carbon atom and an orbital from the appropriate hydrogen atom. The contribution from each hydrogen atom can be taken as a 1s hydrogen AO and these will be symmetric about the appropriate bond axis; however, amongst the AOs of the carbon atom 1s, 2s, $2p_x$, $2p_y$, and $2p_z$ there are not four which are equivalent and symmetric about the four bond axes. We are therefore forced into taking combinations of these primitive orbitals if we wish to have four equivalent and symmetric orbitals; this procedure is called *hybridization* and the combinations are called *hybrid orbitals.* If we restrict ourselves to the broad class of molecules which have a unique central atom A surrounded by a set of other atoms which are bonded to A but not to each other (e.g. mononuclear co-ordination complexes, NO_3^-, SO_4^{2-}, BF_3, PF_5, CH_4, $CHCl_3$, etc.), then the symmetry of the molecule will determine which AOs on atom A should be combined (§ 11-3) and in what proportions (§ 11-5). If there is more than one combination of AOs on atom A having the correct symmetry, and this will usually be the case, then arguments of a more chemical nature will have to be invoked in order to decide which is the most appropriate combination.

A necessary prelude to determining the combinations of AOs which give a hybrid orbital of correct symmetry is the classification of the AOs of the central atom A in terms of the irreducible representations of the point group to which the molecule belongs. This is discussed in § 11-2. In § 11-4 we consider π-bonding systems and in the final section we discuss the relationship between localized and non-localized MO theory.

The reader who is not familiar with the background of this chapter, and it has only been summarized in the preceding paragraphs, is recommended to read C. A. Coulson's excellent book: *Valence.*

11-2. Transformation properties of atomic orbitals

In constructing a localized MO for the bond A—B it is necessary to specify an orbital centred on A (ψ_A) and an orbital centred on B (ψ_B). In principle, provided symmetry about the bond axis is preserved (we are still considering only σ-bonded systems), our choice of ψ_A and ψ_B is not restricted and we could use any well-defined mathematical function or combination of functions. Common sense, however, dictates that the most sensible functions to use for this purpose are the AOs of the free atoms A and B. There are three reasons why this is a sensible choice: one mathematical, one chemical, and one practical.

The mathematical reason is that the AOs of a given free atom form what is known as a *complete set,* that is any function can be produced by taking a combination of them; so we know that it is mathematically possible to replace ψ_A, whatever its form, by a combination of A's AOs. The chemical reason is that the bond A—B is chemically formed by combining atom A with atom B and we expect the electronic distribution, at least close to the nuclei, to be similar in the bond to what it is in the free atoms. The third reason is that we know from atomic calculations the energy order of the AOs and we expect that the lowest energy MOs will be those formed from the lowest energy AOs. This fact can often help us decide which AOs to choose for the construction of an MO when symmetry arguments leave the matter ambiguous.

Having decided to use AOs (or combinations of them) for ψ_A and ψ_B, we will now look at the form these take. They are approximate solutions to the Schrödinger equation for the atom in question. The Schrödinger equation for many-electron atoms is usually solved approximately by writing the total electronic wavefunction as the product of one-electron functions ϕ_i (these are the AOs). Each AO ϕ_i is a function of the polar coordinates r, θ, and ϕ (see Fig. 11-2.1) of a single electron and can be written as

$$\phi_i = R_i(r)Y_i(\theta, \phi).$$

The radial functions $R_i(r)$ will be different for different atoms. Only for the hydrogen atom is the exact analytical form of the $R_i(r)$'s known. For other atoms the $R_i(r)$'s will be approximate and their form will depend on the method used to find them. They might be analytical functions (e.g. Slater orbitals) or tabulated sets of numbers (e.g. numerical Hartree–Fock orbitals).

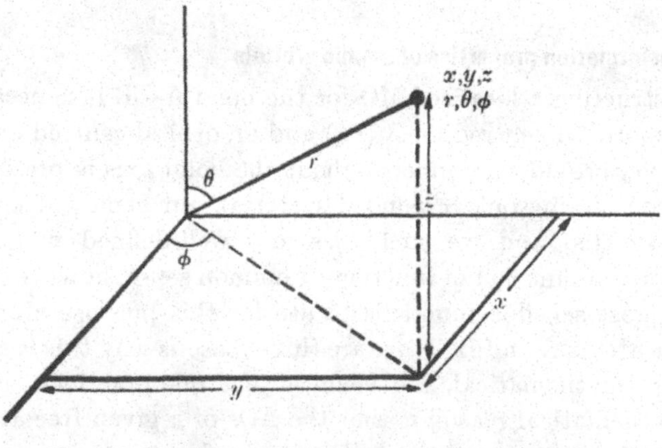

FIG. 11-2.1. The coordinate system for atomic orbitals.

TABLE 11-2.1

Angular functions (un-normalized) for s, p, d, *and* f *orbitals*

Symbol	*Angular function*
s	no angular dependence
p_x	$\sin\theta\cos\phi$
p_y	$\sin\theta\sin\phi$
p_z	$\cos\theta$
$d_{3r^2-z^2}$ or d_{z^2}	$3\cos^2\theta-1$
$d_{x^2-y^2}$	$\sin^2\theta\cos 2\phi$
d_{xy}	$\sin^2\theta\sin 2\phi$
d_{xz}	$\sin\theta\cos\theta\cos\phi$
d_{yz}	$\sin\theta\cos\theta\sin\phi$
$f_{x(5x^2-3r^2)}$ or f_{x^3}	$\sin\theta\cos\phi(5\sin^2\theta\cos^2\phi-3)$
$f_{y(5y^2-3r^2)}$ or f_{y^3}	$\sin\theta\sin\phi(5\sin^2\theta\sin^2\phi-3)$
$f_{z(5z^2-3r^2)}$ or f_{z^3}	$5\cos^3\theta-3\cos\theta$
$f_{x(z^2-y^2)}$	$\sin\theta\cos\phi(\cos^2\theta-\sin^2\theta\sin^2\phi)$
$f_{y(z^2-x^2)}$	$\sin\theta\sin\phi(\cos^2\theta-\sin^2\theta\cos^2\phi)$
$f_{z(x^2-y^2)}$	$\sin^2\theta\cos\theta\cos 2\phi$
f_{xyz}	$\sin^2\theta\cos\theta\sin 2\phi$

The angular functions $Y_l(\theta, \phi)$, called spherical harmonics, are common to all atoms. They are listed in Table 11-2.1 (in this table the normalizing constants have been omitted) together with the well-known symbols for the orbitals to which they correspond, i.e. s, p_x, p_y, p_z, etc. The subscripts in these symbols are directly related to the angular functions; if, for example, the angular function is $\sin^2\theta \sin 2\phi$, then changing to the Cartesian coordinates x, y, and z where: $x = r \sin\theta\cos\phi$, $y = r\sin\theta\sin\phi$ and $z = r\cos\theta$ gives us:

$$\begin{aligned}\sin^2\theta\sin 2\phi &= 2\sin^2\theta\sin\phi\cos\phi\\ &= 2(\sin\theta\cos\phi)(\sin\theta\sin\phi)\\ &= 2(x/r)(y/r)\\ &= (2/r^2)xy\\ &= f(r)xy\end{aligned}$$

and any orbital with this angular dependence is given the subscript xy.

If we restrict ourselves as before to those molecules which have a unique central atom A surrounded by a set of other atoms which are bonded to A, then in order to ascertain which AOs of A can be used to produce a ψ_A which is symmetric about a particular bond axis, it is necessary to know to which irreducible representations of the molecule's point group the AOs of A belong, i.e. for which irreducible representations they form a basis. That they must form the basis of some representation of the molecular point group follows from the fact that

TABLE

The symmetry species of the s, p, *and* d

	$\mathscr{C}_1$	$\mathscr{C}_2$	$\mathscr{C}_3$	$\mathscr{C}_4$	$\mathscr{C}_5$	$\mathscr{C}_6$	$\mathscr{D}_2$	$\mathscr{D}_3$	$\mathscr{D}_4$	$\mathscr{D}_5$	$\mathscr{D}_6$	$\mathscr{C}_{2v}$	$\mathscr{C}_{3v}$	$\mathscr{C}_{4v}$	$\mathscr{C}_{5v}$	$\mathscr{C}_{6v}$	$\mathscr{C}_{2h}$
s	A_g	A	A	A	A	A	A	A_1	A_1	A_1	A_1	A_1	A_1	A_1	A_1	A_1	A_g
p_x	A_u	B					B_3					B_1					B_u
			E	E	E_1	E_1		E	E	E_1	E_1		E	E	E_1	E_1	
p_y	A_u	B					B_2					B_2					B_u
p_z	A_u	A	A	A	A	A	B_1	A_2	A_2	A_2	A_2	A_1	A_1	A_1	A_1	A_1	A_u
d_{z^2}	A_g	A	A	A	A	A	A	A_1	A_1	A_1	A_1	A_1	A_1	A_1	A_1	A_1	A_g
$d_{x^2-y^2}$	A_g	A		B			A		B_1			A_1		B_1			A_g
			E		E_2	E_2		E		E_2	E_2		E		E_2	E_2	
d_{xy}	A_g	A		B			B_1		B_2			A_2		B_2			A_g
d_{xz}	A_g	B					B_2					B_1					B_g
			E	E	E_1	E_1		E	E	E_1	E_1		E	E	E_1	E_1	
d_{yz}	A_g	B					B_3					B_2					B_g

since atom A is unique, it must lie on all the planes and axes of symmetry which the molecule possesses. We have considered this question for p- and d-orbitals using the $\mathscr{D}_{4h}$ point group in § 7-9 (see also § 5-9) and the technique used there can be applied to all orbitals and all point groups. The results are given in Table 11-2.2; they are also incorporated in the character tables in Appendix I at the end of the book. As an example of using Table 11-2.2, consider the phosphorus atom in PF_5. This molecule belongs to the $\mathscr{D}_{3h}$ point group and we immediately see that the phosphorus atomic orbitals belong to the following irreducible representations:

$$\Gamma^{A_1'}: \quad s, d_{z^2}$$
$$\Gamma^{E'}: \quad (p_x, p_y), (d_{x^2-y^2}, d_{xy})$$
$$\Gamma^{A_2''}: \quad p_z$$
$$\Gamma^{E''}: \quad (d_{xz}, d_{yz})$$

where the parentheses indicate that the two orbitals inside them *together* form a basis for the given two-dimensional irreducible representation.

The reader should note that no transformation operator O_R can alter the radial function $R_i(r)$ of an orbital† and consequently the symmetry properties of the AOs are completely defined by the angular functions, $Y_i(\theta, \phi)$. Since these angular functions are the same in all 'one-electron product function' approximations, the orbitals in all these approximations (Slater orbitals, numerical Hartree–Fock orbitals,

† The only exception to this rule is the hydrogen atom, see the footnote on page 155.

11-2.2
orbitals for different point groups

$\mathscr{C}_{3h}$	$\mathscr{C}_{4h}$	$\mathscr{C}_{5h}$	$\mathscr{C}_{6h}$	$\mathscr{D}_{2h}$	$\mathscr{D}_{3h}$	$\mathscr{D}_{4h}$	$\mathscr{D}_{5h}$	$\mathscr{D}_{6h}$	$\mathscr{D}_{2d}$	$\mathscr{D}_{3d}$	$\mathscr{D}_{4d}$	$\mathscr{D}_{5d}$	$\mathscr{D}_{6d}$	$\mathscr{S}_4$	$\mathscr{S}_6$	$\mathscr{T}_d$	$\mathscr{O}_h$
A'	A_g	A'	A_g	A_g	A_1'	A_{1g}	A_1'	A_{1g}	A_1	A_{1g}	A_1	A_{1g}	A_1	A	A_g	A_1	A_{1g}
E'	E_u	E_1'	E_{1u}	B_{3u} B_{2u}	E'	E_u	E_1'	E_{1u}	E	E_u	E_1	E_{1u}	E_1	E	E_u	T_2	T_{1u}
A''	A_u	A''	A_u	B_{1u}	A_2''	A_{2u}	A_2''	A_{2u}	B_2	A_{2u}	B_2	A_{2u}	B_2	B	A_u		
A'	A_g	A'	A_g	A_g	A_1'	A_{1g}	A_1'	A_{1g}	A_1	A_{1g}	A_1	A_{1g}	A_1	A	A_g	E	E_g
E'	B_g B_g	E_2'	E_{2g}	A_g B_{1g}	E'	B_{1g} B_{2g}	E_2'	E_{2g}	B_1 B_2	E_g	E_2	E_{2g}	E_2	B B	E_g		
E''	E_g	E_1''	E_{1g}	B_{2g} B_{3g}	E''	E_g	E_1''	E_{1g}	E	E_g	E_3	E_{1g}	E_5	E	E_g	T_2	T_{2g}

etc.) will belong to the symmetry species designated in Table 11-2.2. Similarly the symmetry species of an orbital will be independent of the atom involved and of the principle quantum number of the shell to which it belongs.

11-3. Hybrid orbitals for σ-bonding systems

A nodal plane or surface is the locus of all points at which a wave-function has zero amplitude as a result of its changing sign on passing from one side of the surface to the other; the probability of finding an electron on such a surface is zero. σ-Bonds and σ-orbitals are defined as those having *no* nodal surface which contains the bond axis; such bonds and orbitals will be symmetric about the bond axis. In this section we consider which AOs of a central atom A (which is bonded to a set of other atoms) can be combined to form a hybrid orbital which is symmetric about the bond axis and therefore capable of σ-bonding.

We will explain how to do this by taking the specific example of methane. Methane has a central carbon atom which is σ-bonded to four hydrogen atoms with each σ-bond pointing to one of the corners of a tetrahedron. We therefore require four hybrid orbitals on the carbon atom which similarly point to the corners of a tetrahedron. Since the four bonds are indistinguishable, the four hybrids must be equivalent, that is to say they must be identical in all respects except for their orientation. For the reasons given in § 11-2, they will be taken to be linear combinations of the atomic orbitals of carbon, which are themselves bases for certain irreducible representations of the point group of the molecule $\mathscr{T}_d$. The hybrid orbitals, as we will see in a moment, form the basis for a *reducible* representation of the $\mathscr{T}_d$ point group and we will give this representation the symbol Γ^{hyb}. Using the equations in § 7-4 we can reduce Γ^{hyb} and find the irreducible representations which it contains. Let us suppose we find

$$\Gamma^{hyb} = \Gamma^1 \oplus \Gamma^2$$

where Γ^1 and Γ^2 are two irreducible representations of $\mathscr{T}_d$. This implies (see Chapter 6) that there are combinations of hybrid orbitals which

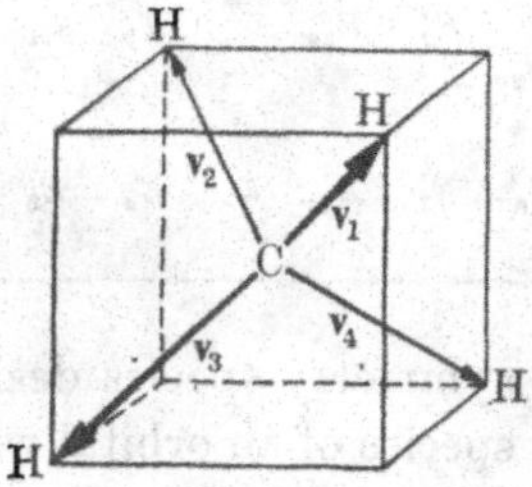

FIG. 11-3.1. A set of vectors v_1, v_2, v_3, and v_4 representing the four σ-hybrid orbitals used by carbon to bond the four hydrogens in methane.

will serve as a basis for the Γ^1 and Γ^2 irreducible representations of $\mathscr{T}_d$. Now we can *reverse* the argument and say that there are combinations of those functions which belong to Γ^1 and Γ^2 which will serve as a basis for Γ^{hyb} and which can be taken as the hybrid functions with the symmetry properties which we desire. We conclude, therefore, that we should take linear combinations of those atomic orbitals which belong to Γ^1 and Γ^2.

First then, for methane, we must obtain Γ^{hyb}. To do this let us associate with each carbon hybrid orbital a vector pointing in the appropriate direction and let us label these vectors v_1, v_2, v_3, v_4 (see Fig. 11-3.1). All of the symmetry properties of the four hybrid orbitals will be identical to those of the four vectors. The reducible representation Γ^{hyb} using these vectors (or hybrids) as a basis can be obtained from

$$Rv_i = \sum_{j=1}^{4} D_{ji}^{hyb}(R)v_j \qquad i = 1, 2, \ldots 4$$

(compare with eqn (9-4.2)), but since we only wish to carry out the reduction of Γ^{hyb}, we only need the character of Γ^{hyb}, which is given by

$$\chi^{hyb}(R) = \sum_{i=1}^{4} D_{ii}^{hyb}(R).$$

It is clear that only if $\mathbf{v}_i$ is unshifted by $\boldsymbol{R}$ will $D_{ii}^{\mathrm{hyb}}(\boldsymbol{R})$ be non-zero and then it will be unity. Consequently $\chi^{\mathrm{hyb}}(\boldsymbol{R})$ is equal to the number of vectors which are unshifted by $\boldsymbol{R}$ (this line of argument is analogous to that used in considering the representation Γ^0 in § 9-6). Proceeding in this way, we find the character below for Γ^{hyb} for the $\mathscr{T}_{\mathrm{d}}$ point

$\mathscr{T}_{\mathrm{d}}$	E	$8C_3$	$3C_2$	$6S_4$	$6\sigma_{\mathrm{d}}$
$\chi^{\mathrm{hyb}}(C_i)$	4	1	0	0	2

group. Using this information together with the character table for $\mathscr{T}_{\mathrm{d}}$ and eqn (7-4.2), we find

$$\Gamma^{\mathrm{hyb}} = \Gamma^{A_1} \oplus \Gamma^{T_2}.$$

Recalling our earlier discussion, this equation means that the four AOs which are combined to make the set of four hybrid orbitals must be chosen so as to include one orbital which belongs to the Γ^{A_1} representation and a set of three orbitals which belong to the Γ^{T_2} representation. Reference to Table 11-2.2 shows that the AOs fall into the categories below for the $\mathscr{T}_{\mathrm{d}}$ point group. So we can combine an s-orbital

Γ^{A_1}	Γ^{T_2}	Γ^{E}
s	$(\mathrm{p}_x, \mathrm{p}_y, \mathrm{p}_z)$ $(\mathrm{d}_{xy}, \mathrm{d}_{xz}, \mathrm{d}_{yz})$	$(\mathrm{d}_{z^2}, \mathrm{d}_{x^2-y^2})$

with the three p-orbitals in four different ways to produce four equivalent hybrids, called sp^3 hybrids, or we can combine an s-orbital with the three d-orbitals $(\mathrm{d}_{xy}, \mathrm{d}_{xz}, \mathrm{d}_{yz})$ in four different ways to form four equivalent sd^3 hybrids. Both sets of hybrids will have the correct symmetry properties and will point in tetrahedral directions.

This is as far as symmetry arguments alone will take us and we can only conclude that the most general solution to the problem would be a set of hybrid orbitals which are linear combinations of both possibilities, namely

$$\psi_{\mathrm{C}} = a(\mathrm{sp}^3) + b(\mathrm{sd}^3)$$

(plus additional terms if f, g,... orbitals are taken into account). The numerical coefficients a and b might be determined by some quantum mechanical technique. However, it is at this stage that our chemical intuition can be used, or rather our knowledge concerning the relative energies of the AOs in a free carbon atom, and we can predict that b is quite small compared with a.

If we assume that the 1s orbital cannot be used on the grounds that it is already occupied by two core (or non-bonding) electrons then the

sp^3 hybrids can be constructed from the 2s- and 2p-orbitals of carbon. The lowest energy, and therefore most stable, d-orbitals available to carbon are the 3d-orbitals, so that the most stable sd^3 hybrids would be constructed from the 2s, $3d_{xy}$, $3d_{xz}$, and $3d_{yz}$ orbitals. However, in the carbon atom the 3d orbitals lie about 230 kcal/mole higher than the 2p-orbitals. Therefore, in order for the bonds formed by using sd^3 hybrids to be more stable than a set using sp^3 hybrids, each sd^3 bond would have to be about $3 \times 230/4 \sim 170$ kcal/mole stronger than each sp^3 bond. This is highly unlikely and even a limited usage of sd^3 hybrids will

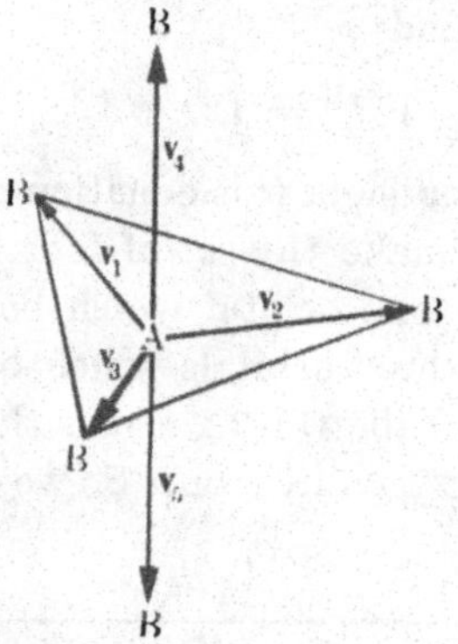

FIG. 11-3.2. A set of vectors v_1, v_2,..., and v_5 representing the five σ-hybrid orbitals used by A to bond the five B atoms which surround A in a trigonal bipyramid structure.

be of very minor importance. We conclude that for CH_4 the hybrids are of sp^3 character. However, for a species like MnO_4^- the sd^3 hybrids will predominate because in this case the lowest available d-orbitals are 3d and the lowest available p-orbitals are 4p and the 3d-orbitals are of lower energy than the 4p-orbitals.

As a second example, let us consider a molecule with the formula AB_5 having the symmetry of a trigonal bipyramid $\mathscr{D}_{3h}$. The vector system is shown in Fig. 11-3.2. The set of five hybrid orbitals (or vectors) on A form a basis for a reducible representation of the $\mathscr{D}_{3h}$ point group, with the following character:

$\mathscr{D}_{3h}$	E	$2C_3$	$3C_2$	σ_h	$2S_3$	$3\sigma_v$
$\chi^{hyb}(C_i)$	5	2	1	3	0	3

From this we deduce that:

$$\Gamma^{hyb} = 2\Gamma^{A_1'} \oplus \Gamma^{A_2''} \oplus \Gamma^{E'}$$

and that the set of five atomic orbitals on A which are combined to produce the set of five hybrid orbitals must be chosen so as to include two orbitals which belong to $\Gamma^{A_1'}$, one orbital which belongs to $\Gamma^{A_2''}$ and a set of two orbitals which together belong to $\Gamma^{E'}$. The atomic orbitals of A fall into the following categories for the $\mathscr{D}_{3h}$ point group (see Table 11-2.2):

$\Gamma^{A_1'}$	$\Gamma^{E'}$	$\Gamma^{A_2''}$	$\Gamma^{E''}$
s	(p_x, p_y)	p_z	(d_{xz}, d_{yz})
d_{z^2}	$(d_{x^2-y^2}, d_{xy})$		

Therefore, any of the following combinations of AOs will produce appropriate (from a symmetry point of view) hybrid orbitals:

(1) ns, $(n+1)$s, p_z, p_x, and p_y,
(2) ns, $(n+1)$s, p_z, d_{xy}, and $d_{x^2-y^2}$,
(3) nd_{z^2}, $(n+1)d_{z^2}$, p_z, p_x, and p_y,
(4) nd_{z^2} $(n+1)d_{z^2}$, p_z, d_{xy}, and $d_{x^2-y^2}$,
(5) s, d_{z^2}, p_z, p_x, and p_y,
(6) s, d_{z^2}, p_z, d_{xy}, and $d_{x^2-y^2}$.

In molecules which are known to have a trigonal bipyramid structure, energy criteria make it unlikely that the first four combinations are important. In PF_5 combination (5) is the most probable and the hybrids are labelled dsp^3. In gaseous $MoCl_5$ it is likely that a combination of (5) and (6) are used, since the 4d and 5p AOs of Mo are close in energy. The hybrid formed from the orbitals in (6) is labelled d^3sp and the most general form for ψ_{Mo} in $MoCl_5$ is

$$\psi_{Mo} = a(dsp^3)+b(d^3sp)$$

with $a \simeq b$.

In Table 11-3.1 hybrid orbitals for a selection of geometries are tabulated.

11-4. Hybrid orbitals for π-bonding systems

In contrast to σ-orbitals and σ-bonds, a π-orbital or π-bond is defined as one which has *one* nodal surface or plane containing the bond axis. (The reader might note that, though we will not deal with them, δ-orbitals and δ-bonds have *two* nodal surfaces which intersect on the bond axis.) If a π-bond is to be formed in MO fashion by combining two orbitals, one on each of the two bonded atoms, it is obviously necessary that each orbital have π-character with respect to the bond axis *and* that their two nodal planes coincide. The formation of a π-type bonding MO from two π-type AOs is shown in Fig. 11-4.1, where the plus and minus signs refer to the sign of the wavefunction.

TABLE 11-3.1
Different hybrid orbitals

Number of equivalent orbitals	*Designation*	*Geometry*
2	sp	linear
	dp	linear
3	sp^2	trigonal plane
	dp^2	trigonal plane
	ds^2	trigonal plane
	d^3	trigonal plane
	p^3	trigonal pyramid
	d^2p	trigonal pyramid
4	sp^3	tetrahedral
	sd^3	tetrahedral
	dsp^2	tetragonal plane
	d^2p^2	tetragonal plane
	d^4	tetragonal pyramid
5	dsp^3	trigonal bipyramid
	d^3sp	trigonal bipyramid
	d^2sp^2	tetragonal pyramid
	d^4s	tetragonal pyramid
	d^2p^3	tetragonal pyramid
	d^4p	tetragonal pyramid
	d^3p^2	pentagonal plane
	d^5	pentagonal pyramid
6	d^2sp^3	octahedron
	d^4sp	trigonal prism
	d^5p	trigonal prism
	d^3p^3	trigonal antiprism

Let us consider a molecule AB_n in which each atom B is bonded to atom A. We will assume that each atom B has available two π-AOs which are orthogonal to each other, i.e. their nodal planes are mutually perpendicular (they might, for example, be a $2p_x$ and a $2p_y$ AO). If we wish to form $2n$ π-bonds, requiring $2n$ π-MOs, then we must furnish $2n$ π-AOs or hybrids on A which match up with respect to their nodal

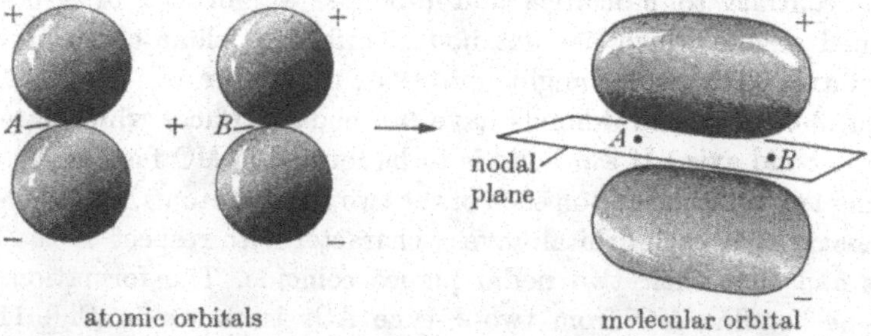

FIG. 11-4.1. The formation of a π-type bonding molecular orbital.

planes with the $2n$ π-AOs of the B's. Each bond will then consist of two shared pairs of electrons (a double bond). In this section we are concerned with the choice of the AOs of A which, when formed into linear combinations, provide the $2n$ hybrid orbitals of the appropriate symmetry.

As an example of an AB_n molecule, we will discuss the planar symmetrical molecule BCl_3 which belongs to the $\mathscr{D}_{3h}$ point group. First we assign to each chlorine atom a pair of mutually perpendicular

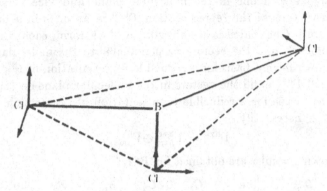

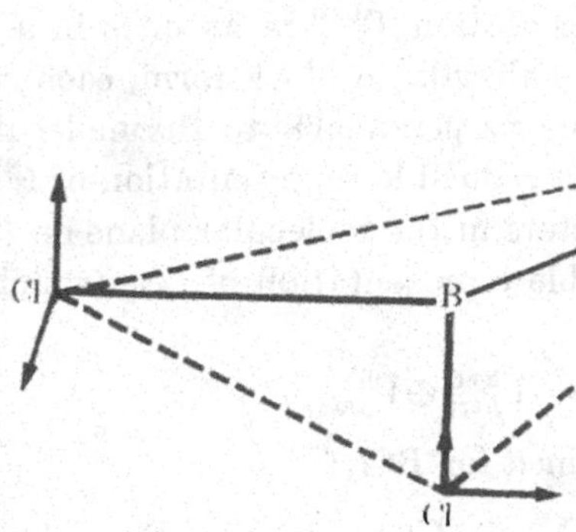

FIG. 11-4.2. Vectors representing π-orbitals on the Cl atoms in BCl_3.

vectors to represent the π-AOs of the chlorine atoms. Each vector points towards the positive lobe of the orbital and, in a pair, one vector is perpendicular to the molecular plane and the other is in the molecular plane and perpendicular to the relevant bond axis (see Fig. 11-4.2). This particular arrangement is chosen simply for convenience and other arrangements are equally valid provided that the two vectors on a given chlorine atom are mutually perpendicular and in a plane perpendicular to the B—Cl axis.

The six necessary hybrid orbitals on the boron atom can also be assigned vectors. If π-bonds are to be formed, these vectors must have the same orientation as the six vectors on the chlorine atoms. If we followed in the footsteps of § 11-3, we would now construct the reducible representation Γ^{hyb} from a consideration of how the six vectors on the boron atom change under the symmetry operations of the $\mathscr{D}_{3h}$ point group. However, it is clear that since the six vectors on the chlorine atoms match the six on the boron atom, exactly the same representation Γ^{hyb} can be found by using these vectors instead. Since it is less confusing to have three pairs of vectors separated in space than six originating from one point, we will take this latter approach.

(We could have done exactly the same thing for methane in § 11-3, it would simply have meant reversing the vectors in Fig. 11-3.1.)

Except for this change, we find $\chi^{\text{hyb}}(R)$ in the same way as before. We note, however, that this time the direction of a vector may be reversed as the result of a symmetry operation and in such a case there will be a contribution of -1 to the character of that operation. Furthermore, we immediately see that in carrying out the different symmetry operations, no vector perpendicular to the molecular plane is ever interchanged with one in the molecular plane and vice versa. This implies two things: the representation Γ^{hyb} is at once in a partially reduced form (the matrices are already in block form, each consisting of two blocks); and the vectors perpendicular to the molecular plane on their own form a basis for a reducible representation of $\mathscr{D}_{3h}$ (which we will call $\Gamma^{\text{hyb}}_{\text{perp}}$) and the vectors in the molecular plane on their own also form a basis for a reducible representation of $\mathscr{D}_{3h}$ (which we will call $\Gamma^{\text{hyb}}_{\text{plane}}$); necessarily

$$\Gamma^{\text{hyb}} = \Gamma^{\text{hyb}}_{\text{perp}} \oplus \Gamma^{\text{hyb}}_{\text{plane}}.$$

The following results are obtained for BCl_3:

$\mathscr{D}_{3h}$	E	$2C_3$	$3C_2$	σ_h	$2S_3$	$3\sigma_v$
$\chi^{\text{hyb}}(C_i)$	6	0	-2	0	0	0
$\chi^{\text{hyb}}_{\text{perp}}(C_i)$	3	0	-1	-3	0	1
$\chi^{\text{hyb}}_{\text{plane}}(C_i)$	3	0	-1	3	0	-1

These characters, using the standard decomposition formula, lead to

$$\Gamma^{\text{hyb}}_{\text{perp}} = \Gamma^{A_2''} \oplus \Gamma^{E''}$$

and

$$\Gamma^{\text{hyb}}_{\text{plane}} = \Gamma^{A_2'} \oplus \Gamma^{E'}.$$

Therefore, in order for boron to form a π-bond perpendicular to the molecular plane to each of the chlorine atoms, it must use three hybrid orbitals constructed from one AO which belongs to $\Gamma^{A_2''}$ and a pair of AOs which belong to $\Gamma^{E''}$. Table 11-2.2 shows that only the following orbitals meet these requirements:

$\Gamma^{A_2''}$	$\Gamma^{E''}$
p_z	(d_{xz}, d_{yz})

and a set of three equivalent hybrids (pd^2 type) using these orbitals is the only one possible for the 'perpendicular' π-bonds. For the hybrids which are in the molecular plane, we require an atomic orbital which belongs to $\Gamma^{A_2'}$ and inspection of Table 11-2.2 shows that no s-, p-, or d-orbital fulfills this requirement, though there are two pairs of atomic orbitals, (p_x, p_y) and $(d_{x^2-y^2}, d_{xy})$, which belong to $\Gamma^{E'}$, the other necessary component. Therefore we cannot form a set of three equivalent π-bonds in the molecular plane. However, this does not mean *no* π-bonds in the plane can be formed or that only two of the chlorine atoms can be π-bonded in the plane. It simply means that there can only be two π-bonds in the plane *shared equally* amongst the three

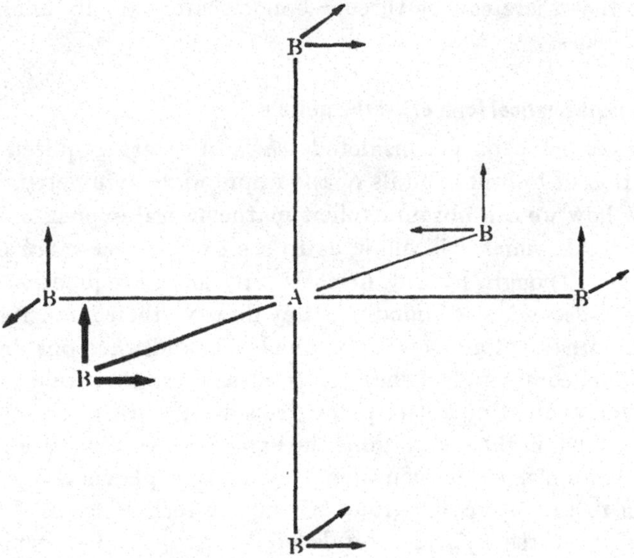

FIG. 11-4.3. Vectors representing π-orbitals on the B atoms in an octahedral AB_6 molecule.

chlorine atoms; these two π-bonds using the $\Gamma^{E'}$ orbitals. This type of situation arises quite often.

Now let us consider the important case of an octahedral AB_6 molecule. If we associate two mutually perpendicular vectors with each atom B as in Fig. 11-4.3, we obtain the following character for Γ^{hyb}:

$\mathcal{O}_h$	E	$8C_3$	$3C_2$	$6C_4$	$6C_2'$	i	$8S_6$	$3\sigma_h$	$6S_4$	$6\sigma_d$
$\chi^{\text{hyb}}(C_i)$	12	0	-4	0	0	0	0	0	0	0

For this class of molecule the vectors do not fall into two categories and we obtain a single reducible representation:

$$\Gamma^{\text{hyb}} = \Gamma^{T_{1g}} \oplus \Gamma^{T_{2g}} \oplus \Gamma^{T_{1u}} \oplus \Gamma^{T_{2u}}.$$

For the irreducible representations in this symbolic equation, inspection of Table 11-2.2 shows that we have the following s-, p-, and d-orbitals:

$\Gamma^{T_{1g}}$	$\Gamma^{T_{2g}}$	$\Gamma^{T_{1u}}$	$\Gamma^{T_{2u}}$
none	(d_{xy}, d_{xz}, d_{yz})	$(p_x, p_y\ p_z)$	none

Since the p-orbitals on A have most likely been used up in σ-bonding, we have only the three T_{2g} d-orbitals for π-bonding. We therefore conclude that there can be three π-bonds shared equally amongst the six A—B pairs.

11-5. The mathematical form of hybrid orbitals

So far we have only considered *which* AOs are required for the construction of hybrid orbitals of the appropriate symmetry. We now will show how we can obtain explicit mathematical expressions for the hybrid orbitals which will allow us to see exactly *how much* each AO contributes. Though hybrid orbitals are most frequently used in qualitative discussions of bonding, they do have their quantitative use when one carries out an exact MO calculation and when one deals with coordination compounds, where it is often necessary to use hybrid orbitals for evaluating overlap integrals which are often related to bond strengths; in these situations the explicit expressions are required.

As an example, let us consider a symmetric planar AB_3 molecule belonging to the $\mathscr{D}_{3h}$ point group. Using the techniques of § 11-3, we find that the three hybrid orbitals of A ψ_1, ψ_2, and ψ_3 which form σ-bonds with the three B atoms, are composed of one AO which belongs to $\Gamma^{A_1'}$ and a pair which belong to $\Gamma^{E'}$. By use of the projection operator technique (see § 7-6) we can project out of ψ_1, ψ_2, and ψ_3, functions which belong to the irreducible representations $\Gamma^{A_1'}$ and $\Gamma^{E'}$ and, by equating these functions with the AOs which are being used, we can obtain equations mathematically linking the hybrids with the AOs.

The projection operator corresponding to the μth irreducible representation is:

$$\boldsymbol{P}^{\mu} = \sum_{R} \chi^{\mu}(R)^{*}\boldsymbol{O}_{R}$$

(see eqn (7-6.6)) and in Table 11-5.1 the results of applying $\boldsymbol{O}_R$ to the three hybrid orbitals are given (the directions of the hybrids are shown in Fig. 11-5.1). From this information we deduce that

$$\begin{aligned} \boldsymbol{P}^{A_1'}\psi_1 &= 4(\psi_1+\psi_2+\psi_3) \\ \boldsymbol{P}^{E'}\psi_1 &= 2(2\psi_1-\psi_2-\psi_3) \\ \boldsymbol{P}^{E'}\psi_2 &= 2(2\psi_2-\psi_1-\psi_3). \end{aligned}$$

TABLE 11-5.1

Transformation of the $\mathscr{D}_{3h}$ hybrids under $\boldsymbol{O}_R$ for all $\boldsymbol{R}$ of $\mathscr{D}_{3h}$

	E	C_3	C_3^2	C_{2a}	C_{2b}	C_{2c}	σ_h	S_3	S_3^2	σ_{va}	σ_{vb}	σ_{vc}
ψ_1	ψ_1	ψ_2	ψ_3	ψ_1	ψ_3	ψ_2	ψ_1	ψ_2	ψ_3	ψ_1	ψ_3	ψ_2
ψ_2	ψ_2	ψ_3	ψ_1	ψ_3	ψ_2	ψ_1	ψ_2	ψ_3	ψ_1	ψ_3	ψ_2	ψ_1
ψ_3	ψ_3	ψ_1	ψ_2	ψ_2	ψ_1	ψ_3	ψ_3	ψ_1	ψ_2	ψ_2	ψ_1	ψ_3

Applying $P^{A_1'}$ to ψ_2 and ψ_3 is not necessary since the same combination as $P^{A_1'}\psi_1$ will be produced. The two combinations obtained from $P^{E'}$ are by inspection, linearly independent and a third combination which can be found by applying $P^{E'}$ to ψ_3, will be a combination of these two. If the combinations are normalized with the assumption that

$$\int \psi_i \psi_j \, d\tau = \delta_{ij},$$

then we have:

$$(\psi_1+\psi_2+\psi_3)/\sqrt{3} \qquad \text{(belonging to } \Gamma^{A_1'}\text{)}$$

and

$$\left.\begin{array}{l}(2\psi_1-\psi_2-\psi_3)/\sqrt{6}\\(2\psi_2-\psi_1-\psi_3)/\sqrt{6}\end{array}\right\} \qquad \text{(belonging to } \Gamma^{E'}\text{).}$$

Under the $\mathscr{D}_{3h}$ point group, an s-orbital belongs to $\Gamma^{A_1'}$ and the pair of p-orbitals p_x and p_y belongs to $\Gamma^{E'}$. If these orbitals have been used to construct the three hybrid orbitals (sp^2 type), then we can

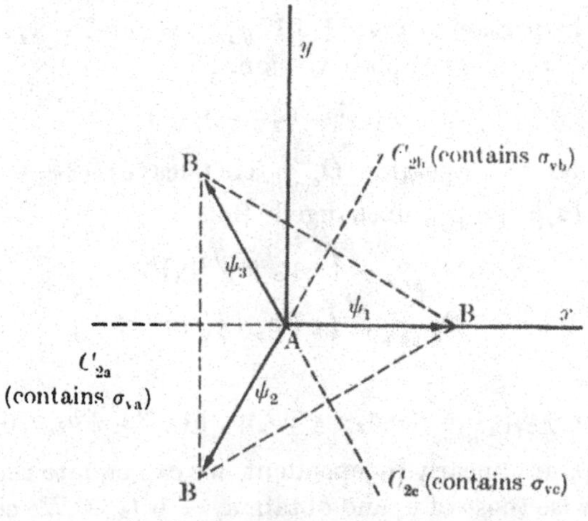

FIG. 11-5.1. Directions of the hybrids and x and y axes for an AB_3 molecule belonging to $\mathscr{D}_{3h}$.

immediately identify, since it is the only one of $\Gamma^{A_1'}$ symmetry,

$$(\psi_1+\psi_2+\psi_3)/\sqrt{3}$$

with the s-orbital, i.e.

$$s = (\psi_1+\psi_2+\psi_3)/\sqrt{3}. \tag{11-5.1}$$

We can also identify the other two combinations $(2\psi_1-\psi_2-\psi_3)/\sqrt{6}$ and $(2\psi_2-\psi_1-\psi_3)/\sqrt{6}$ with two normalized combinations of p_x and p_y orbitals, since both pairs form a basis for $\Gamma^{E'}$, i.e.

$$(2\psi_1-\psi_2-\psi_3)/\sqrt{6} = \pm(a_1^2+b_1^2)^{-\frac{1}{2}}(a_1p_x+b_1p_y)$$
$$(2\psi_2-\psi_1-\psi_3)/\sqrt{6} = \pm(a_2^2+b_2^2)^{-\frac{1}{2}}(a_2p_x+b_2p_y).$$

The x and y axes are shown in Fig. 11-5.1. To establish the values of the coefficients a_1, a_2, b_1, and b_2 it is necessary to investigate the detailed effect of one or more of the transformation operators O_R of $\mathscr{D}_{3h}$ on the pair of combinations.

The operator $O_{\sigma_{va}}$ (see Fig. 11-5.1 for the definition of the σ_{va} plane) leaves $(2\psi_1-\psi_2-\psi_3)$ unchanged, therefore it must leave $(a_1p_x+b_1p_y)$ unchanged:

$$O_{\sigma_{va}}(a_1p_x+b_1p_y) = a_1p_x+b_1p_y. \tag{11-5.2}$$

But $O_{\sigma_{va}}p_x = p_x$ and $O_{\sigma_{va}}p_y = -p_y$, hence:

$$O_{\sigma_{va}}(a_1p_x+b_1p_y) = a_1p_x-b_1p_y \tag{11-5.3}$$

and by comparing eqns (11-5.2) and (11-5.3) we see that $b_1 = 0$. Therefore

$$(2\psi_1-\psi_2-\psi_3)/\sqrt{6} = \pm p_x$$

and since, by inspection of Fig. 11-5.1, ψ_1, $-\psi_2$, and $-\psi_3$ have positive x components, we take the positive sign, i.e.

$$p_x = (2\psi_1-\psi_2-\psi_3)/\sqrt{6}. \tag{11-5.4}$$

Now consider the operator $O_{\sigma_{vb}}$. This leaves $(2\psi_2-\psi_1-\psi_3)$, and consequently $(a_2p_x+b_2p_y)$, unchanged. But

$$O_{\sigma_{vb}}p_x = (-p_x+\sqrt{3}p_y)/2$$

and

$$O_{\sigma_{vb}}p_y = (\sqrt{3}p_x+p_y)/2$$

and hence

$$(a_2p_x+b_2p_y) = \{(-a_2+\sqrt{3}b_2)p_x+(\sqrt{3}a_2+b_2)p_y\}/2.$$

Since p_x and p_y are linearly independent, we can equate the coefficients of p_x and likewise those of p_y and obtain $a_2 = b_2/\sqrt{3}$. We conclude that

$$(2\psi_2-\psi_1-\psi_3)/\sqrt{6} = \pm(p_x+\sqrt{3}p_y)/2.$$

Since ψ_2 and $-\psi_3$ have negative y components (ψ_1 has none), we take the negative sign, i.e.

$$(2\psi_2-\psi_1-\psi_3)/\sqrt{6} = -(p_x+\sqrt{3}p_y)/2.$$

Combining this equation with eqn (11-5.4), we get

$$p_y = (\psi_3-\psi_2)/\sqrt{2}. \tag{11-5.5}$$

Eqns (11-5.1), (11-5.4), and (11-5.5) may be brought together in the single matrix equation

$$\begin{Vmatrix} s \\ p_x \\ p_y \end{Vmatrix} = \begin{Vmatrix} 1/\sqrt{3} & 1/\sqrt{3} & 1/\sqrt{3} \\ 2/\sqrt{6} & -1/\sqrt{6} & -1/\sqrt{6} \\ 0 & -1/\sqrt{2} & 1/\sqrt{2} \end{Vmatrix} \begin{Vmatrix} \psi_1 \\ \psi_2 \\ \psi_3 \end{Vmatrix}$$

which on inversion leads to

$$\begin{Vmatrix} \psi_1 \\ \psi_2 \\ \psi_3 \end{Vmatrix} = \begin{Vmatrix} 1/\sqrt{3} & 1/\sqrt{3} & 1/\sqrt{3} \\ 2/\sqrt{6} & -1/\sqrt{6} & -1/\sqrt{6} \\ 0 & -1/\sqrt{2} & 1/\sqrt{2} \end{Vmatrix}^{-1} \begin{Vmatrix} s \\ p_x \\ p_y \end{Vmatrix}$$

$$= \begin{Vmatrix} 1/\sqrt{3} & 2/\sqrt{6} & 0 \\ 1/\sqrt{3} & -1/\sqrt{6} & -1/\sqrt{2} \\ 1/\sqrt{3} & -1/\sqrt{6} & 1/\sqrt{2} \end{Vmatrix} \begin{Vmatrix} s \\ p_x \\ p_y \end{Vmatrix}.$$

(Note that since the 3×3 matrix is orthogonal, its inverse is simply its transpose.) So we finally achieve the following mathematical expressions for ψ_1, ψ_2, and ψ_3:

$$\psi_1 = (\sqrt{2}s+2p_x)/\sqrt{6},$$
$$\psi_2 = (\sqrt{2}s-p_x-\sqrt{3}p_y)/\sqrt{6},$$
$$\psi_3 = (\sqrt{2}s-p_x+\sqrt{3}p_y)/\sqrt{6}.$$

In this particular example we could have avoided some of the labour involved in finding the combinations of hybrid orbitals which are equal to p_x and p_y, by using the $\mathscr{C}_3$ point group (to which the molecule also belongs). For this point group, the two-dimensional representation, the cause of all the trouble, can be expressed as two complex one-dimensionl representations. The orbitals p_x and p_y are then just as easy to obtain as the s-orbital. Any complex numbers which result are eliminated at the end of the treatment by addition and subtraction of the orbitals formed. This is the technique which was used in § 10-7 to find the π-molecular orbitals of the trivinylmethyl radical. It is, however, of no avail when dealing with point groups which have three-dimensional irreducible representations as in our next example, CH_4.

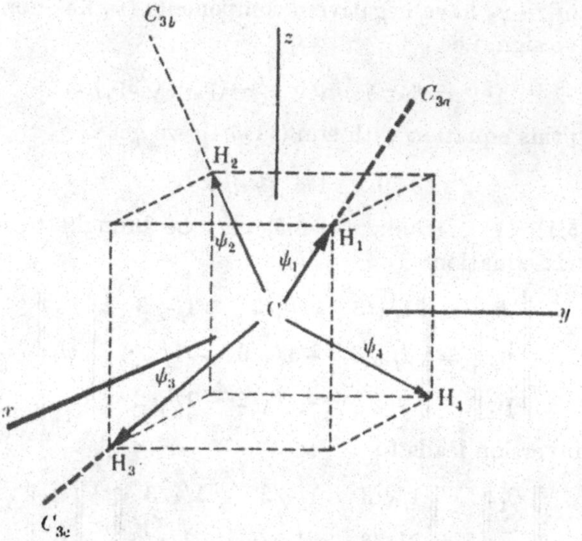

FIG. 11-5.2. Directions of the hybrids and x, y, and z axes for CH_4.

For methane we have seen that there are four hybrid orbitals ψ_1, ψ_2, ψ_3, and ψ_4 (see Fig. 11-5.2), each composed of an s-orbital belonging to Γ^{A_1} and three p-orbitals p_x, p_y, and p_z, belonging to Γ^{T_2}; for the choice of x, y, and z axes see Fig. 11-5.2. Using the relevant projection operators, we obtain the following combinations:

$$\begin{aligned}
P^{A_1}\psi_1 &= 6(\psi_1+\psi_2+\psi_3+\psi_4) \\
P^{T_2}\psi_1 &= 2(3\psi_1-\psi_2-\psi_3-\psi_4) \\
P^{T_2}\psi_2 &= 2(-\psi_1+3\psi_2-\psi_3-\psi_4) \\
P^{T_2}\psi_3 &= 2(-\psi_1-\psi_2+3\psi_3-\psi_4).
\end{aligned}$$

The last three combinations are linearly independent and the normalized combinations are:

$$(\psi_1+\psi_2+\psi_3+\psi_4)/2 \qquad \text{(belonging to } \Gamma^{A_1})$$

$$\left.\begin{aligned}
&(3\psi_1-\psi_2-\psi_3-\psi_4)/\sqrt{12} \\
&(-\psi_1+3\psi_2-\psi_3-\psi_4)/\sqrt{12} \\
&(-\psi_1-\psi_2+3\psi_3-\psi_4)/\sqrt{12}
\end{aligned}\right\} \qquad \text{(belonging to } \Gamma^{T_2}).$$

We can therefore establish that

$$(\psi_1+\psi_2+\psi_3+\psi_4)/2 = s \tag{11-5.6}$$

$$(3\psi_1-\psi_2-\psi_3-\psi_4)/\sqrt{12} = \pm(a_1^2+b_1^2+c_1^2)^{-\frac{1}{2}}(a_1p_x+b_1p_y+c_1p_z) \tag{11-5.7}$$

$$(-\psi_1+3\psi_2-\psi_3-\psi_4)/\sqrt{12} = \pm(a_2^2+b_2^2+c_2^2)^{-\frac{1}{2}}(a_2p_x+b_2p_y+c_2p_z) \tag{11-5.8}$$

$$(-\psi_1-\psi_2+3\psi_3-\psi_4)/\sqrt{12} = \pm(a_3^2+b_3^2+c_3^2)^{-\frac{1}{2}}(a_3p_x+b_3p_y+c_3p_z) \tag{11-5.9}$$

Under the operator $\boldsymbol{O}_{C_{3a}}$ (see Fig. 11-5.2 for the C_{3a} axis), the left hand side of eqn (11-5.7) is unchanged and since

$$\boldsymbol{O}_{C_{3a}}p_x = p_z, \qquad \boldsymbol{O}_{C_{3a}}p_y = p_x, \qquad \boldsymbol{O}_{C_{3a}}p_z = p_y,$$

$(a_1p_x+b_1p_y+c_1p_z)$ becomes $(a_1p_z+b_1p_x+c_1p_y)$. Therefore $a_1 = c_1, b_1 = a_1$, $c_1 = b_1$ and

$$(3\psi_1-\psi_2-\psi_3-\psi_4)/\sqrt{12} = \pm(p_x+p_y+p_z)/\sqrt{3}.$$

Since, by inspection of Fig. 11-5.2, ψ_1 and $-\psi_2$ have positive x components and the x components of ψ_3 and ψ_4 cancel, we take the positive sign, i.e.

$$(3\psi_1-\psi_2-\psi_3-\psi_4)/\sqrt{12} = (p_x+p_y+p_z)/\sqrt{3}. \tag{11-5.10}$$

Under the operator $\boldsymbol{O}_{C_{3b}}$, the left hand side of eqn (11-5.8) is unchanged and since $\boldsymbol{O}_{C_{3b}}p_x = -p_z$, $\boldsymbol{O}_{C_{3b}}p_y = p_x$, $\boldsymbol{O}_{C_{3b}}p_z = -p_y$, we find that $(a_2p_x+b_2p_y+c_2p_z)$ becomes $(-a_2p_z+b_2p_x-c_2p_y)$. Therefore, $a_2 = b_2$, $b_2 = -c_2$, $c_2 = -a_2$ and

$$(-\psi_1+3\psi_2-\psi_3-\psi_4)/\sqrt{12} = \pm(p_x+p_y-p_z)/\sqrt{3}.$$

Inspection of Fig. 11-5.2 justifies the negative sign and we have

$$(-\psi_1+3\psi_2-\psi_3-\psi_4)/\sqrt{12} = -(p_x+p_y-p_z)/\sqrt{3}. \tag{11-5.11}$$

Consideration of eqn (11-5.9) and the effect of operator $\boldsymbol{O}_{C_{3c}}$ on p_x, p_y, and p_z leads to

$$(-\psi_1-\psi_2+3\psi_3-\psi_4)/\sqrt{12} = (p_x-p_y-p_z)/\sqrt{3}. \tag{11-5.12}$$

Eqns (11-5.10) to (11-5.12) can be solved to yield p_x, p_y, and p_z:

$$\begin{aligned} p_x &= (\psi_1-\psi_2+\psi_3-\psi_4)/2 \\ p_y &= (\psi_1-\psi_2-\psi_3+\psi_4)/2 \\ p_z &= (\psi_1+\psi_2-\psi_3-\psi_4)/2 \end{aligned} \tag{11-5.13}$$

and these equations together with eqn (11-5.6) can be written as

$$\left\|\begin{matrix} s \\ p_x \\ p_y \\ p_z \end{matrix}\right\| = \left\|\begin{matrix} \frac{1}{2} & \frac{1}{2} & \frac{1}{2} & \frac{1}{2} \\ \frac{1}{2} & -\frac{1}{2} & \frac{1}{2} & -\frac{1}{2} \\ \frac{1}{2} & -\frac{1}{2} & -\frac{1}{2} & \frac{1}{2} \\ \frac{1}{2} & \frac{1}{2} & -\frac{1}{2} & -\frac{1}{2} \end{matrix}\right\| \left\|\begin{matrix} \psi_1 \\ \psi_2 \\ \psi_3 \\ \psi_4 \end{matrix}\right\|$$

or

$$\begin{Vmatrix}\psi_1\\\psi_2\\\psi_3\\\psi_4\end{Vmatrix} = \begin{Vmatrix}\frac{1}{2} & \frac{1}{2} & \frac{1}{2} & \frac{1}{2}\\ \frac{1}{2} & -\frac{1}{2} & \frac{1}{2} & -\frac{1}{2}\\ \frac{1}{2} & -\frac{1}{2} & -\frac{1}{2} & \frac{1}{2}\\ \frac{1}{2} & \frac{1}{2} & -\frac{1}{2} & -\frac{1}{2}\end{Vmatrix}^{-1} \begin{Vmatrix}s\\p_x\\p_y\\p_z\end{Vmatrix}$$

$$= \begin{Vmatrix}\frac{1}{2} & \frac{1}{2} & \frac{1}{2} & \frac{1}{2}\\ \frac{1}{2} & -\frac{1}{2} & -\frac{1}{2} & \frac{1}{2}\\ \frac{1}{2} & \frac{1}{2} & -\frac{1}{2} & -\frac{1}{2}\\ \frac{1}{2} & -\frac{1}{2} & \frac{1}{2} & -\frac{1}{2}\end{Vmatrix} \begin{Vmatrix}s\\p_x\\p_y\\p_z\end{Vmatrix}$$

(note that the 4×4 matrix is orthogonal, so that its inverse equals its transpose). Hence we obtain the final relationships:

$$\begin{aligned}\psi_1 &= (s+p_x+p_y+p_z)/2,\\ \psi_2 &= (s-p_x-p_y+p_z)/2,\\ \psi_3 &= (s+p_x-p_y-p_z)/2,\\ \psi_4 &= (s-p_x+p_y-p_z)/2.\end{aligned} \tag{11-5.14}$$

We saw in § 11-3 that a set of equivalent tetrahedral hybrid orbitals could also be constructed from a set of s, d_{xy}, d_{xz}, and d_{yz} orbitals. Mathematical expressions for these hybrids (sd^3) can be obtained from eqns (11-5.14) by changing p_x to d_{yz}, p_y to d_{xz}, and p_z to d_{xy}. That these are the correct changes can be deduced from the fact that the operators $O_{C_{3a}}$, $O_{C_{3b}}$, and $O_{C_{3c}}$ acting on the column matrix

$$\begin{Vmatrix}p_x\\p_y\\p_z\end{Vmatrix}$$

produce, in each case, a matrix which is identical with the one obtained when the same operator acts on the column matrix

$$\begin{Vmatrix}d_{yz}\\d_{xz}\\d_{xy}\end{Vmatrix}.$$

For example,

$$O_{C_{3b}}\begin{Vmatrix}p_x\\p_y\\p_z\end{Vmatrix} = \begin{Vmatrix}0 & 0 & -1\\ 1 & 0 & 0\\ 0 & -1 & 0\end{Vmatrix}\begin{Vmatrix}p_x\\p_y\\p_z\end{Vmatrix}$$

and

$$O_{C_{3v}} \begin{Vmatrix} d_{yz} \\ d_{xz} \\ d_{xy} \end{Vmatrix} = \begin{Vmatrix} 0 & 0 & -1 \\ 1 & 0 & 0 \\ 0 & -1 & 0 \end{Vmatrix} \begin{Vmatrix} d_{yz} \\ d_{xz} \\ d_{xy} \end{Vmatrix}.$$

11-6. Relationship between localized and non-localized molecular orbital theory

Localized MO theory using hybrid orbitals is essentially a special case of non-localized MO theory. In the latter theory, if we construct MOs from AOs scattered over the entire molecule, we do not *need* to make any initial assumptions about which AOs are being used to bond particular atoms together, nor for that matter *need* we consider the symmetry properties of the AOs which are used. All these things will be taken care of by the quantum mechanical method used to determine the coefficients of the AOs in the linear combinations: if an AO is of inappropriate symmetry, its coefficient will turn out to be zero and if an AO plays a small role in bonding, its coefficient will turn out to be small. However, it is clear that the major contribution to a non-localized MO will be similar to the localized MO formed from appropriate hybrid orbitals, since both, in the long run, have to describe the same thing: the electronic distribution in the bonding regions. We can therefore utilize our knowledge of hybrid orbitals when we construct non-localized MOs by ignoring those AOs which on the basis of the simpler theory would not contribute significantly. Such an approach not only conforms with our intuitive understanding of bonding, but also, by cutting out that which is irrelevant, saves time in carrying out the calculations.

We will return to the more general MO theory again in the next chapter. The key virtue of this theory is that, unlike the hybridization approach, it affords a mechanism for making direct energy calculations.

As a final note, it should be emphasized that like the phenomenon of resonance, hybridization is not a real physical process (atoms don't hybridize any more than molecules resonate). It is a man-made process for describing an already existing situation, the molecular bond, when the simple model using single AOs fails to work.

PROBLEMS

11.1. Determine the irreducible representations of $\mathcal{T}_d$ to which f-orbitals belong.

11.2. Show that for a molecule of octahedral symmetry the σ-bonding hybrid orbitals on the central atom are composed of six atomic orbitals: s, p_x, p_y, p_z, d_{z^2}, and $d_{x^2-y^2}$.

11.3. Determine what type of π-bonding hybrid orbitals can be formed for the square planar AB_4 molecule which belongs to the $\mathcal{D}_{4h}$ point group.

11.4. Show that for the square planar AB_4 molecule a possible set of four σ-hybrid orbitals on A is composed of the atomic orbitals: s, $d_{x^2-y^2}$, p_x, and p_y. Find explicit expressions for the four hybrid orbitals.

12. Transition-metal chemistry

12-1. Introduction

SINCE the 1950's there has been a dramatic revival of interest in inorganic chemistry. This has largely been due to the synthesis of many transition-metal compounds and complexes and the development of theories to explain their properties. Nearly all of these compounds have a high degree of symmetry and in this chapter we see how the group theoretical rules which we have previously developed can reduce the effort involved in calculating their properties. In some cases the reduction in labour is quite startling.

Although strictly speaking the transition metals are defined as those which, *as elements*, have partly filled d- or f-shells, it is more common also to include in the definition elements which have partly filled d- or f-shells in any of their commonly occurring oxidation states. The atoms or molecules which are bonded to the metal in a transition metal compound are called *ligands*.

In broad terms, there are two very different theories which have been used to interpret the properties of transition-metal compounds; MO theory and crystal field theory.

MO theory is the more general of the two and, in the long run, the one most capable of giving the best results. However, in its application, the calculations which are entailed are very lengthy and only now, with the introduction of high-speed computers, are reliable results being produced. In § 12-2 and § 12-3 we will see how valuable symmetry principles can be in a MO treatment of octahedral MX_6 and tetrahedral MX_4 molecules. We choose these two types of molecule because they occur often in transition-metal chemistry and also because they are prototypes to which many molecules of lower symmetry can be referred. In § 12-4 we construct MOs for ferrocene, which is an example of a sandwich compound. Here the ligands are carbocyclic rings and so we are able to build upon the techniques which were developed in Chapter 10 for such rings.

Crystal field theory has its origins in Hans Bethe's famous 1929 paper *Splitting of terms in crystals*. In that paper Bethe demonstrated what happens to the various states of an ion when it is placed in a crystalline environment of definite symmetry. Later, John Van Vleck showed that the results of that investigation would apply equally well to a transition-metal compound if it could be approximated as a metal ion surrounded by ligands which only interact *electrostatically* with the ion. One of the key elements in the application of crystal field theory to transition-metal compounds is the construction of so-called correlation diagrams. These diagrams relate the splitting of the electronic states of the transition-metal ion under the perturbation of a set of ligands to a parameter which reflects the strength of the *electrostatic* interaction between the ion and the ligands. They resemble the well known correlation diagrams for the formation of a united atom from the separated atoms of a diatomic molecule and like them depend on a non-crossing rule for states of the same symmetry and multiplicity. In § 12-5 and § 12-6 we develop some necessary preliminary steps before constructing, in § 12-7, a correlation diagram for a metal ion which has two d-electrons in the valence shell and is surrounded by an octahedral or a tetrahedral set of ligands. In § 12-8 and § 12-9, the way in which correlation diagrams can predict the spectral and magnetic properties of transition-metal compounds and complexes is described.

It is clear from the start that the approximations involved in applying crystal field theory to transition-metal compounds are extreme and the results of MO theory confirm this. In very few cases can the ligands be considered simply as point charges or point dipoles which interact only electrostatically with the metal. In general the electrons of the ligands will be distributed throughout the molecule and there will be covalent (electron sharing) as well as electrostatic interactions. However, crystal field theory can, to some extent, be adapted to handle this, the real situation, by reinterpreting the parameters which occur. We let these parameters become strictly semi-empirical quantities and give them values which can no longer be ascribed solely to electrostatic interactions. This adaptation is called *ligand field theory*† and it is briefly considered at the end of the chapter.

Two excellent books which offer useful background material to this chapter are: *An introduction to ligand fields* by B. N. Figgis (published by Interscience Publishers) and *Atomic and molecular orbital theory* by P. O'D. Offenhartz (published by the McGraw-Hill Book Co.).

12-2. LCAO MOs for octahedral compounds

Let us consider the construction of molecular orbitals from linear combinations of atomic orbitals (LCAO MOs) for an octahedral molecule

† Some authors have used this term to describe all aspects of the manner in which an ion or atom is influenced by ligands. With this definition, crystal field theory is a special case of ligand field theory and the latter is indistinguishable from MO theory.

or ion MX_6 where M is a transition metal and the ligands X are bonded solely to M. We begin by selecting the AOs to be used. The transition metal in its ground state will have an inner core of electrons with a noble-gas electronic configuration, which we will assume does not participate in bonding, and a number of valence electrons which occupy nd- and $(n+1)$s-AOs. Since, for the transition metals, the $(n+1)$p-orbitals are close in energy to both the nd- and $(n+1)$s-orbitals, we will assume that the metal M contributes five d-, one s-, and three

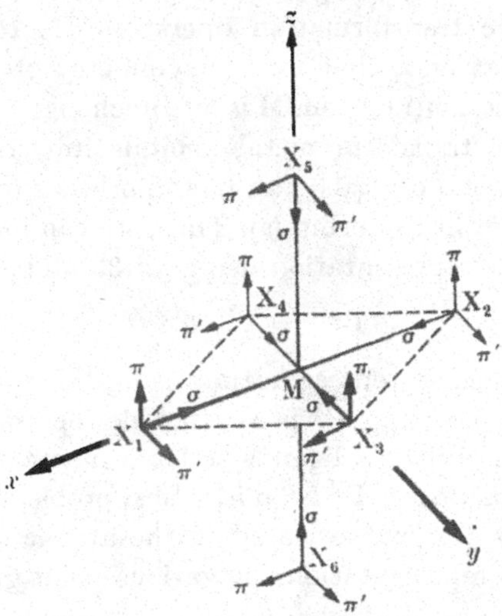

FIG. 12-2.1. p-type ligand orbitals for an octahedral MX_6 compound.

p-orbitals to the MOs. Of course, the selection of AOs of M is made easier by knowing which orbitals participate in the localized molecular orbital bonding scheme (see the previous chapter). We will assume that each ligand X contributes one s-orbital and three p-orbitals. We will distinguish the ligand orbitals from the metal orbitals by a subscript (1, 2,... 6) which indicates the ligand with which they are associated. The three p-orbitals on each X may be given the symbols σ, π, and π' and associated with a set of three mutually perpendicular vectors as shown in Fig. 12-2.1. The σ-type p-orbitals are chosen to have their positive lobes directed towards M.

We therefore have a grand total of 33 atomic orbitals: s, p_x, p_y, p_z, d_{z^2}, $d_{x^2-y^2}$, d_{xy}, d_{xz}, d_{yz}; s_i, σ_i, π_i, π'_i ($i = 1, 2, \ldots 6$). Application of the LCAO MO method *without* applying the principles of symmetry would therefore lead to an equation (see eqn (10-3.5))

$$\det(H^{\text{eff}}_{jk} - \varepsilon S_{jk}) = 0, \qquad (12\text{-}2.1)$$

which involves a 33×33 determinant. Fortunately, by using the techniques we have previously discussed, we can avoid the Herculean task of solving such an equation. What we do is to create linear combinations of the 33 AOs in such a way that they form bases for the irreducible representations of the $\mathcal{O}_h$ point group to which the molecule belongs. Using these symmetry adapted combinations in place of the original 33 AOs immediately leads to the factorization of eqn (12-2.1). (cf. the benzene problem in § 10-6).

To find the symmetry adapted combinations we first consider the application of the transformation operators $\boldsymbol{O_R}$ to the 33 atomic orbitals. We see at once that for all symmetry operations $\boldsymbol{R}$ of the $\mathcal{O}_h$ point group, the central atom M is left unchanged and consequently any $\boldsymbol{O_R}$ will only transform metal orbitals into metal orbitals (or combinations of metal orbitals) and ligand orbitals into ligand orbitals (or combinations of ligand orbitals). Thus, we can immediately reduce Γ^{AO} (the reducible representation using all 33 atomic orbitals) to the form:

$$\Gamma^{AO} = \Gamma^{M} \oplus \Gamma^{X}$$

where Γ_M is a reducible representation using the nine atomic orbitals of M as basis functions and Γ^X is a reducible representation using the 24 atomic orbitals of the six ligands as basis functions.

The further reduction of Γ^M is an identical problem to that discussed in § 11-2. In that section we classified the atomic orbitals according to the irreducible representations of various point groups. For $\mathcal{O}_h$ we found

s	belongs to $\Gamma^{A_{1g}}$,
(p_x, p_y, p_z)	belongs to $\Gamma^{T_{1u}}$,
$(d_{z^2}, d_{x^2-y^2})$	belongs to Γ^{E_g},
(d_{xy}, d_{xz}, d_{yz})	belongs to $\Gamma^{T_{2g}}$.

Hence we can write

$$\Gamma^M = \Gamma^{A_{1g}} \oplus \Gamma^{T_{1u}} \oplus \Gamma^{E_g} \oplus \Gamma^{T_{2g}}.$$

The further reduction of Γ^X is more complicated. In the first place no transformation operator $\boldsymbol{O_R}$ can change a ligand s-orbital into a ligand p-orbital (or combination of ligand p-orbitals), therefore the s-orbitals themselves form a representation Γ^s of $\mathcal{O}_h$. Furthermore, for the particular choice of orientation of the p-orbitals (or vectors) in Fig. 12-2.1, no $\boldsymbol{O_R}$ can change a σ-type p-orbital into a π-type one (or combination of π-type p-orbitals), so we have two further representations Γ^σ and Γ^π and overall we can write

$$\Gamma^X = \Gamma^s \oplus \Gamma^\sigma \oplus \Gamma^\pi.$$

To reduce Γ^s, Γ^σ, and Γ^π we must apply O_R for all R to $(s_1, s_2, \ldots s_6)$, $(\sigma_1, \sigma_2, \ldots \sigma_6)$ and $(\pi_1, \pi_1', \pi_2, \pi_2', \ldots \pi_6, \pi_6')$ respectively and thereby find $\chi^s(R)$, $\chi^\sigma(R)$, $\chi^\pi(R)$ for all R. Using the equations:

$$O_R\phi_i = \sum_{j=1}^{6 \text{ or } 12} D^\alpha_{ji}(R)\phi_j \qquad \text{(or its vector analogue)}$$

and

$$\chi^\alpha(R) = \sum_{i=1}^{6 \text{ or } 12} D^\alpha_{ii}(R),$$

this is a straight forward procedure and we obtain

$\mathcal{O}_h$	E	$8C_3$	$3C_2$	$6C_4$	$6C_2'$	i	$8S_6$	$3\sigma_h$	$6S_4$	$6\sigma_d$
$\chi^s(C_i)$	6	0	2	2	0	0	0	4	0	2
$\chi^\sigma(C_i)$	6	0	2	2	0	0	0	4	0	2
$\chi^\pi(C_i)$	12	0	−4	0	0	0	0	0	0	0

The decomposition rule (eqn (7-4.2)) in conjunction with these results shows that

$$\Gamma^s = \Gamma^{A_{1g}} \oplus \Gamma^{E_g} \oplus \Gamma^{T_{1u}},$$
$$\Gamma^\sigma = \Gamma^{A_{1g}} \oplus \Gamma^{E_g} \oplus \Gamma^{T_{1u}},$$
$$\Gamma^\pi = \Gamma^{T_{1g}} \oplus \Gamma^{T_{2g}} \oplus \Gamma^{T_{1u}} \oplus \Gamma^{T_{2u}}.$$

What do these symbolic equations mean? Consider for example the first one. This shows that it is possible to find a new basis for Γ^s in which the basis functions are linear combinations of $s_1, s_2, \ldots s_6$ such that the matrices of Γ^s, $D^s(R)$, have a completely reduced form. There will be one linear combination which is a basis for $\Gamma^{A_{1g}}$, two linear combinations which form a basis for Γ^{E_g} and three linear combinations which form a basis for $\Gamma^{T_{1u}}$. The other two symbolic equations can be interpreted in the same fashion.

The actual linear combinations which reduce Γ^s, Γ^σ, and Γ^π could be found by the projection operator technique of § 7-6 but because of the large number of symmetry operations contained in $\mathcal{O}_h$ and because of the problems connected with the multi-dimensional representations (similar to those encountered in § 11-5), to do so would require much time and even more patience. Fortunately, we can find the correct combinations by inspection.

Any symmetry operation of $\mathcal{O}_h$ simply permutes the AOs $s_1, s_2, \ldots s_6$ amongst themselves, so that if O_R is applied to $(s_1+s_2+s_3+s_4+s_5+s_6)$ only the order of the orbitals within the bracket is changed and

$$O_R(s_1+s_2+s_3+s_4+s_5+s_6) = (s_1+s_2+s_3+s_4+s_5+s_6)$$

for all R. Consequently the normalized basis function (assuming $\int \phi_i \phi_j \, d\tau = \delta_{ij}$, ϕ_i = any ligand orbital) for the identical representation $\Gamma^{A_{1g}}$ is

$$\psi^{s}_{A_{1g}} = (s_1+s_2+s_3+s_4+s_5+s_6)/\sqrt{6}.$$

To find a pair of linear combinations of $s_1, s_2, \ldots s_6$ that form a basis for Γ^{E_g} (and there are many), we can look for a pair which mirror the Γ^{E_g} AOs of the central atom, i.e. d_{z^2} and $d_{x^2-y^2}$. The d_{z^2}-orbital has the symmetry of the expression $3z^2-r^2 = 2z^2-x^2-y^2$ and it is clear that the s-orbital combination which reflects this is†

$$\psi^{s}_{E_g(1)} = (2s_5+2s_6-s_1-s_2-s_3-s_4)/\sqrt{12}.$$

The $d_{x^2-y^2}$ orbital has positive lobes along the positive and negative x-axes and negative lobes along the positive and negative y-axes, therefore this will be mirrored by the combination:

$$\psi^{s}_{E_g(2)} = (s_1+s_2-s_3-s_4)/2.$$

The three combinations of $s_1, s_2, \ldots s_6$ which form a basis for $\Gamma^{T_{1u}}$ can be chosen to behave like the p_x, p_y, p_z AOs of M (since these also form a basis for $\Gamma^{T_{1u}}$), hence, by inspection,

$$\psi^{s}_{T_{1u}(1)} = (s_1-s_2)/\sqrt{2} \qquad \text{(like } p_x)$$
$$\psi^{s}_{T_{1u}(2)} = (s_3-s_4)/\sqrt{2} \qquad \text{(like } p_y)$$
$$\psi^{s}_{T_{1u}(3)} = (s_5-s_6)/\sqrt{2} \qquad \text{(like } p_z).$$

That these combinations do indeed behave like their central atom counterparts may be verified by applying the operators O_R. One can also confirm that, like the AOs of M, the combinations are orthogonal to each other.

It is clear that the required combinations of $\sigma_1, \sigma_2, \ldots \sigma_6$ can be obtained directly from the previous combinations by replacing s by σ.

For the representations based on the π- and π'-orbitals, we have to find combinations for $\Gamma^{T_{1g}}$, $\Gamma^{T_{2g}}$, $\Gamma^{T_{1u}}$ and $\Gamma^{T_{2u}}$. The $\Gamma^{T_{1u}}$ combinations must mirror the p_x, p_y, and p_z AOs of M. The ligand orbitals π'_3, π'_4, π_5, and π_6 all point in the x direction, π'_1, π'_2, π'_5, and π'_6 all point in the y direction and π_1, π_2, π_3, and π_4 in the z direction. Hence suitable normalized combinations are

$$\psi^{\pi}_{T_{1u}(1)} = (\pi'_3+\pi'_4+\pi_5+\pi_6)/2 \qquad \text{(like } p_x)$$
$$\psi^{\pi}_{T_{1u}(2)} = (\pi'_1+\pi'_2+\pi'_5+\pi'_6)/2 \qquad \text{(like } p_y)$$
$$\psi^{\pi}_{T_{1u}(3)} = (\pi_1+\pi_2+\pi_3+\pi_4)/2 \qquad \text{(like } p_z).$$

† The set of n functions which *together* form a basis for the n-dimensional irreducible representation Γ^{μ} will be labelled $\psi_{\mu(1)}, \psi_{\mu(2)}, \ldots \psi_{\mu(n)}$. This notation does *not* imply that there is an irreducible representation $\Gamma^{\mu(i)}$.

Likewise, comparing the $\Gamma^{T_{2g}}$ combinations with d_{xy}, d_{xz}, and d_{yz}, we find

$$\psi^{\pi}_{T_{2g}(1)} = (\pi_1' - \pi_2' + \pi_3' - \pi_4')/2 \qquad (\text{like } d_{xy})$$
$$\psi^{\pi}_{T_{2g}(2)} = (\pi_1 - \pi_2 + \pi_5 - \pi_6)/2 \qquad (\text{like } d_{xz})$$
$$\psi^{\pi}_{T_{2g}(3)} = (\pi_3 - \pi_4 + \pi_5' - \pi_6')/2 \qquad (\text{like } d_{yz}).$$

For the $\Gamma^{T_{1g}}$ and $\Gamma^{T_{2u}}$ combinations there are no central atom counterparts and consequently, as they therefore do not 'mix' with the central atom orbitals, they do not take part in the bonding. Nonetheless, appropriate $\Gamma^{T_{1g}}$ combinations can be found. The axial vectors R_x, R_y, and R_z (see § 9-6) form a basis for $\Gamma^{T_{1g}}$ and using this piece of information, we find the following $\Gamma^{T_{1g}}$ combinations which mirror R_x, R_y, and R_z:

$$\psi^{\pi}_{T_{1g}(1)} = (\pi_3 - \pi_4 - \pi_5' + \pi_6')/2 \qquad (\text{like } R_x)$$
$$\psi^{\pi}_{T_{1g}(2)} = (\pi_1 - \pi_2 - \pi_5 + \pi_6)/2 \qquad (\text{like } R_y)$$
$$\psi^{\pi}_{T_{1g}(3)} = (\pi_1' - \pi_2' - \pi_3' + \pi_4')/2 \qquad (\text{like } R_z).$$

The three $\Gamma^{T_{2u}}$ combinations must be orthogonal to the previous nine combinations. Such a set is

$$\psi^{\pi}_{T_{2u}(1)} = (\pi_3' + \pi_4' - \pi_5 - \pi_6)/2$$
$$\psi^{\pi}_{T_{2u}(2)} = (\pi_1' + \pi_2' - \pi_5' - \pi_6')/2$$
$$\psi^{\pi}_{T_{2u}(3)} = (\pi_1 + \pi_2 - \pi_3 - \pi_4)/2.$$

What we can now say is that if we replace the initial basis of 33 atomic orbitals by the nine AOs of M and the 24 combinations of ligand AOs ($\psi^{s}_{A_{1g}}$, $\psi^{s}_{E_g(i)}$ $i = 1, 2$, $\psi^{s}_{T_{1u}(i)}$ $i = 1, 2, 3$, $\psi^{\sigma}_{A_{1g}}$, $\psi^{\sigma}_{E_g(i)}$ $i = 1, 2$, $\psi^{\sigma}_{T_{1u}(i)}$, $\psi^{\pi}_{T_{1u}(i)}$, $\psi^{\pi}_{T_{2g}(i)}$, $\psi^{\pi}_{T_{1g}(i)}$, $\psi^{\pi}_{T_{2u}(i)}$ $i = 1, 2, 3$), the determinant of eqn (12-2.1) will be in block form and the equation can be factorized. The reason this happens is the same as in the benzene example (see § 10-6): a combination of the vanishing integral rule (§ 8-4) and the fact that the effective Hamiltonian belongs to Γ^1 (the totally symmetric representation of any point group).

Thus we will have:

(1) A 3×3 block corresponding to $\Gamma^{A_{1g}}$ which will produce three non-degenerate energy levels (a_{1g}-type) with MOs formed from a metal s-orbital, a combination of ligand s-orbitals ($\psi^{s}_{A_{1g}}$) and a combination of ligand σ-type p-orbitals ($\psi^{\sigma}_{A_{1g}}$).

(2) Three equivalent 4×4 blocks corresponding to $\Gamma^{T_{1u}}$. One block will produce four energy levels with MOs formed from a metal p_x-orbital, a combination of ligand s-orbitals ($\psi^{s}_{T_{1u}(1)}$), a combination of ligand σ-type p-orbitals ($\psi^{\sigma}_{T_{1u}(1)}$) and a combination of

ligand π-type p-orbitals ($\psi^{\pi}_{T_{1u}(1)}$). The other two blocks will produce identical energy levels and values for the coefficients of the symmetry adapted basis functions (which are p_y, $\psi^{s}_{T_{1u}(2)}$, $\psi^{\sigma}_{T_{1u}(2)}$, $\psi^{\pi}_{T_{1u}(2)}$ in one case and p_z, $\psi^{s}_{T_{1u}(3)}$, $\psi^{\sigma}_{T_{1u}(3)}$, $\psi^{\pi}_{T_{1u}(3)}$, in the other). Consequently we need only solve eqn (12-2.1) for one of the blocks. Overall, we will have four triply-degenerate energy levels (t_{1u}-type).

(3) Two equivalent 3×3 blocks corresponding to Γ^{E_g}, one will involve d_{z^2}, $\psi^{s}_{E_g(1)}$, and $\psi^{\sigma}_{E_g(1)}$ and the other $d_{x^2-y^2}$, $\psi^{s}_{E_g(2)}$, and $\psi^{\sigma}_{E_g(2)}$. They will both produce identical energy levels and identical values for the coefficients of the symmetry-adapted basis functions in the final MOs. Again, it is necessary to solve eqn (12-2.1) for only one of the blocks. Overall we will have three doubly-degenerate energy levels (e_g-type).

(4) Three equivalent 2×2 blocks corresponding to $\Gamma^{T_{2g}}$, one will involve d_{xy} and $\psi^{\pi}_{T_{2g}(1)}$, another d_{xz} and $\psi^{\pi}_{T_{2g}(2)}$, and a third d_{yz} and $\psi^{\pi}_{T_{2g}(3)}$. These three blocks will lead to identical energy levels and coefficients. Putting the results together we will have two triply-degenerate energy levels (t_{2g}-type).

(5) Six 1×1 blocks, three corresponding to $\Gamma^{T_{1g}}$ and three corresponding to $\Gamma^{T_{2u}}$. These will provide one triply-degenerate level of each type (t_{1g} and t_{2u}) and the final MOs will only involve ligand π-type p-orbitals and will simply be the symmetry adapted combinations themselves ($\psi^{\pi}_{T_{1g}(i)}$ and $\psi^{\pi}_{T_{2u}(i)}$, $i = 1, 2, 3$).

The upshot of all this is quite astounding: a problem which initially involved a 33×33 determinant has been reduced to one involving one 4×4, two 3×3 and one 2×2 determinants.

This is as far as we can go with symmetry arguments alone, the next step involves calculating, or estimating in a semi-empirical fashion, the required matrix elements H^{eff}_{jk} and S_{jk}. This is beyond the scope of the present book, however, once it is done eqn (12-2.1) can be solved and the energy levels determined. Radical approximations are often made in finding H^{eff}_{jk} and S_{jk}, e.g. the off-diagonal elements H^{eff}_{jk} ($j \neq k$) are frequently made simply proportional to the corresponding overlap integrals S_{jk}, and therefore the reader is advised to treat such LCAO MO results critically. Only recently have *ab initio* LCAO MO calculations been attempted for transition-metal compounds.†

The energy-level diagram shown in Fig. 12-2.2 is a summary of the

† See, for example, the calculation on NiF_6^{4-} by J. W. Moskowitz, C. Hollister, C. J Hornback, and H. Basch (*Journal of chemical physics*, **53**, 2570 (1970)).

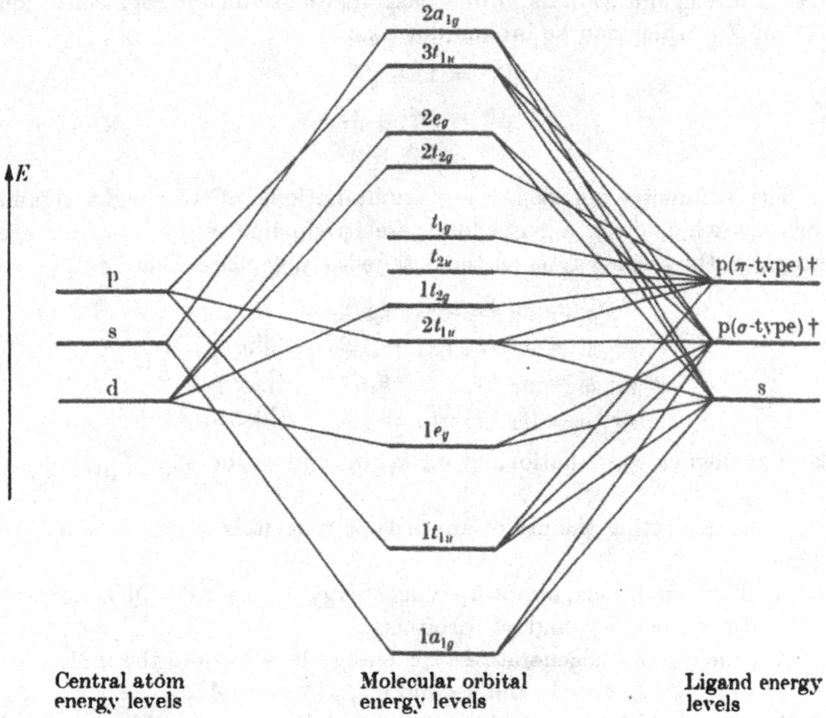

FIG. 12-2.2. The energy level diagram for an octahedral MX_6 molecule or ion.

conclusions given above. Though this diagram is only schematic, the *order* of the energy levels is that accepted by most inorganic chemists.

12-3. LCAO MOs for tetrahedral compounds

We will not consider the LCAO MO treatment of the tetrahedral molecule MX_4 in detail, since the steps involved are essentially the same as those for the MX_6 case. The central atom M can use s-, p- and d-orbitals and for the $\mathscr{T}_d$ point group these are classified as follows:

$$\begin{array}{ll} \text{s} & \text{belong to } \Gamma^{A_1}, \\ (d_{z^2}, d_{x^2-y^2}) & \text{belong to } \Gamma^{E}, \\ (p_x, p_y, p_z) & \text{belong to } \Gamma^{T_2}, \\ (d_{xy}, d_{xz}, d_{yz}) & \text{belong to } \Gamma^{T_2}. \end{array}$$

If we restrict ourselves to only σ-bonding, the four X ligands will each use a s-orbital and a σ-type p-orbital (one which points towards

† These two levels are only separated for visual convenience.

M). These ligand orbitals form a basis for the reducible representation Γ^{X} of $\mathscr{T}_d$, which can be broken down as:

$$\Gamma^{X} = \Gamma^{s} \oplus \Gamma^{\sigma}$$

where

$$\Gamma^{s} = \Gamma^{A_1} \oplus \Gamma^{T_2}$$
$$\Gamma^{\sigma} = \Gamma^{A_1} \oplus \Gamma^{T_2}.$$

The symmetry adapted linear combinations of the eight ligand orbitals which form a basis for these irreducible representations are found in the same way as in the octahedral MX_6 case. They are:

$$\psi^{s}_{A_1} = (s_1+s_2+s_3+s_4)/2,$$
$$\psi^{s}_{T_2(1)} = (s_1-s_2+s_3-s_4)/2 \qquad \text{(like } p_x\text{)},$$
$$\psi^{s}_{T_2(2)} = (s_1-s_2-s_3+s_4)/2 \qquad \text{(like } p_y\text{)},$$
$$\psi^{s}_{T_2(3)} = (s_1+s_2-s_3-s_4)/2 \qquad \text{(like } p_z\text{)},$$

and analogous combinations of σ_1, σ_2, σ_3, and σ_4 for $\psi^{\sigma}_{A_1}$, $\psi^{\sigma}_{T_2(1)}$, $\psi^{\sigma}_{T_2(2)}$, and $\psi^{\sigma}_{T_2(3)}$.†

Thus, neglecting the use of any π-type p-orbitals on the ligands, we have:

(1) Three non-degenerate a_1-type energy levels with MOs formed from the s, $\psi^{s}_{A_1}$ and $\psi^{\sigma}_{A_1}$ orbitals.

(2) One doubly-degenerate e-type energy level, where the molecular orbitals are simply the d_{z^2} and $d_{x^2-y^2}$ AOs of M.

(3) Four triply-degenerate t_2-type energy levels, where for each level the three degenerate MOs are composed of combinations of (a) p_x, d_{yz}, $\psi^{s}_{T_2(1)}$ and $\psi^{\sigma}_{T_2(1)}$, (b) p_y, d_{xz}, $\psi^{s}_{T_2(2)}$ and $\psi^{\sigma}_{T_2(2)}$, (c) p_z, d_{xy}, $\psi^{s}_{T_2(3)}$ and $\psi^{\sigma}_{T_2(3)}$.

There have been several *ab inito* LCAO MO SCF calculations on tetrahedral transition-metal complexes: for example, see the work of Hillier and Saunders (*Molecular physics* **22**, 1025 (1970)), and Demuynck and Veillard (*Theoretica chimica acta* **28**, 241 (1973)) on nickel carbonyl $Ni(CO)_4$.

12-4. LCAO MOs for sandwich compounds

An important class of organo-metallic compounds are the so-called metal-sandwich compounds. These compounds have the formula $(C_nH_n)_2M$ and consist of a transition metal atom M sandwiched symmetrically between two parallel carbocyclic ring systems, e.g. ferrocene, or, to give it its proper name, dicyclopentadienyliron $[(C_5H_5)_2Fe]$, and

† The X atoms are numbered in the same sense as the hydrogens in Fig. 11-5.2. The reader should be able to see for himself why the set of equations for $\psi^{s}_{A_1}$, $\psi^{s}_{T_2(1)}$, $\psi^{s}_{T_2(2)}$, and $\psi^{s}_{T_2(3)}$ is identical in *form* with the set of equations given by eqn (11-5.6) and eqn (11-5.13).

dibenzenechromium $[(C_6H_6)_2Cr]$.† As an example of the construction of MOs for this type of compound we shall consider ferrocene.

Ferrocene, if we assume the two rings to be staggered,‡ belongs to the point group $\mathscr{D}_{5d}$ (see Fig. 3-6.4) and we begin by forming linear combinations of the ten π-type p-orbitals of the two carbocyclic rings which belong to the irreducible representations of $\mathscr{D}_{5d}$. We then combine these π-MOs with the metal AOs, matching the symmetries as we do so, in order to create MOs for the entire system. We exclude from our treatment any possible bonding between the iron atom and the rings through use of the σ-electrons of the latter.

The ten π-type p-orbitals form the basis for a reducible representation Γ^{π} of the $\mathscr{D}_{5d}$ point group with the following character:

$\mathscr{D}_{5d}$	E	$2C_5$	$2C_5^2$	$5C_2'$	i	$2S_{10}^3$	$2S_{10}$	$5\sigma_d$
$\chi^{\pi}(C_i)$	10	0	0	0	0	0	0	2

and hence, using the standard reduction procedure:

$$\Gamma^{\pi} = \Gamma^{A_{1g}} \oplus \Gamma^{A_{2u}} \oplus \Gamma^{E_{1g}} \oplus \Gamma^{E_{1u}} \oplus \Gamma^{E_{2g}} \oplus \Gamma^{E_{2u}}.$$

The exact form of the linear combinations which belong to these six irreducible representations can be found by using the techniques which were applied to benzene in § 10-6. If the π-type p-orbitals are labelled as $\phi_1, \phi_2, \ldots \phi_5, \phi_1', \phi_2', \ldots \phi_5'$ (see Fig. 12-4.1) and have their positive

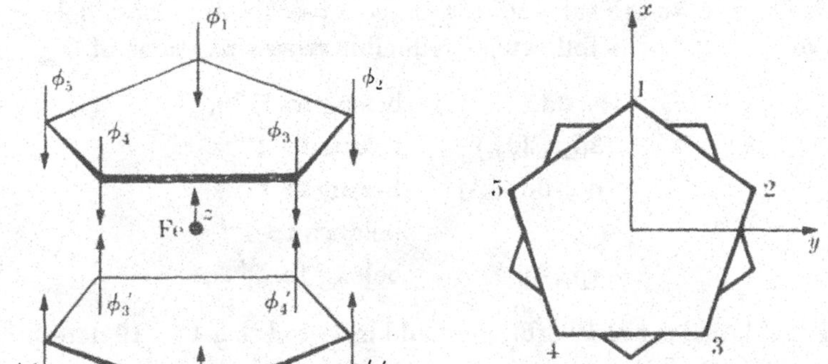

FIG. 12-4.1. Directions and labels of the π-type p-orbitals of ferrocene and a set of x, y, and z axes.

† Dibenzenechromium was discovered by Hein in 1919 but not recognized as a sandwich compound until the 1950's.

‡ Experimentally it is not clear whether the staggered or eclipsed form is the more stable; they are very close in energy. Most derivatives of ferrocene show the eclipsed conformation in the solid state, but there is evidence that ferrocene itself is staggered.

lobes directed towards the opposite ring, then the normalized π-MOs are:

$$\psi_{A_{1g}} = (\phi_1+\phi_2+\phi_3+\phi_4+\phi_5+\phi_1'+\phi_2'+\phi_3'+\phi_4'+\phi_5')/\sqrt{10},$$
$$\psi_{A_{2u}} = \{(\phi_1+\phi_2+\phi_3+\phi_4+\phi_5)-(\phi_1'+\phi_2'+\phi_3'+\phi_4'+\phi_5')\}/\sqrt{10},$$
$$\psi_{E_{1g}(1)} = \psi_1+\psi_1',$$
$$\psi_{E_{1g}(2)} = \psi_2+\psi_2',$$
$$\psi_{E_{1u}(1)} = \psi_1-\psi_1',$$
$$\psi_{E_{1u}(2)} = \psi_2-\psi_2',$$
$$\psi_{E_{2g}(1)} = \psi_3+\psi_3',$$
$$\psi_{E_{2g}(2)} = \psi_4+\psi_4',$$
$$\psi_{E_{2u}(1)} = \psi_3-\psi_3',$$
$$\psi_{E_{2u}(2)} = \psi_4-\psi_4',$$

where

$$\psi_1 = \{\phi_1+(\cos w)\phi_2+(\cos 2w)\phi_3+(\cos 2w)\phi_4+(\cos w)\phi_5\}/\sqrt{5}$$
$$\psi_2 = \{(\sin w)\phi_2+(\sin 2w)\phi_3-(\sin 2w)\phi_4-(\sin w)\phi_5\}/\sqrt{5}$$
$$\psi_3 = \{\phi_1+(\cos 2w)\phi_2+(\cos w)\phi_3+(\cos w)\phi_4+(\cos 2w)\phi_5\}/\sqrt{5}$$
$$\psi_4 = \{(\sin 2w)\phi_2-(\sin w)\phi_3+(\sin w)\phi_4-(\sin 2w)\phi_5\}/\sqrt{5}$$

and $w = 2\pi/5$. ψ_n' is obtained from ψ_n by replacing ϕ_i by ϕ_i', $i = 1, 2, \ldots 5$.

For the metal atom, iron, the valence orbitals are the five 3d-orbitals, the 4s-orbital and the three 4p-orbitals. They belong (see Table 11-2.2) to the following irreducible representations of $\mathscr{D}_{5d}$

4s, $3d_{z^2}$	belong to $\Gamma^{A_{1g}}$,
$(3d_{xz}, 3d_{yz})$	belong to $\Gamma^{E_{1g}}$,
$(d_{xy}, 3d_{x^2-y^2})$	belong to $\Gamma^{E_{2g}}$,
$4p_z$	belongs to $\Gamma^{A_{2u}}$,
$(4p_x, 4p_y)$	belong to $\Gamma^{E_{1u}}$.

The original set of 19 orbitals would have led to a 19×19 determinant in eqn (12-2.1), but now instead, by using the equally valid set of symmetry adapted orbitals, we have:

(1) A 3×3 determinant corresponding to $\Gamma^{A_{1g}}$ which will produce three non-degenerate energy levels (a_{1g}-type) and a corresponding set of MOs formed from combinations of the 4s, $3d_{z^2}$, and $\psi_{A_{1g}}$ orbitals.

(2) Two equivalent 2×2 determinants corresponding to $\Gamma^{E_{1g}}$. One determinant will produce two energy levels and two MOs formed

from the $3d_{xz}$ and $\psi_{E_{1g}(1)}$ orbitals (that this is the correct MO of the $\Gamma^{E_{1g}}$ pair to match up with $3d_{xz}$ is verified by inspection of Fig. 12-4.2, i.e. $\psi_{E_{1g}(1)}$ is positive (negative) where $3d_{xz}$ is positive (negative)). The other determinant will produce an identical set of energy levels with molecular orbitals formed from the $3d_{yz}$ and $\psi_{E_{1g}(2)}$ orbitals (the values of the coefficients of these component functions will be identical with those obtained from the first determinant). Together, there will be two doubly-degenerate energy levels of the e_{1g}-type.

(3) Two equivalent 2×2 determinants corresponding to $\Gamma^{E_{2g}}$, one 'mixing' the $3d_{x^2-y^2}$ and $\psi_{E_{2g}(1)}$ orbitals and the other 'mixing' the $3d_{xy}$ and $\psi_{E_{2g}(2)}$ orbitals (see Fig. 12-4.2 for the matching). These will provide two doubly-degenerate energy levels of e_{2g}-type.

(4) One 2×2 determinant corresponding to $\Gamma^{A_{2u}}$. The two MOs formed will be mixtures of the $4p_z$ and $\psi_{A_{2u}}$ orbitals and the two energy levels will be non-degenerate and of the a_{2u}-type.

(5) Two equivalent 2×2 determinants corresponding to $\Gamma^{E_{1u}}$, one 'mixing' the $4p_x$ and $\psi_{E_{1u}(1)}$ orbitals (see Fig. 12-4.2) and the other 'mixing' the $4p_y$ and $\psi_{E_{1u}(2)}$ orbitals. These will lead to two doubly-degenerate energy levels of the e_{1u}-type.

(6) Two equivalent 1×1 determinants corresponding to $\Gamma^{E_{2u}}$ which will produce a doubly-degenerate energy level of the e_{2u}-type. The MOs will be the pure ligand MOs $\psi_{E_{2u}(1)}$ and $\psi_{E_{2u}(2)}$ (there are no metal orbitals of $\Gamma^{E_{2u}}$ symmetry) and consequently they do not participate in the bonding of the iron atom to the rings.

If certain assumptions are made about the matrix elements H_{jk}^{eff} and S_{jk} in these determinants, then the energy levels for the valence electrons in ferrocene can be calculated. An energy level diagram, based on the results of such a calculation, is shown in Fig. 12-4.3. This diagram implies that the electronic configuration for the 18 bonding electrons of ferrocene is $1a_{1g}^2\, 1a_{2u}^2\, 1e_{1u}^4\, 1e_{1g}^4\, 2a_{1g}^2\, 1e_{2g}^4$, each individual MO accommodating two electrons of opposite spin. The reader is warned that there is much disagreement about the exact order of the MO energy levels in ferrocene since they depend rather critically on the assumptions made about H_{jk}^{eff} and S_{jk}. In 1972, however, Veillard and co-workers† carried out a strictly *ab initio* calculation and made no such assumptions. Their results are likely to be more reliable than the previous ones.

† M.- M. Coutière, J. Demuynck and A. Veillard, *Theoretica chimica acta* **27**, 281 (1972).

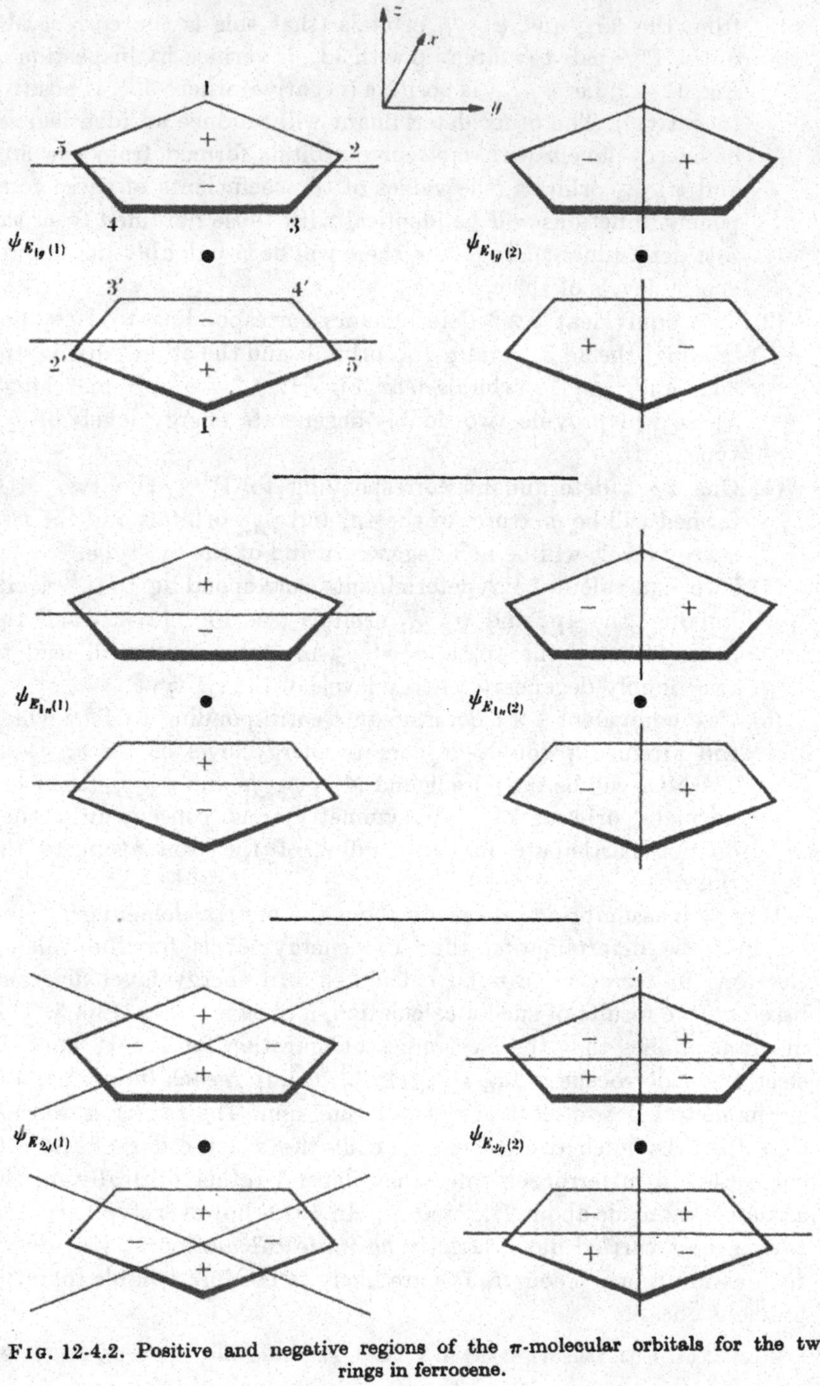

Fig. 12-4.2. Positive and negative regions of the π-molecular orbitals for the two rings in ferrocene.

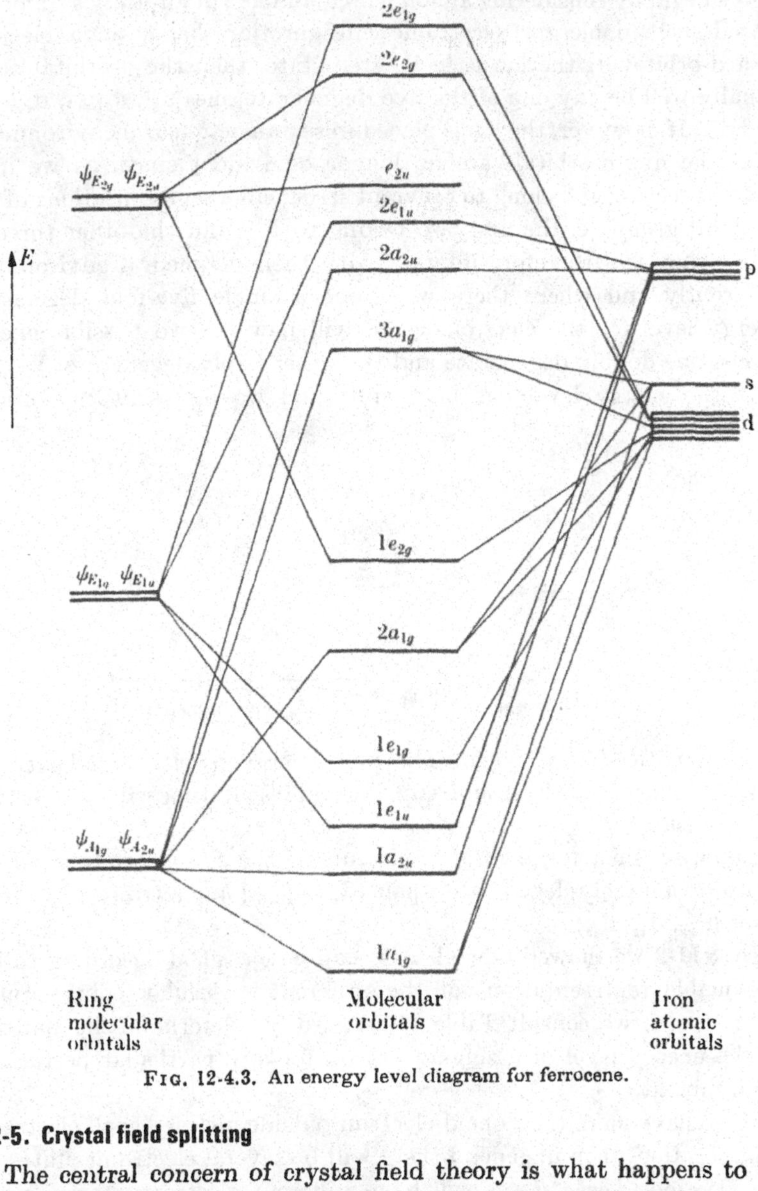

FIG. 12-4.3. An energy level diagram for ferrocene.

12-5. Crystal field splitting

The central concern of crystal field theory is what happens to the electronic states of an ion when it is placed in some perturbing symmetric environment. If we assume that a set of ligands placed symmetrically about a transition-metal ion interact only electrostatically with the electrons of that ion, then the answer to this question is relevant to our study of transition–metal compounds.

We begin by considering an ion which, outside of a totally symmetric closed shell (noble gas) electronic configuration, has a single electron in a d-orbital. If the ion is in its free state, then the d-orbital could equally well be any one of the five degenerate ones d_{z^2}, $d_{x^2-y^2}$, d_{xy}, d_{xz}, and d_{yz}. If, however, the ion is placed in say an octahedral environment, then the five d-orbitals are no longer equivalent since, as we have already seen, they belong to different irreducible representations of the $\mathcal{O}_h$ point group i.e. the first two belong to Γ^{E_g} and the other three to $\Gamma^{T_{2g}}$. They will therefore interact with their octahedral environment differently and where there was once a single five-fold degenerate energy level for the electron, there will now be two possible energy levels, one doubly degenerate and the other triply degenerate. We say that the energy level has been split and we represent the process

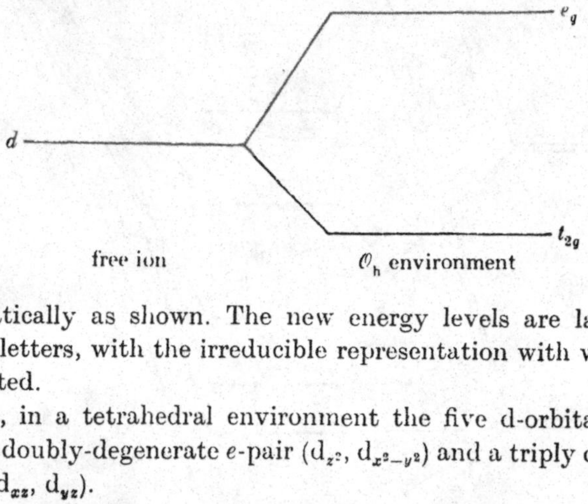

diagrammatically as shown. The new energy levels are labelled, in lower case letters, with the irreducible representation with which they are associated.

Likewise, in a tetrahedral environment the five d-orbitals will be split into a doubly-degenerate e-pair (d_{z^2}, $d_{x^2-y^2}$) and a triply degenerate t_2-set (d_{xy}, d_{xz}, d_{yz}).

In § 11-2 we showed how all AOs can be classified according to the irreducible representations of the different molecular point groups. Therefore, if we consult Table 11-2.2, we can determine the splitting of the energy level of a single electron for any particular perturbing environment.

If we have more than one d-electron outside a closed shell electronic configuration, then in general there will be several electronic states for the free ion. These states will have different electronic energies and will be characterized by their term symbols: ${}^{2S+1}T_J$ (see Appendix A.12-1 for a brief review of atomic-spectroscopic notation). In a given term-symbol, T will be S, P, D, F, or G etc. depending on whether the total electronic orbital angular-momentum quantum number L is 0, 1,

2, 3, or 4 etc. Just as there are, for example, for a single electron with an orbital angular-momentum quantum number of $l = 2$, five degenerate d-orbitals, so there are five degenerate total electronic wavefunctions for a many-electron atom or ion in a D ($L = 2$) state.† Furthermore, it can be shown that the symmetry properties of the five total-electronic

TABLE 12-5.1

Splitting of electronic states by different symmetrical environments†

Free ion states	Split states‡ $\mathscr{O}_h$	$\mathscr{T}_d$	$\mathscr{D}_{4h}$	$\mathscr{D}_3$	$\mathscr{D}_{2d}$
S	A_{1g}	A_1	A_{1g}	A_1	A_1
P	T_{1g}	T_1	A_{2g}, E_g	A_2, E	A_2, E
D	E_g, T_{2g}	E, T_2	$A_{1g}, B_{1g}, B_{2g}, E_g$	$A_1, E(2)$	A_1, B_1, B_2, E
F	A_{2g}, T_{1g}, T_{2g}	A_2, T_1, T_2	$A_{2g}, B_{1g}, B_{2g}, E_g(2)$	$A_1, A_2(2), E(2)$	$B_1, A_2, B_2, E(2)$
G	$A_{1g}, E_g, T_{1g}, T_{2g}$	A_1, E, T_1, T_2	$A_{1g}(2), A_{2g}, B_{1g}, B_{2g}, E_g(2)$	$A_1(2), A_2, E(3)$	$A_1(2), A_2, B_1, B_2, E(2)$
H	$E_g, T_{1g}(2), T_{2g}$	$E, T_1(2), T_2$	$A_{1g}, A_{2g}(2), B_{1g}, B_{2g}, E_g(3)$	$A_1, A_2(2), E(4)$	$A_1, A_2(2), B_1, B_2, E(3)$
I	$A_{1g}, A_{2g}, E_g, T_{1g}, T_{2g}(2)$	$A_1, A_2, E, T_1, T_2(2)$	$A_{1g}(2), A_{2g}, B_{1g}(2), B_{2g}(2), E_g(3)$	$A_1(3), A_2(2), E(4)$	$A_1(2), A_2, B_1(2), B_2(2), E(3)$

† The states are derived from d^n configurations only and hence for $\mathscr{O}_h$ and $\mathscr{D}_{4h}$ the split states all have g character.

‡ The numbers in parentheses refer to the number of states of the given symmetry, e.g. the 13 degenerate total wavefunctions for the state I form a reducible representation for $\mathscr{D}_3$ which can be decomposed as

$$\Gamma^I = 3\Gamma^{A_1} \oplus 2\Gamma^{A_2} \oplus 4\Gamma^E.$$

wavefunctions belonging to a D state are the same as those of the five d-orbitals. Necessarily then, a D state will split under an $\mathscr{O}_h$ perturbing environment into two states. The new states will again be labelled by the relevant irreducible representations, but this time upper case letters will be used. In Table 12-5.1 the states into which the states of a free ion are split under a variety of environments are listed.

Since a chemical environment does not normally interact directly with electron spins, the spin multiplicity of a state is unaffected by the splitting and the split states will have the same multiplicity as the parent free ion state. The quantum number J also remains unaltered. For this reason, multiplicities and J values are left out in Table 12-5.1.

† Each wavefunction will correspond to a different value of the quantum number M_L (the equivalent to m_l for a single electron) and will be a linear combination of Slater determinants, composed of atomic orbitals, which have the same value of M_L.

12-6. Order of orbital energy levels in crystal field theory

Our object in discussing crystal field theory is to construct energy level diagrams (*correlation diagrams*) which show how the energies of the various electronic states into which the free-ion terms split depend on the strength of the interaction of the ion with its environment. Before we can do this, however, it is necessary to have some idea about the *relative* order of the energy levels of the d-orbitals in tetrahedral and octahedral surroundings (our discussion of correlation diagrams will be restricted to ions with a d^n configuration and environments with a $\mathcal{T}_d$ or $\mathcal{O}_h$ symmetry). Such relative order concerns quantum mechanics rather than symmetry or group theory and a rigorous treatment is beyond the scope of this book. Nonetheless, if we are willing to forego rigour, a simple model using crystal field principles will establish all we need to know for the purposes of constructing correlation diagrams.

For an octahedral or tetrahedral environment we have already seen that five, originally degenerate, d-orbitals are split into two sets which have different energies: one set is d_{z^2} and $d_{x^2-y^2}$ (called e_g or e orbitals, depending on the environment) and the other is d_{xy}, d_{xz}, and d_{yz} (called t_{2g} or t_2 orbitals). To investigate the relative order of the two new energy levels, it is only necessary to look at the effect of the environment on one orbital of each set, say $d_{x^2-y^2}$ and d_{xy}.

In crystal field theory each ligand is approximated as a point charge or a point dipole and each metal–ligand interaction is taken to be purely electrostatic. So our problem is reduced to one of investigating the effect of point charges (or point dipoles) arranged tetrahedrally or octahedrally about an electron in a $d_{x^2-y^2}$ or a d_{xy} orbital.

Since the electron has a negative charge and since we will assume the ligands to be anions or dipolar molecules with their negative ends pointed towards the central ion, an octahedral arrangement of ligands will clearly result in greater electrostatic repulsive forces for an electron in a $d_{x^2-y^2}(e_g)$ orbital than for an electron in a $d_{xy}(t_{2g})$ orbital (compare the two diagrams in Fig. 12-6.1). The t_{2g} orbitals should therefore be more stable than the e_g orbitals in an octahedral environment. The exact magnitude of the difference in energies will depend, amongst other things, on the nature of the ligands and it is commonly given the symbol Δ_o or $10Dq$ (see Fig. 12-6.2).

The difference between the effect of a tetrahedral arrangement of ligands on an electron in a $d_{x^2-y^2}$ (e-type) orbital and on an electron in a d_{xy} (t_2-type) orbital is perhaps not quite so obvious as the previous case. However, a study of Fig. 12-6.3 suggests that an electron in a

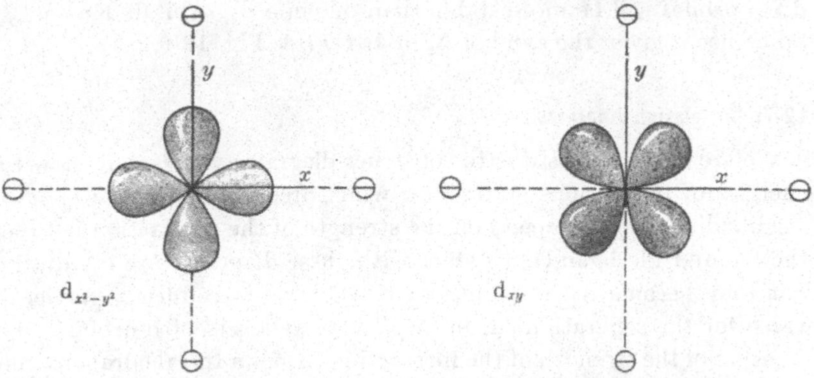

FIG. 12-6.1. $d_{x^2-y^2}$ and d_{xy} orbitals surrounded by an octahedral arrangement of ligands. The ligands which are directly above and below the origin of the axes are not shown.

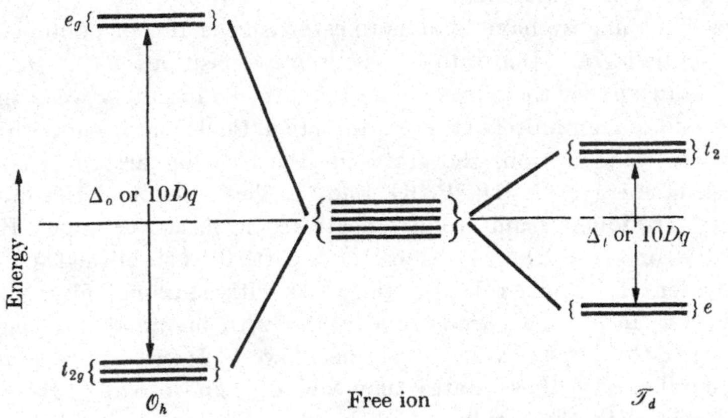

FIG. 12-6.2. The relative energies of d-orbitals in octahedral $\mathcal{O}_h$ and tetrahedral $\mathcal{T}_d$ environments.

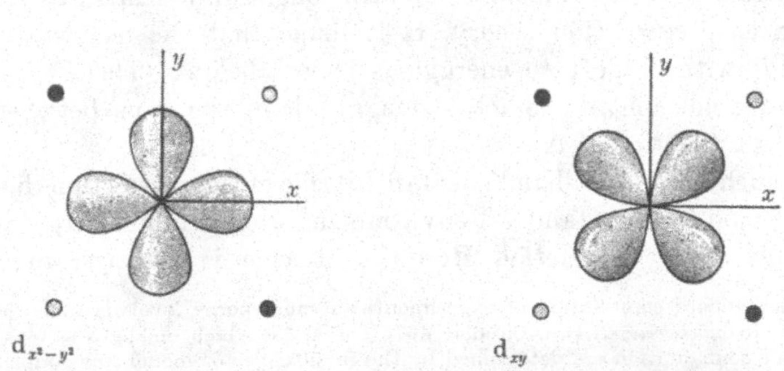

FIG. 12-6.3. $d_{x^2-y^2}$ and d_{xy} orbitals surrounded by a tetrahedral arrangement of ligands. The black circles indicate ligands below the xy plane and the shaded circles indicate ligands above it. See Fig. 11-5.2 for the relations between the axes and the ligands.

$d_{x^2-y^2}$ orbital will be more stable than one in a d_{xy} orbital. The energy difference is given the symbol Δ_t or $10Dq$ (see Fig. 12-6.2).

12-7. Correlation diagrams

We are now in a position to construct diagrams which show how the energies of the various states into which free ion states are split by surrounding ligands depend on the strength of the interaction between the ion and the ligands. To start with these diagrams are qualitative but they become more quantitative when we introduce a particular value for the separation, Δ_t or Δ_o, of the two sets of d-orbitals as a measure of the strength of the interaction (Δ_t for a tetrahedral environment and Δ_o for an octahedral environment). We will restrict our discussion to ions having a d^n electronic configuration and surroundings having a $\mathscr{T}_d$ or $\mathscr{O}_h$ symmetry.

The first thing we have to do is to establish the anchor points of the correlation diagram, that is to say the relative positions of the spectroscopic states under the extreme conditions of an infinitely weak interaction and of an infinitely strong interaction. Consider an ion with a d^2 electronic configuration, the states of the free ion are, in order of increasing energy, ^3F, ^1D, ^3P, ^1G, and ^1S. This order and the precise energies are found from atomic spectroscopic measurements.† However the order of the states can be partially established by the application of Hund's rule: (*a*) the state with maximal multiplicity lies lowest, (*b*) if there are several states with maximal multiplicity that with the largest value of L lies lowest. If we apply a *weak* octahedral field to these states they will be split into new states in accordance with the results which have been given in Table 12-5.1. Since the perturbation to the ion is small, the relative energies of the new states may be calculated by the quantum-mechanical method known as perturbation theory. It is found that the new states are ordered, with respect to energy, as shown below. (The ^1S and ^3P states are not split.) The actual magnitude of the gaps between the states is a function of Δ_o.

So much for the left-hand (weak-field) side of the correlation diagram for a d^2 ion in an octahedral environment. Now we must consider an infinitely–strong interaction. Here the situation is directly contrary to

† The standard compilation of experimental atomic energy levels is C. E. Moore's *Atomic Energy Levels*, N.B.S. Circular 467. The states which emanate from a given electronic configuration are determined by the familiar vector model rule coupled with the Pauli exclusion principle. The reader might note that the vector model rule itself can be derived from group theoretical principles. However, the derivation involves the full rotation group which is beyond the scope of this book.

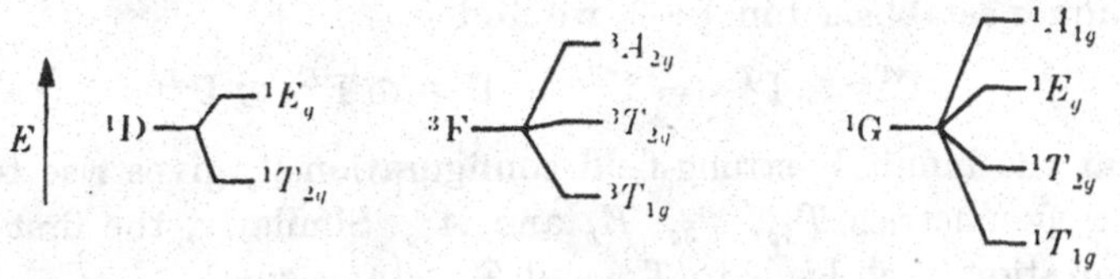

the earlier one. We now assume that the interelectronic repulsions in the ion are small compared with the infinitely-strong crystal field. In other words we have a splitting of the free-ion states which is very large compared to the separation between them. The wavefunctions which are constructed for the free ion and are based on relatively strong interelectronic repulsions (see the footnote on page 259) are quite unsuitable for this new situation and instead infinitely-strong-field configurations become the natural choice. An infinitely-strong-field configuration is obtained by assigning each d-electron to either a t_{2g} or an e_g orbital. For a d^2 ion in an octahedral environment, both the electrons go into the t_{2g} orbital to form the lowest lying configuration: t_{2g}^2. The next highest configuration will be $t_{2g}^1 e_g^1$ and after that e_g^2. The energy separation between these configurations will be Δ_o or $10Dq$ (see Fig. 12-6.2).

Now let us consider what happens as we weaken this infinitely–strong interaction with the environment to just a strong interaction. The interelectronic repulsions, in a relative sense, start to increase as the electrons start to influence each other. As they begin to couple, a set of states of the entire configuration will be produced i.e. there is a splitting into the strong-field states. The new states and their energies can be found by treating the interelectronic repulsions as perturbations on the infinitely-strong-field configurations. The qualitative picture, however, can be obtained by the use of group theory in the following way. If the states which arise from a given configuration are known, then the states which arise from the configuration with one additional electron must have the symmetry labels of the irreducible representations which are contained in the direct product (see § 8-3) between the symmetry labels (i.e. irreducible representations) of the initial states and the added electron. With only one d-electron, there are no interelectronic repulsions, so the symmetry labels of the states of d^1 must be the same as those of the electron. Thus the t_{2g} and e_g electrons of d^1 in $\mathcal{O}_h$ give rise to ${}^2T_{2g}$ and ${}^2E_g$ states, respectively.

For a d^2 configuration, to obtain the states of t_{2g}^2 it is necessary to take the direct product between T_{2g} (the symmetry label of the state of t_{2g}^1) and T_{2g} (the symmetry label of the added electron). Using the techniques established in § 8-3, we find

$$\Gamma^{T_{2g}} \otimes \Gamma^{T_{2g}} = \Gamma^{T_{1g}} \oplus \Gamma^{T_{2g}} \oplus \Gamma^{E_g} \oplus \Gamma^{A_{1g}}$$

so that the infinitely-strong-field configuration t_{2g}^2 gives rise to states having symmetries T_{1g}, T_{2g}, E_g and A_{1g}. Similarly, the first excited configuration, $t_{2g}^1e_g^1$, leads to T_{1g} and T_{2g} states since

$$\Gamma^{T_{2g}} \otimes \Gamma^{E_g} = \Gamma^{T_{1g}} \oplus \Gamma^{T_{2g}}$$

and the second excited configuration, e_g^2, leads to E_g, A_{1g}, and A_{2g} states since

$$\Gamma^{E_g} \otimes \Gamma^{E_g} = \Gamma^{E_g} \oplus \Gamma^{A_{1g}} \oplus \Gamma^{A_{2g}}.$$

As well as the symmetry labels of these strong-field states, we also require the multiplicities. The completely general method of determining these is beyond the scope of this book, so we will confine ourselves to consideration of a case which can be resolved on the basis of some simple arguments. Consider first the configuration t_{2g}^2 and let the three t_{2g} orbitals be represented by three boxes. In Fig. 12-7.1 it is shown that, if an electron with spin quantum number $m_s = \frac{1}{2}$ is represented by an arrow pointing upwards and one with spin quantum number $m_s = -\frac{1}{2}$ by an arrow pointing downwards, then the number of ways of arranging the arrows in the boxes is 15. This corresponds to the number of distinct wavefunctions for t_{2g}^2. As the field strength is decreased this total degeneracy must remain at 15. We now recall that a T-type state is of three-fold degeneracy, an E-type state of two-fold degeneracy, an A-type state is non-degenerate and also that only triplet or singlet multiplicities can arise from two electrons. Therefore, if the required multiplicities a, b, c, and d are attached to the states in the following way:

$$^aT_{2g},\ ^bT_{1g},\ ^cE_g,\ ^dA_{1g};$$

and if we require that the degeneracy for the strong-field case remain at 15, then

$$3a+3b+2c+d = 15,$$

with a, b, c, and d each equal to either 1 or 3. This equation has three possible solutions:

	a	b	c	d
(1)	3	1	1	1
(2)	1	3	1	1
(3)	1	1	3	3.

That solution (2) is the correct one, we will discover only when we finally set up the correlation diagram.

For the configuration $t_{2g}^1e_g^1$ it is possible to write 24 wavefunctions (for each of the six ways of putting an electron in the t_{2g} set, there are

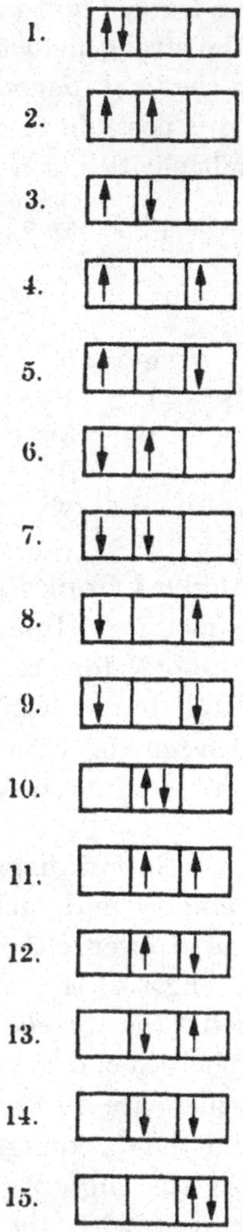

FIG. 12-7.1. Symbolic wavefunctions for the t_{2g}^2 configuration.

four ways of putting one in the e_g set). Hence, if a and b are the multiplicities of the states T_{1g} and T_{2g} respectively, we have

$$3a + 3b = 24$$

which is satisfied, for example, by $a = 4$ and $b = 4$. But since we have already stated that the multiplicities are restricted to 1 and 3, this result is unacceptable. We can extract ourselves from this dilemma by assuming that we have in fact four states $^3T_{1g}$, $^3T_{2g}$, $^1T_{1g}$, and $^1T_{2g}$.

[Choosing eight singlet states is ruled out on the grounds that we would then have more states in the strong field than in the weak field (this will become apparent when the final diagram is set up).]

For the e_g^2 configuration it is possible to construct six wavefunctions and if a, b, and c are the multiplicities of E_g, A_{1g}, and A_{2g} respectively, then

$$2a+b+c = 6$$

for which there are two solutions:

	a	b	c
(1)	1	1	3
(2)	1	3	1

Again the correlation diagram itself will dictate that solution (1) is the correct one.

The order of the states derived from a given infinitely–strong-field configuration is given by a modified Hund's rule: (1) states with the highest multiplicity lie lowest, (2) for states with equal multiplicity, the ones with highest orbital degeneracy ($T > E > A$) tend to lie lower. Any ambiguities which remain after the application of this rule, can only be resolved by recourse to detailed quantum mechanical calculations.

Once the two sides of a correlation diagram have been established, the states of the same symmetry and multiplicity are connected by straight lines in such a way as to observe the non-crossing rule: identical states cannot cross as the strength of the interaction is changed. When this is done we have completed the correlation diagram.

The assignment of multiplicities can now be settled. For a d^2 ion in an octahedral environment there are no $^3A_{1g}$ states in the weak crystal field and thus solution (3) for the t_{2g}^2 configuration is ruled out since it includes such a state. Also the highest of the $^3T_{1g}$ states in the weak crystal-field must connect with the highest $^3T_{1g}$ state in the strong field, namely the one arising from the $t_{2g}^1e_g^1$ configuration, this leaves the other weak crystal field $^3T_{1g}$ state with only the possibility of connecting with the T_{1g} state from t_{2g}^2, thus this state must be a triplet and solution (2) is the correct one. Finally, the fact that the only A_{2g} state in the weak crystal field is a triplet requires that we accept solution (1) for the e_g^2 configuration.

A correlation diagram for a d^2 ion (e.g. V^{3+}) in an octahedral environment is shown in Fig. 12-7.2. What this diagram does is to demonstrate how the energy levels of the free ion behave as a function of the strength (Δ_o) of the ion's interaction with a set of octahedrally disposed ligands.

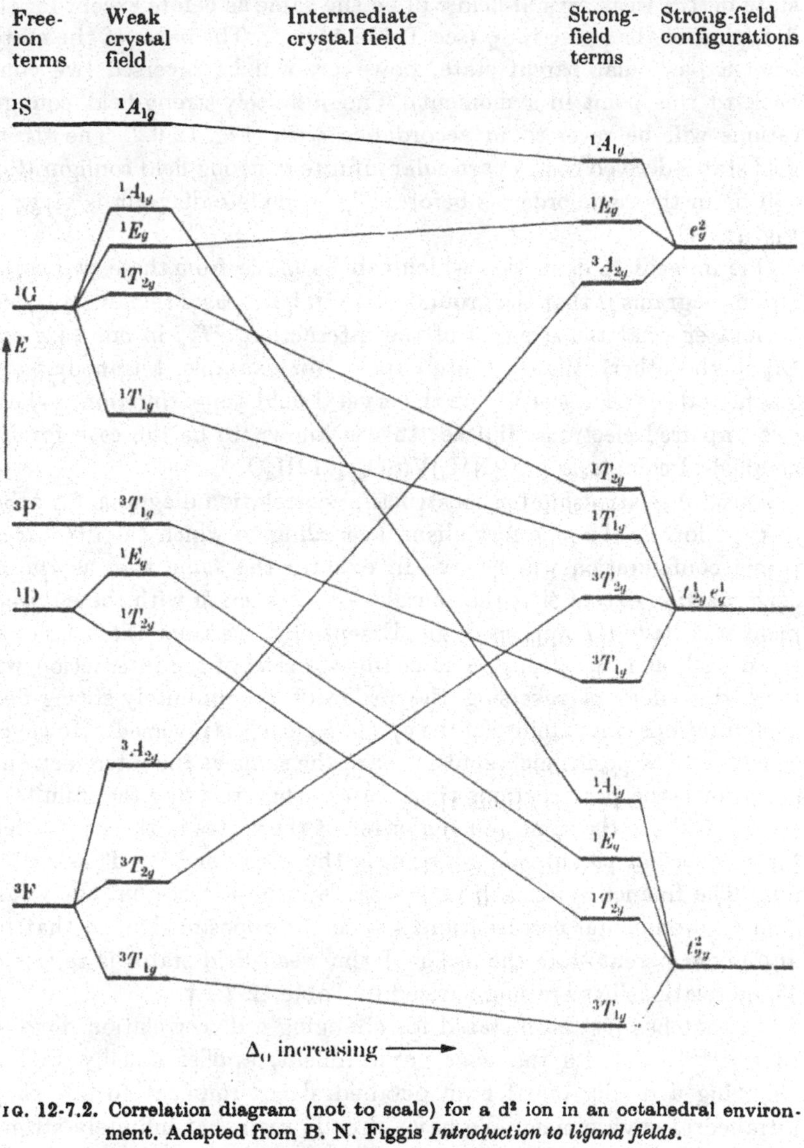

FIG. 12-7.2. Correlation diagram (not to scale) for a d^2 ion in an octahedral environment. Adapted from B. N. Figgis *Introduction to ligand fields*.

If Δ_0 is known for a particular ion and set of ligands, then a correlation diagram will immediately predict the order of the ion's energy levels.

For a d^2 ion in a tetrahedral environment, exactly the same procedure can be carried out. The free ion states will be the same as in the octahedral case. The *type* of states produced from a particular free-ion state by the weak crystal-field will be the same as before except for the dropping of the subscript g (see Table 11-2.2). The *order* of the states from a particular parent state, however, will be reversed (we come back to this point in a moment). The infinitely-strong-field configurations will be reversed in accordance with Fig. 12-6.2. The strong field states derived from a particular infinitely-strong field configuration will be in the same order as before. The complete diagram is given in Fig. 12-7.3.

One immediate deduction which can be made from these two correlation diagrams is that the ground state in both cases remains a triplet no matter what the strength of the interaction ($^3T_{1g}$ in one case and $^3A_2$ in the other). We therefore expect, for example, tetrahedral and octahedral complexes of V^{3+}, in the crystal field approximation, to have two unpaired electrons. Indeed, this is known to be the case for the octahedral complexes e.g. $(NH_4)V(SO_4)_2.12H_2O$.

A useful relationship for constructing correlation diagrams for other d^n-type ions is the hole formalism, according to which the d^{10-n} electronic configuration will behave in exactly the same way as the d^n configuration except that the energies of interaction with the environment will have the opposite sign. Essentially, we treat the n holes in the d shell as n 'positrons'. The change of sign of the interaction will have the effect of reversing the order of the infinitely-strong-field configurations (the stability of the e_g and t_{2g} levels is reversed). However, since the 'interpositronic' repulsions are the same as the interelectronic repulsions, the perturbations these cause when relaxing the infinitely-strong-field are the same and the order of the states in the strong field for a *particular* parent configuration is the same for both d^n and d^{10-n} ions. The free-ion states will be the same in both cases but the weak-field environmental perturbations will be of opposite sign, so that for any given parent state the order of the weak-field states is reversed. These relationships are summarized in Table 12-7.1.†

All that has just been stated for changing a d^n correlation diagram to a d^{10-n} one with the *same* environment, applies equally well to changing a d^n diagram for an octahedral environment to one for a tetrahedral environment. We have already seen that infinitely-strong-field configurations are reversed by such an environmental change (Fig. 12-6.2) and, if we assume therefore that the environmental perturbation in going from the free-ion to the weak-field case is also reversed, then we can conclude that the order of states emanating

† A precise and formal discussion of the hole formalism is given by J. S. Griffith: *The theory of transition-metals ions*, Cambridge University Press, 1961.

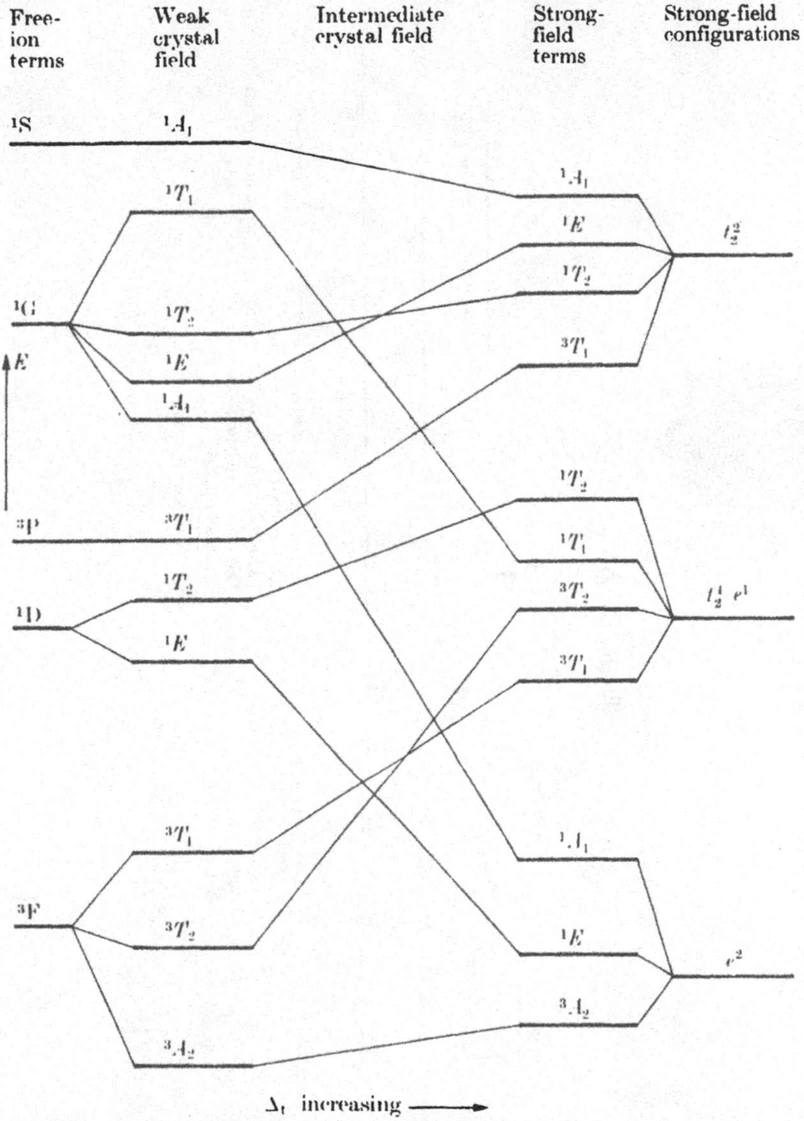

FIG. 12-7.3. Correlation diagram (not to scale) for a d^2 ion in a tetrahedral environment. Adapted from B. N. Figgis *Introduction to ligand fields*.

from a particular parent free-ion state will, in the weak field, be the opposite in a $\mathscr{T}_d$ environment to that in an $\mathscr{O}_h$ environment. The free ion states and the interelectronic-repulsion perturbation are the same in both cases. Hence, Table 12-7.1 applies also to the $\mathscr{O}_h \leftrightarrow \mathscr{T}_d$ change.

It should now be clear that if we change both the configuration, $d^n \leftrightarrow d^{10-n}$ (i.e. change the ion), *and* the environmental symmetry,

TABLE 12-7.1

Change	Free ion	Environmental perturbation	Weak field	Strong field	Perturbation due to interelectronic or 'interpositronic' repulsions	Infinitely-strong field
$d^n \leftrightarrow d^{10-n}$ or $\mathcal{O}_h \leftrightarrow \mathcal{T}_d$	No change		For a given parent free ion state the order of states is reversed	For a given parent infinitely-strong field configuration the order of states is unchanged		The order of configurations is reversed

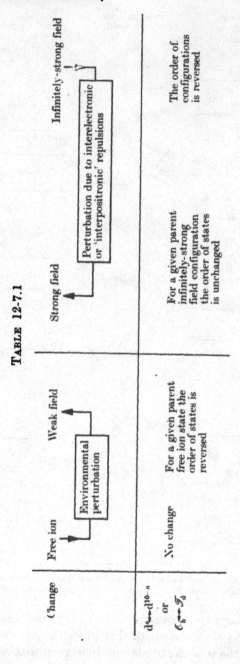

$\mathscr{T}_d \leftrightarrow \mathscr{O}_h$, then the correlation diagram is unaltered (except for the obvious and minor change of adding or dropping the subscript g on various symbols). We can express this result by

$$d^n(\text{oct}) \equiv d^{10-n}(\text{tetr})$$

and

$$d^n(\text{tetr}) \equiv d^{10-n}(\text{oct}).$$

These relationships show that far fewer individual correlation diagrams need be constructed from scratch than might have first been anticipated.

12-8. Spectral properties

One of the most important applications of correlation diagrams concerns the interpretation of the spectral properties of transition-metal complexes. The visible and near ultra-violet spectra of transition-metal complexes can generally be assigned to transitions from the ground state to the excited states of the metal ion (mainly d–d transitions). There are two selection rules for these transitions: the spin selection rule and the Laporte rule.

The spin selection rule states that no transition can occur between states of different multiplicity i.e. $\Delta S = 0$. Transitions which violate this rule are generally so weak that they can usually be ignored.

The Laporte rule states that transitions between states of the same parity, u or g, are forbidden i.e. $u \rightarrow g$ and $g \rightarrow u$ but $g \nleftrightarrow g$ and $u \nleftrightarrow u$. This rule follows from the symmetry of the environment and the invoking of the Born–Oppenheimer approximation. But since, due to vibrations, the environment will not always be strictly symmetrical, these forbidden transitions *will* in fact occur, though rather weakly (oscillator strengths of the order of 10^{-4}). All the states of a transition-metal ion in an octahedral environment are g states, so that it will be these weak symmetry forbidden transitions (called d–d transitions) that will be of most interest to us when we study the spectra of octahedral complexes.

We will exclude from our discussion the so-called charge-transfer bands. These relate to the transfer of electrons from the surrounding ligands to the metal ion or vice versa. They may be fully allowed and hence have greater intensities than the d–d transitions. They usually, though not always, occur at high enough energies and with such high intensities that they are not too easily confused with the d–d bands. A third type of transition, transitions occurring within the ligands, will also be ignored.

By consulting the appropriate correlation diagram, it is possible to see what kind of d–d spectrum a transition–metal ion in a given environment should have. For qualitative predictions we can use diagrams of the kind which were developed in the last section. However, for quantitative predictions it is necessary to use the so-called Tanabe–Sugano correlation diagrams (*J. phys. soc. Japan* **9**, 753 (1954)).

These diagrams are based on proper quantum-mechanical calculations of the energy levels of a d^n system in the presence of *both* interelectronic repulsions and crystal fields of medium strength. Such calculations are very difficult to carry out and we will simply discuss the form of the results. It turns out that the energy of each state depends not only on the field strength (as measured by Δ_o or Δ_t) but also on two electronic-repulsion parameters B and C called Racah parameters. (B and C are related to the Slater–Condon parameters F_2 and F_4 by the equations: $B = F_2 - 5F_4$, $C = 35F_4$.) In Tanabe–Sugano diagrams it is assumed that C is directly proportional to B with a proportionality constant which has a fixed value for each diagram (the diagrams are apparently not too sensitive to the value of this proportionality constant). Furthermore, the diagrams are made independent of B by plotting E/B against Δ_o/B (or Δ_t/B) rather than E, the energy, against Δ_o (or Δ_t). Consequently, to obtain from a given diagram the relative energies of the states of a metal ion–ligand system, it is necessary to specify both B and Δ_o (or Δ_t). This is usually done by using *two* pieces of experimental data, e.g. by fitting two d–d transitions to the appropriate Tanabe–Sugano diagram.

Now let us consider some particular cases. $[V(H_2O)_6]^{3+}$ is a d^2 ion in an octahedral environment and the pertinent qualitative correlation diagram, Fig. 12-7.2, shows that there should be three spin-allowed transitions: from the $^3T_{1g}(F)$ ground state to the excited states $^3T_{2g}(F)$, $^3T_{1g}(P)$ and $^3A_{2g}(F)$; the symbol in brackets, in each case, denotes the parent state of the free ion. Experimentally, aqueous solutions of trivalent vanadium salts show two absorption bands, one at 17 200 cm^{-1} and the other at 25 600 cm^{-1}; these give rise to the green colour of such solutions. If we specify the complex (i.e. determine Δ_o and B) by fitting the transitions $^3T_{2g}(F) \leftarrow {}^3T_{1g}(F)$ and $^3T_{1g}(P) \leftarrow {}^3T_{1g}(F)$ to 17 200 and 25 600 cm^{-1} respectively, then Δ_o is found to be 18 600 cm^{-1} and B to be 665 cm^{-1}. With these values the transition $^3A_{2g}(F) \leftarrow T_{1g}(F)$ is predicted to lie in the region of 36 000 cm^{-1}. Unfortunately this cannot be verified as there is a very strong charge-transfer band in the same region. However, in the solid state, particularly for V^{3+} in Al_2O_3, where charge transfer occurs at a higher energy, a weak band at about the right position has been found. Since the oxygen ligand atoms in the Al_2O_3 structure are known to produce about the same value of Δ_o as water molecules, this can be considered as partial experimental confirmation of the assignments for $[V(H_2O)_6]^{3+}$. An aqueous solution of a

V^{3+} salt also shows some *very* weak bands ($f \approx 10^{-7}$) in the 20 000–30 000 cm^{-1} region; these are thought to be due to spin-forbidden transitions to excited singlet states.

A very well studied group of complexes are those with d^3 configurations in an octahedral environment. We have not shown the correlation diagram for this case, but the important features of such a diagram are a $^4A_{2g}(F)$ ground state and three other excited quartet states which, in order of increasing energy, are $^4T_{2g}(F)$, $^4T_{1g}(F)$, and $^4T_{1g}(P)$; furthermore, none of these states cross each other as the strength of the interaction changes. As an example, we take the case of $[Cr(H_2O)_6]^{3+}$. The aqueous solutions of salts of trivalent chromium are green in colour as a result of absorption bands at 17 000, 24 000, and 37 000 cm^{-1} (there are also two very weak spin forbidden bands at 15 000 and 22 000 cm^{-1}). If the complex is specified by fitting the transitions $^4T_{2g}(F) \leftarrow {}^4A_{2g}(F)$ and $^4T_{1g}(F) \leftarrow {}^4A_{2g}(F)$ to 17 000 and 24 000 cm^{-1} respectively, then Δ_o has to be 17 000 cm^{-1} and B has to be 695 cm^{-1}. The transition $^4T_{1g}(P) \leftarrow {}^4A_{2g}(F)$ is then predicted to lie at 37 000 cm^{-1} which is in excellent agreement with the observed spectrum.

The same correlation diagram can be used for the tetrahedral complexes of Co^{2+} (d^7) and, as an example, we consider $[CoCl_4]^{2-}$. The spectrum of $[CoCl_4]^{2-}$ in HCl solutions shows bands at 5800 and 15 000 cm^{-1} and some weak absorption in the 17 000–23 000 cm^{-1} region. If the $^4T_2(F) \leftarrow {}^4A_2(F)$ transition† is assigned to the 5800 cm^{-1} band, then it is found to be impossible to fit the remaining bands. However, the assignment:

$$^4T_1(F) \leftarrow {}^4A_2(F) \qquad 5800 \text{ cm}^{-1}$$
$$^4T_1(P) \leftarrow {}^4A_2(F) \qquad 15\,000 \text{ cm}^{-1}$$

can be fitted to the correlation diagram and this leads to a value of 3200 cm^{-1} for Δ_t and 730 cm^{-1} for B. With these values, the transition $^4T_2(F) \leftarrow {}^4A_2(F)$ is predicted to lie at 3300 cm^{-1}. Studies in the infrared region show a band at 3500 cm^{-1} which can probably be assigned to this transition. There are no other spin-allowed transitions, so the bands in the 17 000–23 000 cm^{-1} region must be assigned to spin-forbidden transitions.

12-9. Magnetic properties

The effective magnetic moment of a transition metal ion μ_{eff} is defined by the equation

$$\mu_{eff} = 2{\cdot}84(\chi T)^{\frac{1}{2}} \qquad \text{Bohr magnetons}$$

† Note that the g subscript is not applicable in the tetrahedral case.

where χ is the molar magnetic susceptibility and T the temperature. Different formulae can be derived which relate μ_{eff} to the ion's angular momentum quantum numbers L, S, and J (definitions of L, S, and J are given in Appendix A.12-1). The particular formula to be used depends on how far the excited states of the ion lie above the ground state in comparison with kT.

If the excited states are separated from the ground state by an amount much larger than kT, then

$$\mu_{eff} = g\{J(J+1)\}^{\frac{1}{2}}$$

where g, the Landé splitting factor, is given by

$$g = 1+\{(S(S+1)-L(L+1)+J(J+1)\}/2J(J+1).$$

Hence, for a system in which the ion has no orbital angular momentum: $L = 0$, $J = S$, $g = 2$ and

$$\mu_{eff} = 2\{S(S+1)\}^{\frac{1}{2}}.$$

Also, under these circumstances, if n is the number of electrons in the ion with unpaired spin, then $S = n/2$ and

$$\mu_{eff} = \{n(n+2)\}^{\frac{1}{2}}.$$

This is known as the *spin-only* formula. Many ions which in the free state have orbital angular momentum ($L \neq 0$), lose it, completely or partially, when incorporated into a complex. This phenomenon is called orbital angular momentum *quenching*. It can be shown that for A states the quenching is complete and that for T and E states it is incomplete.† Because of this, the spin-only formula applies to more situations than might have been expected.

The number of unpaired spins n for an ion in its ground state is obviously determined by the multiplicity, $2S+1$, of that state and vice versa. The multiplicity is written in the upper left hand corner of the symbol for the state. Thus the ground state symbol can be read off from the appropriate correlation diagram and, if spin-only conditions apply, the effective magnetic moment can be immediately determined.

For octahedral d^1, d^2, d^3, d^8, and d^9 cases, the ground state is derived from the lowest term of the free ion for all values of the crystal field strength (defined by Δ_o). Hence the multiplicity, the number of unpaired spins and, if the spin-only formula applies, the effective magnetic moment must all be the same as that of the free ion, no matter how strong the interaction between the ion and the ligands. For octahedral d^4, d^5, d^6, and d^7 cases the ground state is derived from

†See P.J. Stiles, *Mol. Phys.*, *15*, 405 (1968).

TABLE 12-9.1

Magnetic properties of some transition-metal ions

No. of d electrons	*Ion*	*Compound*	*Ground state*	*n*	$\{n(n+2)\}^{\frac{1}{2}}$	μ_{eff} (expt) 300 K
1	Ti^{3+}	$CsTi(SO_4)_2.12H_2O$	$^2T_{2g}$	1	1·73	1·84
2	V^{3+}	$(NH_4)V(SO_4)_2.12H_2O$	$^3T_{1g}$	2	2·83	2·80
3	Cr^{3+}	$KCr(SO_4)_2.12H_2O$	$^4A_{2g}$	3	3·87	3·84
	Mo^{3+}	K_3MoCl_6	$^4A_{2g}$	3	3·87	3·79
	Mn^{4+}	$BaMnF_6$	$^4A_{2g}$	3	3·87	3·80
	Re^{4+}	Cs_2ReCl_6	$^4A_{2g}$	3	3·87	3·35
4	Cr^{2+}	$Cr(SO_4).6H_2O$	^5E_g	4	4·90	4·82
	Mn^{3+}	$Mn(acac)_3$	^5E_g	4	4·90	4·86
		$K_3Mn(CN)_6$	$^3T_{1g}$	2	2·83	3·50
	Ru^{4+}	K_2RuCl_6	$^3T_{1g}$	2	2·83	2·96
	Os^{4+}	K_2OsCl_6	$^3T_{1g}$	2	2·83	1·50
5	Mn^{2+}	$K_2Mn(SO_4)_2.6H_2O$	$^6A_{1g}$	5	5.92	5.92
		$K_4Mn(CN)_6.3H_2O$	$^2T_{2g}$	1	1·73	2·18
	Fe^{3+}	$KFe(SO_4)_2.12H_2O$	$^6A_{1g}$	5	5·92	5·89
		$K_3Fe(CN)_6$	$^2T_{2g}$	1	1·73	2·25
	Ru^{3+}	$Ru(NH_3)_6.Cl_3$	$^2T_{2g}$	1	1·73	2·13
	Os^{3+}	$Os(NH_3)_6.Cl_3$	$^2T_{2g}$	1	1·73	1·81
6	Fe^{2+}	$K_4Fe(CN)_6$	$^1A_{1g}$	0	0·0	0·35
		$(NH_4)_2Fe(SO_4)_2.6H_2O$	$^5T_{2g}$	4	4·90	5·47
	Co^{3+}	$Co(NH_3)_6.Cl_3$	$^1A_{1g}$	0	0·0	0·46
	Rh^{3+}	$Rh(NH_3)_6.Cl_3$	$^1A_{1g}$	0	0·0	0·35
	Ir^{3+}	$K_3IrCl_6.3H_2O$	$^1A_{1g}$	0	0·0	0·0
	Pt^{4+}	K_2PtCl_6	$^1A_{1g}$	0	0·0	0·0
7	Co^{2+}	$K_2BaCo(NO_2)_6$	^2E_g	1	1·73	1·81
		$(NH_4)_2Co(SO_4)_2.6H_2O$	$^4T_{1g}$	3	3·87	5·10
8	Ni^{2+}	$(NH_4)_2Ni(SO_4)_2.6H_2O$	$^3A_{2g}$	2	2·83	3·23
		$[(C_2H_5)_4N]_2NiCl_4$	$^3T_1$	2	2·83	3·89
9	Cu^{2+}	$K_2Cu(SO_4)_2.6H_2O$	^2E_g	1	1·73	1·91

the lowest free-ion state only out to a certain critical Δ_o value, beyond which a state of lower multiplicity, originating in a higher free-ion state, drops below it and hence becomes the ground state.† So in these cases the multiplicity, the number of unpaired spins and the effective magnetic moment will depend on Δ_o and therefore on the nature of the ligands. For strong interactions between the ion and its environment (Δ_o large) there will be fewer unpaired spins than for weak interactions (Δ_o small). Similar predictions can be made for ions in tetrahedral environments.

In Table 12-9.1 calculated (spin-only formula) and experimental effective magnetic moments are listed for a number of ions, they are in accord with the previous discussion.

† Studies of the cross-over point have been made by E. König; see, for example, *Theor. Chim. Acta* *26*, 311 (1972).

12-10 Ligand field theory

In the introduction to this chapter we stated that the approximations made in applying crystal field theory to most transition-metal complexes and compounds are extreme. The question which arises is: can we modify the theory so as to take account of its known defects? The answer is a qualified 'yes'. Essentially, what we must do is to drop the assumption that the metal ion's partially-filled shell consists solely of its d- or f-orbitals and allow for the overlap between the orbitals of the ion and those of the ligands (MO calculations show that there invariably is such an overlap). Doing this has two consequences. We can no longer consider the crystal field parameters Δ_o or Δ_t (and, if Tanabe–Sugano diagrams are used, B) within the framework of simple electrostatics and they lose their initial significance and become quite arbitrary parameters to be adjusted in any way necessary.† In other words, the corrections due to the approximations are assimilated in these parameters. Further, in the construction of the correlation diagrams, the separations of the energies of the free-ion states become adjustable and are not taken as the observed values given by atomic spectroscopy.

With the exception of these changes, the practical development of ligand field theory and crystal field theory are the same.

Appendix

A.12-1. Spectroscopic states and term symbols for many-electron atoms or ions

So far in this book we have only discussed non-relativistic Hamiltonian operators but when atomic or molecular spectra are considered it is necessary to account for relativistic effects. These lead to additional terms in the Hamiltonian operator which can be related to the following phenomena:

(1) The coupling of spin and orbital angular momenta among the electrons.
(2) The coupling of spin angular momenta among the electrons.
(3) Interactions among the orbital magnetic moments of the electrons.
(4) The coupling of spin angular momenta among the nuclei.
(5) The coupling of spin angular momenta of the electrons with spin angular momenta of the nuclei.
(6) The coupling of nuclear-spin angular momenta with electron-orbital angular momenta.
(7) Nuclear electric-quadrupole-moment interactions.

As well as these additional terms there will also be changes to the Hamiltonian operator due to the relativistic change of electron mass with velocity. In ordinary optical spectroscopy the first two phenomena, (1) and (2), are the most important, leading to changes to the non-relativistic energy levels which are observable (effects (4) and (5) are important in n.m.r. and e.s.r. spectroscopy).

† For example, the parameters can be adjusted so as to reproduce the experimental d–d spectral transitions. This, in fact, was what was done in §12-8.

For these reasons the electronic energies, and therefore the electronic states, of a many-electron atom or ion will depend upon the electronic spins and how the spin angular momenta are coupled with the orbital angular momenta. The coupling scheme which is most appropriate for our purposes is known as *L–S* (or Russell–Saunders) coupling. It first couples the electronic spin angular momenta together, then the electronic orbital angular momenta together and finally couples these total momenta together. Like all coupling schemes, it is an approximation. Associated with the spin and orbital angular momenta of a single electron are quantum numbers l and s, respectively, and for a n-electron system there are equivalent quantum numbers L and S. The quantum number L defines the total orbital angular momentum and its allowed values are

$$L = l_1+l_2+ \dots l_n, l_1+l_2+ \dots l_n-1, \dots, -(l_1+l_2+ \dots l_n)$$

where l_i is the orbital quantum number of the ith electron. Capital letter symbols are assigned to states having different L values as follows:

$L =$	0	1	2	3	4	5	6
symbol =	S	P	D	F	G	H	I.

The quantum number S defines the total electronic spin angular momentum and its allowed values are

$$S = n/2, (n/2)-1, \dots, 1/2 \qquad \text{(if } n \text{ is odd)}$$
$$S = n/2, (n/2)-1, \dots, 0 \qquad \text{(if } n \text{ is even).}$$

In *L–S* coupling, the total electronic angular momentum (spin and orbital) is defined by the quantum number J whose allowed values are

$$L+S, L+S-1, \dots, |L-S|.$$

TABLE A.12-1.1

Spectroscopic terms arising from equivalent electronic configurations in L–S coupling

configuration	*L–S terms*†
s^2	^1S
p or p^5	^2P
p^2 or p^4	^1S, ^1D, ^3P
p^3	^2P, ^2D, ^4S
d or d^9	^2D
d^2 or d^8	^1S, ^1D, ^1G, ^3P, ^3F
d^3 or d^7	$^2D(2)$, ^2P, ^2F, ^2G, ^2H, ^4P, ^4F
d^4 or d^6	$^1S(2)$, $^1D(2)$, ^1F, $^1G(2)$, ^1I, $^3P(2)$, ^3D, $^3F(2)$, ^3G, ^3H, ^5D
d^5	^2S, ^2P, $^2D(3)$, $^2F(2)$, $^2G(2)$, ^2H, ^2I, ^4P, ^4D, ^4F, ^4G, ^6S

† The number in parentheses is the number of distinct terms with the same L and S quantum numbers. For each distinct term there will be different states corresponding to the different possible J values.

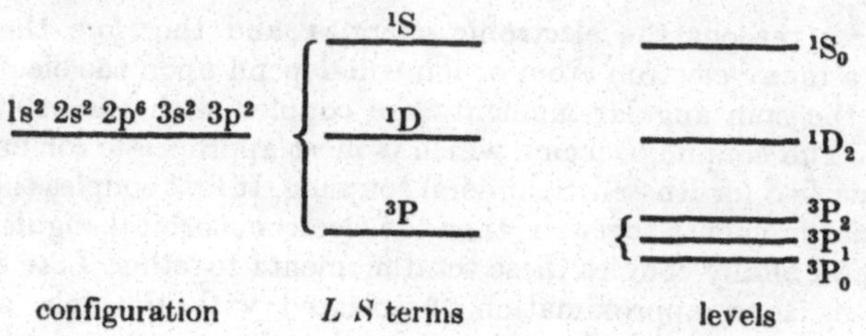

FIG. A.12-1.1. Levels for the silicon atom.

J therefore can have $2S+1$ values if $L \geqslant S$ and $2L+1$ values if $L < S$. The number $2S+1$ is called the multiplicity.

As the electronic energy of an atom or ion will depend on the quantum numbers L, S, and J, we designate the various energy states which may arise from a given electronic configuration by what is known as a spectroscopic *term symbol:*

$$^{2S+1}T_J$$

where $T =$ S, P, D,... as $L = 0, 1, 2, \ldots$. When all the electrons have different principal quantum numbers there are no restrictions on the combinations of L and S, but, if this is not so, some combinations will be excluded by virtue of the Pauli Principle. In Table A.12-1.1 the spectroscopic states of common configurations of electrons with the same principal quantum number are shown. The reader should note that we are only concerned with that part of an electronic configuration which is outside of any closed shells (noble-gas structures). The latter are spherically symmetrical and do not play any role in the effects which are of interest to us in this chapter. In Fig. A.12-1.1, as an example of the above notation, the hierarchy of levels for the ground state configuration of silicon is shown.

PROBLEMS

12.1. Determine the qualitative form of the molecular orbitals for the square-planar complex $Ni(CN)_4^{2-}$. (Assume that each CN ligand provides one σ-type and two π-type orbitals to the system.)

12.2. Determine the qualitative form of the molecular orbitals for the tetrahedral molecule MnO_4^-. [Assume that each oxygen atom provides just three p-orbitals (set these up so that one points towards the Mn and the other two are perpendicular to each other and to the Mn—O axis) and that the Mn atom provides 4s and 3d orbitals.] You will be on the right track if you find that

$$\Gamma^{\sigma} = \Gamma^{A_1} \oplus \Gamma^{T_2}$$
$$\Gamma^{\pi} = \Gamma^{E} \oplus \Gamma^{T_1} \oplus \Gamma^{T_2}.$$

12.3. Determine the qualitative form of the molecular orbitals for the eclipsed conformation of ferrocene.

12.4. For an octahedral environment the d-orbitals are split into two sets (d_{e_g} and $d_{t_{2g}}$); how would they be split for a square-planar environment?

12.5. Set up a qualitative correlation diagram for the d^3 configuration in an octahedral environment.

Appendix I: Character Tables

The x, y, z axes referred to in these tables are a set of three mutually perpendicular axes chosen as follows:

(1) $\mathscr{C}_s$: the z axis is perpendicular to the reflection plane.

(2) Groups with one main axis of symmetry: the z axis points along the main axis of symmetry and, where applicable, the x axis lies in one of the σ_v planes or coincides with one of the C_2 axes. For $\mathscr{D}_2$ and $\mathscr{D}_{2h}$ the x, y, z axes coincide with the three equivalent two-fold axes.

(3) $\mathscr{T}_d$ and $\mathscr{O}_h$: see Figs. 3-6.2 and 3-6.3 respectively.

I: Groups $\mathscr{C}_s$, $\mathscr{C}_i$, and $\mathscr{C}_n$ ($n = 2, 3, 4, 5, 6$)

$\mathscr{C}_s$	E	σ_h		
A'	1	1	$x; y; R_z$	$x^2; y^2; z^2; xy$
A''	1	−1	$z; R_x; R_y$	$xz; yz$

$\mathscr{C}_i$	E	i		
A_g	1	1	$R_x; R_y; R_z$	$x^2; y^2; z^2; xy; xz; yz$
A_u	1	−1	$x; y; z$	

$\mathscr{C}_2$	E	C_2		
A	1	1	$z; R_z$	$x^2; y^2; z^2; xy$
B	1	−1	$x; y; R_x; R_y$	$xz; yz$

$\varepsilon = \exp(2\pi i/3)$

$\mathscr{C}_3$	E	C_3	C_3^2		
A	1	1	1	$z; R_z$	$x^2+y^2; z^2$
E	1	ε	ε^*	$(x, y); (R_x, R_y)$	$(x^2-y^2, xy); (xz, yz)$
	1	ε^*	ε		

$\mathscr{C}_4$	E	C_4	C_2	C_4^3		
A	1	1	1	1	$z; R_z$	$x^2+y^2; z^2$
B	1	−1	1	−1		$x^2-y^2; xy$
E	1	i	−1	−i	$(x, y); (R_x, R_y)$	(xz, yz)
	1	−i	−1	i		

$\varepsilon = \exp(2\pi i/5)$

$\mathscr{C}_5$	E	C_5	C_5^2	C_5^3	C_5^4		
A	1	1	1	1	1	$z; R_z$	$x^2+y^2; z^2$
E_1	1	ε	ε^2	ε^{2*}	ε^*	$(x, y); (R_x, R_y)$	(xz, yz)
	1	ε^*	ε^{2*}	ε^2	ε		
E_2	1	ε^2	ε^*	ε	ε^{2*}		(x^2-y^2, xy)
	1	ε^{2*}	ε	ε^*	ε^2		

$\varepsilon = \exp(2\pi i/6)$

$\mathscr{C}_6$	E	C_6	C_3	C_2	C_3^2	C_6^5		
A	1	1	1	1	1	1	$z; R_z$	$x^2+y^2; z^2$
B	1	−1	1	−1	1	−1		
E_1	1	ε	$-\varepsilon^*$	−1	$-\varepsilon$	ε^*	$(x, y); (R_x, R_y)$	(xz, yz)
	1	ε^*	$-\varepsilon$	−1	$-\varepsilon^*$	ε		
E_2	1	$-\varepsilon^*$	$-\varepsilon$	1	$-\varepsilon^*$	$-\varepsilon$		(x^2-y^2, xy)
	1	$-\varepsilon$	$-\varepsilon^*$	1	$-\varepsilon$	$-\varepsilon^*$		

II: Groups $\mathscr{D}_n$ $(n = 2, 3, 4, 5, 6)$

$\mathscr{D}_2 = \mathscr{V}$	E	$C_2(z)$	$C_2(y)$	$C_2(x)$		
A	1	1	1	1		$x^2; y^2; z^2$
B_1	1	1	-1	-1	$z; R_z$	xy
B_2	1	-1	1	-1	$y; R_y$	xz
B_3	1	-1	-1	1	$x; R_x$	yz

$\mathscr{D}_3$	E	$2C_3$	$3C_2'$		
A_1	1	1	1		$x^2+y^2; z^2$
A_2	1	1	-1	$z; R_z$	
E	2	-1	0	$(x, y); (R_x, R_y)$	$(x^2-y^2, xy); (xz, yz)$

$\mathscr{D}_4$	E	$2C_4$	C_2	$2C_2'$	$2C_2''$		
A_1	1	1	1	1	1		$x^2+y^2; z^2$
A_2	1	1	1	-1	-1	$z; R_z$	
B_1	1	-1	1	1	-1		x^2-y^2
B_2	1	-1	1	-1	1		xy
E	2	0	-2	0	0	$(x, y); (R_x, R_y)$	(xz, yz)

$\mathscr{D}_5$	E	$2C_5$	$2C_5^2$	$5C_2'$		$\alpha = 72°$
A_1	1	1	1	1		$x^2+y^2; z^2$
A_2	1	1	1	-1	$z; R_z$	
E_1	2	$2\cos\alpha$	$2\cos 2\alpha$	0	$(x, y); (R_x, R_y)$	(xz, yz)
E_2	2	$2\cos 2\alpha$	$2\cos\alpha$	0		(x^2-y^2, xy)

$\mathscr{D}_6$	E	$2C_6$	$2C_3$	C_2	$3C_2'$	$3C_2''$		
A_1	1	1	1	1	1	1		x^2+y^2, z^2
A_2	1	1	1	1	-1	-1	$z; R_z$	
B_1	1	-1	1	-1	1	-1		
B_2	1	-1	1	-1	-1	1		
E_1	2	1	-1	-2	0	0	$(x, y); (R_x, R_y)$	(xz, yz)
E_2	2	-1	-1	2	0	0		(x^2-y^2, xy)

III: Groups $\mathscr{C}_{nv}$ ($n = 2, 3, 4, 5, 6$)

$\mathscr{C}_{2v}$	E	C_2	$\sigma_v(xz)$	$\sigma_v(yz)$		
A_1	1	1	1	1	z	$x^2; y^2; z^2$
A_2	1	1	−1	−1	R_z	xy
B_1	1	−1	1	−1	$x; R_y$	xz
B_2	1	−1	−1	1	$y; R_x$	yz

$\mathscr{C}_{3v}$	E	$2C_3$	$3\sigma_v$		
A_1	1	1	1	z	$x^2+y^2; z^2$
A_2	1	1	−1	R_z	
E	2	−1	0	$(x, y); (R_x, R_y)$	$(x^2-y^2, xy); (xz, yz)$

$\mathscr{C}_{4v}$	E	$2C_4$	C_2	$2\sigma_v$	$2\sigma_d$		
A_1	1	1	1	1	1	z	$x^2+y^2; z^2$
A_2	1	1	1	−1	−1	R_z	
B_1	1	−1	1	1	−1		x^2-y^2
B_2	1	−1	1	−1	1		xy
E	2	0	−2	0	0	$(x, y); (R_x, R_y)$	(xz, yz)

$\mathscr{C}_{5v}$	E	$2C_5$	$2C_5^2$	$5\sigma_v$		$\alpha = 72°$
A_1	1	1	1	1	z	$x^2+y^2; z^2$
A_2	1	1	1	−1	R_z	
E_1	2	$2\cos\alpha$	$2\cos 2\alpha$	0	$(x, y); (R_x, R_y)$	(xz, yz)
E_2	2	$2\cos 2\alpha$	$2\cos\alpha$	0		(x^2-y^2, xy)

$\mathscr{C}_{6v}$	E	$2C_6$	$2C_3$	C_2	$3\sigma_v$	$3\sigma_d$		
A_1	1	1	1	1	1	1	z	$x^2+y^2; z^2$
A_2	1	1	1	1	−1	−1	R_z	
B_1	1	−1	1	−1	1	−1		
B_2	1	−1	1	−1	−1	1		
E_1	2	1	−1	−2	0	0	$(x, y); (R_x, R_y)$	(xz, yz)
E_2	2	−1	−1	2	0	0		(x^2-y^2, xy)

IV: Groups $\mathscr{C}_{nh}$ ($n = 2, 3, 4, 5, 6$)

$\mathscr{C}_{2h}$	E	C_2	i	σ_h		
A_g	1	1	1	1	R_z	$x^2; y^2; z^2; xy$
B_g	1	−1	1	−1	$R_x; R_y$	$xz; yz$
A_u	1	1	−1	−1	z	
B_u	1	−1	−1	1	$x; y$	

$\mathscr{C}_{3h}$	E	C_3	C_3^2	σ_h	S_3	S_3^5		$\varepsilon = \exp(2\pi i/3)$
A'	1	1	1	1	1	1	R_z	$x^2+y^2; z^2$
E'	1	ε	ε^*	1	ε	ε^*	(x, y)	(x^2-y^2, xy)
	1	ε^*	ε	1	ε^*	ε		
A''	1	1	1	−1	−1	−1	z	
E''	1	ε	ε^*	−1	$-\varepsilon$	$-\varepsilon^*$	(R_x, R_y)	(xz, yz)
	1	ε^*	ε	−1	$-\varepsilon^*$	$-\varepsilon$		

$\mathscr{C}_{4h}$	E	C_4	C_2	C_4^3	i	S_4^3	σ_h	S_4		
A_g	1	1	1	1	1	1	1	1	R_z	$x^2+y^2; z^2$
B_g	1	−1	1	−1	1	−1	1	−1		$x^2-y^2; xy$
E_g	1	i	−1	−i	1	i	−1	−i	(R_x, R_y)	(xz, xy)
	1	−i	−1	i	1	−i	−1	i		
A_u	1	1	1	1	−1	−1	−1	−1	z	
B_u	1	−1	1	−1	−1	1	−1	1		
E_u	1	i	−1	−i	−1	−i	1	i	(x, y)	
	1	−i	−1	i	−1	i	1	−i		

$\mathscr{C}_{5h}$	E	C_5	C_5^2	C_5^3	C_5^4	σ_h	S_5	S_5^7	S_5^3	S_5^9		$\varepsilon = \exp(2\pi i/5)$
A'	1	1	1	1	1	1	1	1	1	1	R_z	$x^2+y^2; z^2$
E_1'	1	ε	ε^2	ε^{2*}	ε^*	1	ε	ε^2	ε^{2*}	ε^*	(x, y)	
	1	ε^*	ε^{2*}	ε^2	ε	1	ε^*	ε^{2*}	ε^2	ε		
E_2'	1	ε^2	ε^*	ε	ε^{2*}	1	ε^2	ε^*	ε	ε^{2*}		(x^2-y^2, xy)
	1	ε^{2*}	ε	ε^*	ε^2	1	ε^{2*}	ε	ε^*	ε^2		
A''	1	1	1	1	1	−1	−1	−1	−1	−1	z	
E_1''	1	ε	ε^2	ε^{2*}	ε^*	−1	$-\varepsilon$	$-\varepsilon^2$	$-\varepsilon^{2*}$	$-\varepsilon^*$	(R_x, R_y)	(xz, yz)
	1	ε^*	ε^{2*}	ε^2	ε	−1	$-\varepsilon^*$	$-\varepsilon^{2*}$	$-\varepsilon^2$	$-\varepsilon$		
E_2''	1	ε^2	ε^*	ε	ε^{2*}	−1	$-\varepsilon^2$	$-\varepsilon^*$	$-\varepsilon$	$-\varepsilon^{2*}$		
	1	ε^{2*}	ε	ε^*	ε^2	−1	$-\varepsilon^{2*}$	$-\varepsilon$	$-\varepsilon^*$	$-\varepsilon^2$		

$\mathscr{C}_{6h}$	E	C_6	C_3	C_2	C_3^2	C_6^5	i	S_3^5	S_6^5	σ_h	S_6	S_3		$\varepsilon = \exp(2\pi i/6)$
A_g	1	1	1	1	1	1	1	1	1	1	1	1	R_z	$x^2+y^2; z^2$
B_g	1	−1	1	−1	1	−1	1	−1	1	−1	1	−1		
E_{1g}	1	ε	$-\varepsilon^*$	−1	$-\varepsilon$	ε^*	1	ε	$-\varepsilon^*$	−1	$-\varepsilon$	ε^*	(R_x, R_y)	(xz, yz)
	1	ε^*	$-\varepsilon$	−1	$-\varepsilon^*$	ε	1	ε^*	$-\varepsilon$	−1	$-\varepsilon^*$	ε		
E_{2g}	1	$-\varepsilon^*$	$-\varepsilon$	1	$-\varepsilon^*$	$-\varepsilon$	1	$-\varepsilon^*$	$-\varepsilon$	1	$-\varepsilon^*$	$-\varepsilon$		(x^2-y^2, xy)
	1	$-\varepsilon$	$-\varepsilon^*$	1	$-\varepsilon$	$-\varepsilon^*$	1	$-\varepsilon$	$-\varepsilon^*$	1	$-\varepsilon$	$-\varepsilon^*$		
A_u	1	1	1	1	1	1	−1	−1	−1	−1	−1	−1	z	
B_u	1	−1	1	−1	1	−1	−1	1	−1	1	−1	1		
E_{1u}	1	ε	$-\varepsilon^*$	−1	$-\varepsilon$	ε^*	−1	$-\varepsilon$	ε^*	1	ε	$-\varepsilon^*$	(x, y)	
	1	ε^*	$-\varepsilon$	−1	$-\varepsilon^*$	ε	−1	$-\varepsilon^*$	ε	1	ε^*	$-\varepsilon$		
E_{2u}	1	$-\varepsilon^*$	$-\varepsilon$	1	$-\varepsilon^*$	$-\varepsilon$	−1	ε^*	ε	−1	ε^*	ε		
	1	$-\varepsilon$	$-\varepsilon^*$	1	$-\varepsilon$	$-\varepsilon^*$	−1	ε	ε^*	−1	ε	ε^*		

V: Groups $\mathscr{D}_{nh}$ $(n = 2, 3, 4, 5, 6)$

$\mathscr{D}_{2h} = \mathscr{V}_h$	E	$C_2(z)$	$C_2(y)$	$C_2(x)$	i	$\sigma(xy)$	$\sigma(xz)$	$\sigma(yz)$		
A_g	1	1	1	1	1	1	1	1		$x^2; y^2; z^2$
B_{1g}	1	1	−1	−1	1	1	−1	−1	R_z	xy
B_{2g}	1	−1	1	−1	1	−1	1	−1	R_y	xz
B_{3g}	1	−1	−1	1	1	−1	−1	1	R_x	yz
A_u	1	1	1	1	−1	−1	−1	−1		
B_{1u}	1	1	−1	−1	−1	−1	1	1	z	
B_{2u}	1	−1	1	−1	−1	1	−1	1	y	
B_{3u}	1	−1	−1	1	−1	1	1	−1	x	

$\mathscr{D}_{3h}$	E	$2C_3$	$3C_2$	σ_h	$2S_3$	$3\sigma_v$		
A_1'	1	1	1	1	1	1		$x^2+y^2; z^2$
A_2'	1	1	−1	1	1	−1	R_z	
E'	2	−1	0	2	−1	0	(x, y)	(x^2-y^2, xy)
A_1''	1	1	1	−1	−1	−1		
A_2''	1	1	−1	−1	−1	1	z	
E''	2	−1	0	−2	1	0	(R_x, R_y)	(xz, yz)

$\mathscr{D}_{4h}$	E	$2C_4$	C_2	$2C_2'$	$2C_2''$	i	$2S_4$	σ_h	$2\sigma_v$	$2\sigma_d$		
A_{1g}	1	1	1	1	1	1	1	1	1	1		$x^2+y^2; z^2$
A_{2g}	1	1	1	−1	−1	1	1	1	−1	−1	R_z	
B_{1g}	1	−1	1	1	−1	1	−1	1	1	−1		x^2-y^2
B_{2g}	1	−1	1	−1	1	1	−1	1	−1	1		xy
E_g	2	0	−2	0	0	2	0	−2	0	0	(R_x, R_y)	(xz, yz)
A_{1u}	1	1	1	1	1	−1	−1	−1	−1	−1		
A_{2u}	1	1	1	−1	−1	−1	−1	−1	1	1	z	
B_{1u}	1	−1	1	1	−1	−1	1	−1	−1	1		
B_{2u}	1	−1	1	−1	1	−1	1	−1	1	−1		
E_u	2	0	−2	0	0	−2	0	2	0	0	(x, y)	

$\mathscr{D}_{5h}$	E	$2C_5$	$2C_5^2$	$5C_2$	σ_h	$2S_5$	$2S_5^3$	$5\sigma_v$	$\alpha = 72°$	
A_1'	1	1	1	1	1	1	1	1		$x^2+y^2; z^2$
A_2'	1	1	1	−1	1	1	1	−1	R_z	
E_1'	2	$2\cos\alpha$	$2\cos 2\alpha$	0	2	$2\cos\alpha$	$2\cos 2\alpha$	0	(x, y)	
E_2'	2	$2\cos 2\alpha$	$2\cos\alpha$	0	2	$2\cos 2\alpha$	$2\cos\alpha$	0		(x^2-y^2, xy)
A_1''	1	1	1	1	−1	−1	−1	−1		
A_2''	1	1	1	−1	−1	−1	−1	1	z	
E_1''	2	$2\cos\alpha$	$2\cos 2\alpha$	0	−2	$-2\cos\alpha$	$-2\cos 2\alpha$	0	(R_x, R_y)	(xz, yz)
E_2''	2	$2\cos 2\alpha$	$2\cos\alpha$	0	−2	$-2\cos 2\alpha$	$-2\cos\alpha$	0		

$\mathscr{D}_{6h}$	E	$2C_6$	$2C_3$	C_2	$3C_2'$	$3C_2''$	i	$2S_3$	$2S_6$	σ_h	$3\sigma_d$	$3\sigma_v$		
A_{1g}	1	1	1	1	1	1	1	1	1	1	1	1		$x^2+y^2; z^2$
A_{2g}	1	1	1	1	−1	−1	1	1	1	1	−1	−1	R_z	
B_{1g}	1	−1	1	−1	1	−1	1	−1	1	−1	1	−1		
B_{2g}	1	−1	1	−1	−1	1	1	−1	1	−1	−1	1		
E_{1g}	2	1	−1	−2	0	0	2	1	−1	−2	0	0	(R_x, R_y)	(xz, yz)
E_{2g}	2	−1	−1	2	0	0	2	−1	−1	2	0	0		(x^2-y^2, xy)
A_{1u}	1	1	1	1	1	1	−1	−1	−1	−1	−1	−1		
A_{2u}	1	1	1	1	−1	−1	−1	−1	−1	−1	1	1	z	
B_{1u}	1	−1	1	−1	1	−1	−1	1	−1	1	−1	1		
B_{2u}	1	−1	1	−1	−1	1	−1	1	−1	1	1	−1		
E_{1u}	2	1	−1	−2	0	0	−2	−1	1	2	0	0	(x, y)	
E_{2u}	2	−1	−1	2	0	0	−2	1	1	−2	0	0		

VI: Groups $\mathscr{D}_{nd}$ ($n = 2, 3, 4, 5, 6$)

$\mathscr{D}_{2d} = \mathscr{V}_d$	E	$2S_4$	C_2	$2C_2'$	$2\sigma_d$		
A_1	1	1	1	1	1		$x^2+y^2; z^2$
A_2	1	1	1	−1	−1	R_z	
B_1	1	−1	1	1	−1		x^2-y^2
B_2	1	−1	1	−1	1	z	xy
E	2	0	−2	0	0	(x, y); (R_x, R_y)	(xz, yz)

$\mathscr{D}_{3d}$	E	$2C_3$	$3C_2'$	i	$2S_6$	$3\sigma_d$		
A_{1g}	1	1	1	1	1	1		$x^2+y^2; z^2$
A_{2g}	1	1	−1	1	1	−1	R_z	
E_g	2	−1	0	2	−1	0	(R_x, R_y)	(x^2-y^2, xy); (xz, yz)
A_{1u}	1	1	1	−1	−1	−1		
A_{2u}	1	1	−1	−1	−1	1	z	
E_u	2	−1	0	−2	1	0	(x, y)	

$\mathscr{D}_{4d}$	E	$2S_8$	$2C_4$	$2S_8^3$	C_2	$4C_2'$	$4\sigma_d$		
A_1	1	1	1	1	1	1	1		$x^2+y^2; z^2$
A_2	1	1	1	1	1	−1	−1	R_z	
B_1	1	−1	1	−1	1	1	−1		
B_2	1	−1	1	−1	1	−1	1	z	
E_1	2	$\sqrt{2}$	0	$-\sqrt{2}$	−2	0	0	(x, y)	
E_2	2	0	−2	0	2	0	0		(x^2-y^2, xy)
E_3	2	$-\sqrt{2}$	0	$\sqrt{2}$	−2	0	0	(R_x, R_y)	(xz, yz)

$\mathscr{D}_{5d}$	E	$2C_5$	$2C_5^2$	$5C_2'$	i	$2S_{10}^3$	$2S_{10}$	$5\sigma_d$		$\alpha = 72°$
A_{1g}	1	1	1	1	1	1	1	1		$x^2+y^2; z^2$
A_{2g}	1	1	1	−1	1	1	1	−1	R_z	
E_{1g}	2	$2\cos\alpha$	$2\cos 2\alpha$	0	2	$2\cos\alpha$	$2\cos 2\alpha$	0	(R_x, R_y)	(xz, yz)
E_{2g}	2	$2\cos 2\alpha$	$2\cos\alpha$	0	2	$2\cos 2\alpha$	$2\cos\alpha$	0		(x^2-y^2, xy)
A_{1u}	1	1	1	1	−1	−1	−1	−1		
A_{2u}	1	1	1	−1	−1	−1	−1	1	z	
E_{1u}	2	$2\cos\alpha$	$2\cos 2\alpha$	0	−2	$-2\cos\alpha$	$-2\cos 2\alpha$	0	(x, y)	
E_{2u}	2	$2\cos 2\alpha$	$2\cos\alpha$	0	−2	$-2\cos 2\alpha$	$-2\cos\alpha$	0		

$\mathscr{D}_{6d}$	E	$2S_{12}$	$2C_6$	$2S_4$	$2C_3$	$2S_{12}^5$	C_2	$6C_2'$	$6\sigma_d$		
A_1	1	1	1	1	1	1	1	1	1		$x^2+y^2; z^2$
A_2	1	1	1	1	1	1	1	−1	−1	R_z	
B_1	1	−1	1	−1	1	−1	1	1	−1		
B_2	1	−1	1	−1	1	−1	1	−1	1	z	
E_1	2	$\sqrt{3}$	1	0	−1	$-\sqrt{3}$	−2	0	0	(x, y)	
E_2	2	1	−1	−2	−1	1	2	0	0		(x^2-y^2, xy)
E_3	2	0	−2	0	2	0	−2	0	0		
E_4	2	−1	−1	2	−1	−1	2	0	0		
E_5	2	$-\sqrt{3}$	1	0	−1	$\sqrt{3}$	−2	0	0	(R_x, R_y)	(xz, yz)

VII: Groups $\mathscr{S}_n$ ($n = 4, 6, 8$)

$\mathscr{S}_4$	E	S_4	C_2	S_4^3		
A	1	1	1	1	R_z	x^2+y^2; z^2
B	1	−1	1	−1	z	x^2-y^2; xy
E	1	i	−1	−i	(x, y); (R_x, R_y)	(xz, yz)
	1	−i	−1	i		

$\mathscr{S}_6$	E	C_3	C_3^2	i	S_6^5	S_6		$\varepsilon = \exp(2\pi i/3)$
A_g	1	1	1	1	1	1	R_z	x^2+y^2; z^2
E_g	1	ε	ε^*	1	ε	ε^*	(R_x, R_y)	(x^2-y^2, xy); (xz, yz)
	1	ε^*	ε	1	ε^*	ε		
A_u	1	1	1	−1	−1	−1	z	
E_u	1	ε	ε^*	−1	$-\varepsilon$	$-\varepsilon^*$	(x, y)	
	1	ε^*	ε	−1	$-\varepsilon^*$	$-\varepsilon$		

$\mathscr{S}_8$	E	S_8	C_4	S_8^3	C_2	S_8^5	C_4^3	S_8^7	$\varepsilon = \exp(2\pi i/8)$	
A	1	1	1	1	1	1	1	1	R_z	x^2+y^2; z^2
B	1	−1	1	−1	1	−1	1	−1	z	
E_1	1	ε	i	$-\varepsilon^*$	−1	$-\varepsilon$	−i	ε^*	(x, y);	
	1	ε^*	−i	$-\varepsilon$	−1	$-\varepsilon^*$	i	ε	(R_x, R_y)	
E_2	1	i	−1	−i	1	i	−1	−i		(x^2-y^2, xy)
	1	−i	−1	i	1	−i	−1	i		
E_3	1	$-\varepsilon^*$	−i	ε	−1	ε^*	i	$-\varepsilon$		(xz, yz)
	1	$-\varepsilon$	i	ε^*	−1	ε	−i	$-\varepsilon^*$		

VIII: Group $\mathcal{T}_d$, $\mathcal{O}$ *and* $\mathcal{O}_h$

$\mathcal{T}_d$	E	$8C_3$	$3C_2$	$6S_4$	$6\sigma_d$		
A_1	1	1	1	1	1		$x^2+y^2+z^2$
A_2	1	1	1	−1	−1		
E	2	−1	2	0	0		$(2z^2-x^2-y^2, x^2-y^2)$
T_1, F_1	3	0	−1	1	−1	(R_x, R_y, R_z)	
T_2, F_2	3	0	−1	−1	1	(x, y, z)	(xy, xz, yz)

$\mathcal{O}$	E	$8C_3$	$3C_2$	$6C_4$	$6C_2'$		
A_1	1	1	1	1	1		$x^2+y^2+z^2$
A_2	1	1	1	−1	−1		
E	2	−1	2	0	0		$(2z^2-x^2-y^2, x^2-y^2)$
T_1, F_1	3	0	−1	1	−1	(R_x, R_y, R_z); (x, y, z)	
T_2, F_2	3	0	−1	−1	1		(xy, xz, yz)

$\mathcal{O}_h$	E	$8C_3$	$3C_2$	$6C_4$	$6C_2'$	i	$8S_6$	$3\sigma_h$	$6S_4$	$6\sigma_d$		
A_{1g}	1	1	1	1	1	1	1	1	1	1		$x^2+y^2+z^2$
A_{2g}	1	1	1	−1	−1	1	1	1	−1	−1		
E_g	2	−1	2	0	0	2	−1	2	0	0		$(2z^2-x^2-y^2, x^2-y^2)$
T_{1g}, F_{1g}	3	0	−1	1	−1	3	0	−1	1	−1	(R_x, R_y, R_z)	
T_{2g}, F_{2g}	3	0	−1	−1	1	3	0	−1	−1	1		(xy, xz, yz)
A_{1u}	1	1	1	1	1	−1	−1	−1	−1	−1		
A_{2u}	1	1	1	−1	−1	−1	−1	−1	1	1		
E_u	2	−1	2	0	0	−2	1	−2	0	0		
T_{1u}, F_{1u}	3	0	−1	1	−1	−3	0	1	−1	1	(x, y, z)	
T_{2u}, F_{2u}	3	0	−1	−1	1	−3	0	1	1	−1		

IX: Groups $\mathscr{C}_{\infty v}$ and $\mathscr{D}_{\infty h}$

$\mathscr{C}_{\infty v}$	E	$2C(\phi)$	...	$\infty\sigma_v$		
$A_1 = \Sigma^+$	1	1	...	1	z	x^2+y^2; z^2
$A_2 = \Sigma^-$	1	1	...	-1	R_z	
$E_1 = \Pi$	2	$2\cos\phi$	...	0	(x, y); (R_x, R_y)	(xz, yz)
$E_2 = \Delta$	2	$2\cos 2\phi$	...	0		(x^2-y^2, xy)
$E_3 = \Phi$	2	$2\cos 3\phi$	...	0		
...	...	...	...	...		

$\mathscr{D}_{\infty h}$	E	$2C(\phi)$	...	$\infty\sigma_v$	i	$2S(\phi)$	..	$\infty C_2'$		
Σ_g^+	1	1	..	1	1	1	...	1		x^2+y^2; z^2
Σ_g^-	1	1	...	-1	1	1	...	-1	R_z	
Π_g	2	$2\cos\phi$	...	0	2	$-2\cos\phi$	...	0	(R_x, R_y)	(xz, yz)
Δ_g	2	$2\cos 2\phi$	..	0	2	$2\cos 2\phi$	...	0		(x^2-y^2, xy)
...	...	...	...	...	..	...	...	.		
Σ_u^+	1	1	...	1	-1	-1	...	-1	z	
Σ_u^-	1	1	.	-1	-1	-1	...	1		
Π_u	2	$2\cos\phi$	..	0	-2	$2\cos\phi$	...	0	(x, y)	
Δ_u	2	$2\cos 2\phi$	...	0	-2	$-2\cos 2\phi$	...	0		
...	...	...	...	...	...	...	...	...		

Bibliography

Bethe: *Splitting of terms in crystals*. Annalen der Physik **3**, 133 (1929).
Cotton: *Chemical applications of group theory* (John Wiley).
Ferraro and Ziomek: *Introductory group theory* (Plenum Press).
Hall: *Group theory and symmetry in chemistry* (McGraw-Hill).
Hamermesh: *Group theory* (Addison–Wesley).
Heine: *Group theory in quantum mechanics* (Pergamon Press).
Hochstrasser: *Molecular aspects of symmetry* (Benjamin).
Hollingsworth: *Vectors, matrices, and group theory* (McGraw-Hill).
Jaffé and Orchin: *Symmetry in chemistry* (John Wiley).
Levine: *Quantum chemistry*, Vols. I & II, (Allyn and Bacon).
McWeeny: *Symmetry—an introduction to group theory* (Pergamon Press).
Schonland: *Molecular symmetry* (Van Nostrand).
Tinkham: *Group theory and quantum mechanics* (McGraw-Hill).
Weyl: *Theory of groups and quantum mechanics* (Dover Publications).
Wigner: *Group theory* (Academic Press).

Answers to selected problems

Chapter 2

2.2 CH_4 and SF_6

2.3 *trans*-$Co(en)_2(NH_3)_2^{3+}$

Chapter 3

3.1 (a) $\mathcal{C}_s$
(b) $\mathcal{C}_{3v}$
(c) $\mathcal{D}_{3h}$
(d) $\mathcal{D}_{2d}$
(e) $\mathcal{D}_{3h}$
(f) $\mathcal{D}_{2h}$
(g) $\mathcal{C}_s$

3.3 (a) $\mathcal{D}_{3d}$
(b) $\mathcal{C}_{2v}$
(c) $\mathcal{D}_{3d}$
(d) $\mathcal{D}_{3h}$
(e) $\mathcal{D}_3$

3.6 $\mathcal{D}_3$

Chapter 4

4.7 (a)

$$\left\| \begin{matrix} (a - ib)/D & -(c + id)/D \\ -(-c + id)/D & (a + ib)/D \end{matrix} \right\|$$

$$D = a^2 + b^2 + c^2 + d^2$$

(c) $$\begin{Vmatrix} 0 & 0 & c^{-1} \\ 0 & b^{-1} & 0 \\ a^{-1} & 0 & 0 \end{Vmatrix}$$

(f) $$\begin{Vmatrix} 5 & -3 & \frac{1}{2} \\ -3 & 2 & -\frac{1}{2} \\ 0 & 0 & \frac{1}{2} \end{Vmatrix}$$

4.10 (c) $\lambda_1 = 1;$ $c_1 = 0, c_2 = 0, c_3 = 1$

$\lambda_2 = \cos\theta + i\sin\theta;$ $c_1 = 1/\sqrt{2}, c_2 = i/\sqrt{2}, c_3 = 0$

$\lambda_3 = \cos\theta - i\sin\theta;$ $c_1 = 1/\sqrt{2}, c_2 = -i/\sqrt{2}, c_3 = 0$

4.12 (c) $\lambda_1 = -15;$ $c_1 = (4 - i)/\sqrt{306}, c_2 = -17/\sqrt{306}$

$\lambda_2 = 3;$ $c_1 = (4 - i)/\sqrt{18}, c_2 = 1/\sqrt{18}$

Chapter 5

5.2 $$D(E) = \begin{Vmatrix} 1 & 0 & 0 & 0 \\ 0 & 1 & 0 & 0 \\ 0 & 0 & 1 & 0 \\ 0 & 0 & 0 & 1 \end{Vmatrix}$$

$$D(C_4) = \begin{Vmatrix} 0 & 0 & 0 & 1 \\ 1 & 0 & 0 & 0 \\ 0 & 1 & 0 & 0 \\ 0 & 0 & 1 & 0 \end{Vmatrix}$$

5.4 (a) All matrices are diagonal, the three diagonal elements are:

$\mathbf{R} = \mathbf{E}$;	1, 1, 1
$\mathbf{R} = \mathbf{C}_2(z)$;	-1, -1, 1
$\mathbf{R} = \mathbf{C}_2(y)$;	-1, 1, -1
$\mathbf{R} = \mathbf{C}_2(x)$;	1, -1, -1
$\mathbf{R} = \mathbf{i}$;	-1, -1, -1
$\mathbf{R} = \sigma(xy)$;	1, 1, -1
$\mathbf{R} = \sigma(xz)$;	1, -1, 1
$\mathbf{R} = \sigma(yz)$;	-1, 1, 1

Chapter 7

7.2 $\Gamma^{red} = \Gamma^{A_2} \oplus \Gamma^{B_1} \oplus \Gamma^{E}$

$f^{A_2} = (f_1 + f_2 + f_3 + f_4)/2$
$f^{B_1} = (f_1 - f_2 + f_3 - f_4)/2$
$f_1^E = (f_1 - f_3)/\sqrt{2}$
$f_2^E = (f_2 - f_4)/\sqrt{2}$

7.4 $\Gamma^{red} = 3\Gamma^{A_1} \oplus \Gamma^{A_2} \oplus 2\Gamma^{B_1} \oplus 3\Gamma^{B_2}$

7.6 (a) For $\mathcal{D}_4$, p_3 belongs to Γ^{A_2}
p_1 and p_2 to Γ^E
For $\mathcal{D}_{2h}$, p_1 belongs to $\Gamma^{B_{3u}}$
p_2 belongs to $\Gamma^{B_{2u}}$
p_3 belongs to $\Gamma^{B_{1u}}$

Chapter 8

8.1 (c) $\Gamma^{E_1} \otimes \Gamma^{E_1} = \Gamma^{A_1} \oplus \Gamma^{A_2} \oplus \Gamma^{E_2}$
$\Gamma^{E_1} \otimes \Gamma^{E_2} = \Gamma^{E_1} \oplus \Gamma^{E_2}$
$\Gamma^{E_2} \otimes \Gamma^{E_2} = \Gamma^{A_1} \oplus \Gamma^{A_2} \oplus \Gamma^{E_1}$

8.2 (b) $\sigma = B_{1g}, B_{2g}$ or E_{1g}

$$D(C_4^3) = \begin{Vmatrix} 0 & 1 & 0 & 0 \\ 0 & 0 & 1 & 0 \\ 0 & 0 & 0 & 1 \\ 1 & 0 & 0 & 0 \end{Vmatrix}$$

$$D(C_2) = \begin{Vmatrix} 0 & 0 & 1 & 0 \\ 0 & 0 & 0 & 1 \\ 1 & 0 & 0 & 0 \\ 0 & 1 & 0 & 0 \end{Vmatrix}$$

$$D(C'_{2a}) = \begin{Vmatrix} 0 & 0 & 0 & -1 \\ 0 & 0 & -1 & 0 \\ 0 & -1 & 0 & 0 \\ -1 & 0 & 0 & 0 \end{Vmatrix}$$

$$D(C'_{2b}) = \begin{Vmatrix} 0 & -1 & 0 & 0 \\ -1 & 0 & 0 & 0 \\ 0 & 0 & 0 & -1 \\ 0 & 0 & -1 & 0 \end{Vmatrix}$$

$$D(C''_{2a}) = \begin{Vmatrix} -1 & 0 & 0 & 0 \\ 0 & 0 & 0 & -1 \\ 0 & 0 & -1 & 0 \\ 0 & -1 & 0 & 0 \end{Vmatrix}$$

$$D(C''_{2b}) = \begin{Vmatrix} 0 & 0 & -1 & 0 \\ 0 & -1 & 0 & 0 \\ -1 & 0 & 0 & 0 \\ 0 & 0 & 0 & -1 \end{Vmatrix}$$

Chapter 9

9.2 For $\mathcal{D}_{4d}$, the five B_2 and eight E_1 modes are infra-red active; the six A_1, nine E_2 and eight E_3 modes are Raman active.

For $\mathcal{C}_{2v}$, the three A_1 and two B_2 modes are infra-red active; the three A_1, one A_2 and two B_2 modes are Raman active.

9.5 (a) $\Gamma^0 = 3\Gamma^{A_1} \oplus \Gamma^{A_2} \oplus 4\Gamma^{E}$

(c) $\Gamma^0 = \Gamma^{A_{1g}} \oplus \Gamma^{A_{2g}} \oplus \Gamma^{B_{1g}} \oplus \Gamma^{B_{2g}} \oplus \Gamma^{E_g} \oplus 2\Gamma^{A_{2u}} \oplus \Gamma^{B_{2u}} \oplus 3\Gamma^{E_u}$

Chapter 10

10.1 (a) $\Psi_1(a_u) = 0.526(\phi_1 + \phi_4)/\sqrt{2} + 0.851(\phi_2 + \phi_3)/\sqrt{2}$

$\Psi_2(a_u) = 0.851(\phi_1 + \phi_4)/\sqrt{2} - 0.526(\phi_2 + \phi_3)/\sqrt{2}$

$\Psi_3(b_g) = 0.851(\phi_1 - \phi_4)/\sqrt{2} + 0.526(\phi_2 - \phi_3)/\sqrt{2}$

$\Psi_4(b_g) = 0.526(\phi_1 - \phi_4)/\sqrt{2} - 0.851(\phi_2 - \phi_3)/\sqrt{2}$

$\varepsilon_1 = \alpha + 1.618\beta$

$\varepsilon_2 = \alpha - 0.618\beta$

$\varepsilon_3 = \alpha + 0.618\beta$

$\varepsilon_4 = \alpha - 1.618\beta$

(d) $\Psi_1\,(a_2'') = \sqrt{1/5}\;(\phi_1 + \phi_2 + \phi_3 + \phi_4 + \phi_5)$

$\Psi_2\,(e_1'') = \sqrt{2/5}\;(\phi_1 + c_1\phi_2 + c_2\phi_3 + c_2\phi_4 + c_1\phi_5)$

$\Psi_3\,(e_1'') = \sqrt{2/5}\;(c_1\phi_1 + \phi_2 + c_1\phi_3 + c_2\phi_4 + c_2\phi_5)$

$\Psi_4\,(e_2'') = \sqrt{2/5}\;(\phi_1 + c_2\phi_2 + c_1\phi_3 + c_1\phi_4 + c_2\phi_5)$

$\Psi_5\,(e_2'') = \sqrt{2/5}\;(c_2\phi_1 + \phi_2 + c_2\phi_3 + c_1\phi_4 + c_1\phi_5)$

$c_1 = \cos(2\pi/5)$ and $c_2 = \cos(4\pi/5)$

$\varepsilon_1 = \alpha + 2\beta$

$\varepsilon_2 = \alpha + 0.618\beta$

$\varepsilon_4 = \alpha - 1.618\beta$

Chapter 11

11.1

$$\left.\begin{array}{l} f_{x(5x^2-3r^2)} \\ f_{y(5y^2-3r^2)} \\ f_{z(5z^2-3r^2)} \end{array}\right\} \Gamma^{T_2}$$

$$\left.\begin{array}{l} f_{x(z^2-y^2)} \\ f_{y(z^2-x^2)} \\ f_{z(x^2-y^2)} \end{array}\right\} \Gamma^{T_1}$$

$$f_{xyz} \qquad \Gamma^{A_1}$$

Chapter 12

12.1 a_{1g}: mixtures of s, d_{z^2} and $\Psi^{\sigma}_{A_{1g}}$ (three non-degenerate MOs)

a_{2g}: $\Psi^{\pi}_{A_{2g}}$ (one non-degenerate ligand MO)

b_{1g}: mixtures of $d_{x^2-y^2}$ and $\Psi^{\sigma}_{B_{1g}}$ (two non-degenerate MOs)

b_{2g}: mixtures of d_{xy} and $\Psi^{\pi}_{B_{2g}}$ (two non-degenerate MOs)

e_g: mixtures of d_{xz} and $\Psi^{\pi}_{E_g}$ (1) (two non-degenerate MOs) and degenerate with these two mixtures of d_{yz} and $\Psi^{\pi}_{E_g}$ (2)

a_{2u}: mixtures of p_z and $\Psi^{\pi}_{A_{2u}}$ (two non-degenerate MOs)

b_{2u}: $\Psi^{\pi}_{B_{2u}}$ (one non-degenerate ligand MO)

e_u: mixtures of p_x, $\Psi^{\sigma}_{E_u}$ (1) and $\Psi^{\pi}_{E_u}$ (1) (three non-degenerate MOs) and degenerate with these three mixtures of p_y, $\Psi^{\sigma}_{E_u}$ (2) and $\Psi^{\pi}_{E_u}$ (2).

Index

A CATALOG OF SELECTED
DOVER BOOKS
IN SCIENCE AND MATHEMATICS

Astronomy

BURNHAM'S CELESTIAL HANDBOOK, Robert Burnham, Jr. Thorough guide to the stars beyond our solar system. Exhaustive treatment. Alphabetical by constellation: Andromeda to Cetus in Vol. 1; Chamaeleon to Orion in Vol. 2; and Pavo to Vulpecula in Vol. 3. Hundreds of illustrations. Index in Vol. 3. 2,000pp. 6⅛ x 9¼.
Vol. I: 0-486-23567-X
Vol. II: 0-486-23568-8
Vol. III: 0-486-23673-0

EXPLORING THE MOON THROUGH BINOCULARS AND SMALL TELESCOPES, Ernest H. Cherrington, Jr. Informative, profusely illustrated guide to locating and identifying craters, rills, seas, mountains, other lunar features. Newly revised and updated with special section of new photos. Over 100 photos and diagrams. 240pp. 8¼ x 11. 0-486-24491-1

THE EXTRATERRESTRIAL LIFE DEBATE, 1750–1900, Michael J. Crowe. First detailed, scholarly study in English of the many ideas that developed from 1750 to 1900 regarding the existence of intelligent extraterrestrial life. Examines ideas of Kant, Herschel, Voltaire, Percival Lowell, many other scientists and thinkers. 16 illustrations. 704pp. 5⅜ x 8½. 0-486-40675-X

THEORIES OF THE WORLD FROM ANTIQUITY TO THE COPERNICAN REVOLUTION, Michael J. Crowe. Newly revised edition of an accessible, enlightening book re-creates the change from an earth-centered to a sun-centered conception of the solar system. 242pp. 5⅜ x 8½. 0-486-41444-2

ARISTARCHUS OF SAMOS: The Ancient Copernicus, Sir Thomas Heath. Heath's history of astronomy ranges from Homer and Hesiod to Aristarchus and includes quotes from numerous thinkers, compilers, and scholasticists from Thales and Anaximander through Pythagoras, Plato, Aristotle, and Heraclides. 34 figures. 448pp. 5⅜ x 8½.
0-486-43886-4

A COMPLETE MANUAL OF AMATEUR ASTRONOMY: TOOLS AND TECHNIQUES FOR ASTRONOMICAL OBSERVATIONS, P. Clay Sherrod with Thomas L. Koed. Concise, highly readable book discusses: selecting, setting up and maintaining a telescope; amateur studies of the sun; lunar topography and occultations; observations of Mars, Jupiter, Saturn, the minor planets and the stars; an introduction to photoelectric photometry; more. 1981 ed. 124 figures. 25 halftones. 37 tables. 335pp. 6½ x 9¼. 0-486-42820-8

AMATEUR ASTRONOMER'S HANDBOOK, J. B. Sidgwick. Timeless, comprehensive coverage of telescopes, mirrors, lenses, mountings, telescope drives, micrometers, spectroscopes, more. 189 illustrations. 576pp. 5⅝ x 8¼. (Available in U.S. only.)
0-486-24034-7

STAR LORE: Myths, Legends, and Facts, William Tyler Olcott. Captivating retellings of the origins and histories of ancient star groups include Pegasus, Ursa Major, Pleiades, signs of the zodiac, and other constellations. "Classic."—Sky & Telescope. 58 illustrations. 544pp. 5⅜ x 8½. 0-486-43581-4

Chemistry

THE SCEPTICAL CHYMIST: THE CLASSIC 1661 TEXT, Robert Boyle. Boyle defines the term "element," asserting that all natural phenomena can be explained by the motion and organization of primary particles. 1911 ed. viii+232pp. $5^3/_8$ x $8^1/_2$. 0-486-42825-7

RADIOACTIVE SUBSTANCES, Marie Curie. Here is the celebrated scientist's doctoral thesis, the prelude to her receipt of the 1903 Nobel Prize. Curie discusses establishing atomic character of radioactivity found in compounds of uranium and thorium; extraction from pitchblende of polonium and radium; isolation of pure radium chloride; determination of atomic weight of radium; plus electric, photographic, luminous, heat, color effects of radioactivity. ii+94pp. $5^3/_8$ x $8^1/_2$. 0-486-42550-9

CHEMICAL MAGIC, Leonard A. Ford. Second Edition, Revised by E. Winston Grundmeier. Over 100 unusual stunts demonstrating cold fire, dust explosions, much more. Text explains scientific principles and stresses safety precautions. 128pp. $5^3/_8$ x $8^1/_2$. 0-486-67628-5

MOLECULAR THEORY OF CAPILLARITY, J. S. Rowlinson and B. Widom. History of surface phenomena offers critical and detailed examination and assessment of modern theories, focusing on statistical mechanics and application of results in mean-field approximation to model systems. 1989 edition. 352pp. $5^3/_8$ x $8^1/_2$. 0-486-42544-4

CHEMICAL AND CATALYTIC REACTION ENGINEERING, James J. Carberry. Designed to offer background for managing chemical reactions, this text examines behavior of chemical reactions and reactors; fluid-fluid and fluid-solid reaction systems; heterogeneous catalysis and catalytic kinetics; more. 1976 edition. 672pp. $6^1/_8$ x $9^1/_4$. 0-486-41736-0 $31.95

ELEMENTS OF CHEMISTRY, Antoine Lavoisier. Monumental classic by founder of modern chemistry in remarkable reprint of rare 1790 Kerr translation. A must for every student of chemistry or the history of science. 539pp. $5^3/_8$ x $8^1/_2$. 0-486-64624-6

MOLECULES AND RADIATION: An Introduction to Modern Molecular Spectroscopy. Second Edition, Jeffrey I. Steinfeld. This unified treatment introduces upper-level undergraduates and graduate students to the concepts and the methods of molecular spectroscopy and applications to quantum electronics, lasers, and related optical phenomena. 1985 edition. 512pp. $5^3/_8$ x $8^1/_2$. 0-486-44152-0

A SHORT HISTORY OF CHEMISTRY, J. R. Partington. Classic exposition explores origins of chemistry, alchemy, early medical chemistry, nature of atmosphere, theory of valency, laws and structure of atomic theory, much more. 428pp. $5^3/_8$ x $8^1/_2$. (Available in U.S. only.) 0-486-65977-1

GENERAL CHEMISTRY, Linus Pauling. Revised 3rd edition of classic first-year text by Nobel laureate. Atomic and molecular structure, quantum mechanics, statistical mechanics, thermodynamics correlated with descriptive chemistry. Problems. 992pp. $5^3/_8$ x $8^1/_2$. 0-486-65622-5

ELECTRON CORRELATION IN MOLECULES, S. Wilson. This text addresses one of theoretical chemistry's central problems. Topics include molecular electronic structure, independent electron models, electron correlation, the linked diagram theorem, and related topics. 1984 edition. 304pp. $5^3/_8$ x $8^1/_2$. 0-486-45879-2

Engineering

DE RE METALLICA, Georgius Agricola. The famous Hoover translation of greatest treatise on technological chemistry, engineering, geology, mining of early modern times (1556). All 289 original woodcuts. 638pp. $6^3/_4$ x 11. 0-486-60006-8

FUNDAMENTALS OF ASTRODYNAMICS, Roger Bate et al. Modern approach developed by U.S. Air Force Academy. Designed as a first course. Problems, exercises. Numerous illustrations. 455pp. $5^3/_8$ x $8^1/_2$. 0-486-60061-0

DYNAMICS OF FLUIDS IN POROUS MEDIA, Jacob Bear. For advanced students of ground water hydrology, soil mechanics and physics, drainage and irrigation engineering and more. 335 illustrations. Exercises, with answers. 784pp. $6^1/_8$ x $9^1/_4$. 0-486-65675-6

THEORY OF VISCOELASTICITY (SECOND EDITION), Richard M. Christensen. Complete consistent description of the linear theory of the viscoelastic behavior of materials. Problem-solving techniques discussed. 1982 edition. 29 figures. xiv+364pp. $6^1/_8$ x $9^1/_4$. 0-486-42880-X

MECHANICS, J. P. Den Hartog. A classic introductory text or refresher. Hundreds of applications and design problems illuminate fundamentals of trusses, loaded beams and cables, etc. 334 answered problems. 462pp. $5^3/_8$ x $8^1/_2$. 0-486-60754-2

MECHANICAL VIBRATIONS, J. P. Den Hartog. Classic textbook offers lucid explanations and illustrative models, applying theories of vibrations to a variety of practical industrial engineering problems. Numerous figures. 233 problems, solutions. Appendix. Index. Preface. 436pp. $5^3/_8$ x $8^1/_2$. 0-486-64785-4

STRENGTH OF MATERIALS, J. P. Den Hartog. Full, clear treatment of basic material (tension, torsion, bending, etc.) plus advanced material on engineering methods, applications. 350 answered problems. 323pp. $5^3/_8$ x $8^1/_2$. 0-486-60755-0

A HISTORY OF MECHANICS, René Dugas. Monumental study of mechanical principles from antiquity to quantum mechanics. Contributions of ancient Greeks, Galileo, Leonardo, Kepler, Lagrange, many others. 671pp. $5^3/_8$ x $8^1/_2$. 0-486-65632-2

STABILITY THEORY AND ITS APPLICATIONS TO STRUCTURAL MECHANICS, Clive L. Dym. Self-contained text focuses on Koiter postbuckling analyses, with mathematical notions of stability of motion. Basing minimum energy principles for static stability upon dynamic concepts of stability of motion, it develops asymptotic buckling and postbuckling analyses from potential energy considerations, with applications to columns, plates, and arches. 1974 ed. 208pp. $5^3/_8$ x $8^1/_2$. 0-486-42541-X

BASIC ELECTRICITY, U.S. Bureau of Naval Personnel. Originally a training course; best nontechnical coverage. Topics include batteries, circuits, conductors, AC and DC, inductance and capacitance, generators, motors, transformers, amplifiers, etc. Many questions with answers. 349 illustrations. 1969 edition. 448pp. $6^1/_2$ x $9^1/_4$. 0-486-20973-3

Mathematics

FUNCTIONAL ANALYSIS (Second Corrected Edition), George Bachman and Lawrence Narici. Excellent treatment of subject geared toward students with background in linear algebra, advanced calculus, physics and engineering. Text covers introduction to inner-product spaces, normed, metric spaces, and topological spaces; complete orthonormal sets, the Hahn-Banach Theorem and its consequences, and many other related subjects. 1966 ed. 544pp. $6^1/_8$ x $9^1/_4$. 0-486-40251-7

DIFFERENTIAL MANIFOLDS, Antoni A. Kosinski. Introductory text for advanced undergraduates and graduate students presents systematic study of the topological structure of smooth manifolds, starting with elements of theory and concluding with method of surgery. 1993 edition. 288pp. $5^3/_8$ x $8^1/_2$. 0-486-46244-7

VECTOR AND TENSOR ANALYSIS WITH APPLICATIONS, A. I. Borisenko and I. E. Tarapov. Concise introduction. Worked-out problems, solutions, exercises. 257pp. $5^3/_8$ x $8^1/_4$. 0-486-63833-2

AN INTRODUCTION TO ORDINARY DIFFERENTIAL EQUATIONS, Earl A. Coddington. A thorough and systematic first course in elementary differential equations for undergraduates in mathematics and science, with many exercises and problems (with answers). Index. 304pp. $5^3/_8$ x $8^1/_2$. 0-486-65942-9

FOURIER SERIES AND ORTHOGONAL FUNCTIONS, Harry F. Davis. An incisive text combining theory and practical example to introduce Fourier series, orthogonal functions and applications of the Fourier method to boundary-value problems. 570 exercises. Answers and notes. 416pp. $5^3/_8$ x $8^1/_2$. 0-486-65973-9

COMPUTABILITY AND UNSOLVABILITY, Martin Davis. Classic graduate-level introduction to theory of computability, usually referred to as theory of recurrent functions. New preface and appendix. 288pp. $5^3/_8$ x $8^1/_2$. 0-486-61471-9

AN INTRODUCTION TO MATHEMATICAL ANALYSIS, Robert A. Rankin. Dealing chiefly with functions of a single real variable, this text by a distinguished educator introduces limits, continuity, differentiability, integration, convergence of infinite series, double series, and infinite products. 1963 edition. 624pp. $5^3/_8$ x $8^1/_2$. 0-486-46251-X

METHODS OF NUMERICAL INTEGRATION (SECOND EDITION), Philip J. Davis and Philip Rabinowitz. Requiring only a background in calculus, this text covers approximate integration over finite and infinite intervals, error analysis, approximate integration in two or more dimensions, and automatic integration. 1984 edition. 624pp. $5^3/_8$ x $8^1/_2$. 0-486-45339-1

INTRODUCTION TO LINEAR ALGEBRA AND DIFFERENTIAL EQUATIONS, John W. Dettman. Excellent text covers complex numbers, determinants, orthonormal bases, Laplace transforms, much more. Exercises with solutions. Undergraduate level. 416pp. $5^3/_8$ x $8^1/_2$. 0-486-65191-6

RIEMANN'S ZETA FUNCTION, H. M. Edwards. Superb, high-level study of landmark 1859 publication entitled "On the Number of Primes Less Than a Given Magnitude" traces developments in mathematical theory that it inspired. xiv+315pp. $5^3/_8$ x $8^1/_2$. 0-486-41740-9

INTRODUCTORY REAL ANALYSIS, A.N. Kolmogorov, S. V. Fomin. Translated by Richard A. Silverman. Self-contained, evenly paced introduction to real and functional analysis. Some 350 problems. 403pp. $5^3/_8 \times 8^1/_2$. 0-486-61226-0

APPLIED ANALYSIS, Cornelius Lanczos. Classic work on analysis and design of finite processes for approximating solution of analytical problems. Algebraic equations, matrices, harmonic analysis, quadrature methods, much more. 559pp. $5^3/_8 \times 8^1/_2$. 0-486-65656-X

AN INTRODUCTION TO ALGEBRAIC STRUCTURES, Joseph Landin. Superb self-contained text covers "abstract algebra": sets and numbers, theory of groups, theory of rings, much more. Numerous well-chosen examples, exercises. 247pp. $5^3/_8 \times 8^1/_2$. 0-486-65940-2

QUALITATIVE THEORY OF DIFFERENTIAL EQUATIONS, V. V. Nemytskii and V.V. Stepanov. Classic graduate-level text by two prominent Soviet mathematicians covers classical differential equations as well as topological dynamics and ergodic theory. Bibliographies. 523pp. $5^3/_8 \times 8^1/_2$. 0-486-65954-2

THEORY OF MATRICES, Sam Perlis. Outstanding text covering rank, nonsingularity and inverses in connection with the development of canonical matrices under the relation of equivalence, and without the intervention of determinants. Includes exercises. 237pp. $5^3/_8 \times 8^1/_2$. 0-486-66810-X

INTRODUCTION TO ANALYSIS, Maxwell Rosenlicht. Unusually clear, accessible coverage of set theory, real number system, metric spaces, continuous functions, Riemann integration, multiple integrals, more. Wide range of problems. Undergraduate level. Bibliography. 254pp. $5^3/_8 \times 8^1/_2$. 0-486-65038-3

MODERN NONLINEAR EQUATIONS, Thomas L. Saaty. Emphasizes practical solution of problems; covers seven types of equations. ". . . a welcome contribution to the existing literature. . . ."—*Math Reviews*. 490pp. $5^3/_8 \times 8^1/_2$. 0-486-64232-1

MATRICES AND LINEAR ALGEBRA, Hans Schneider and George Phillip Barker. Basic textbook covers theory of matrices and its applications to systems of linear equations and related topics such as determinants, eigenvalues and differential equations. Numerous exercises. 432pp. $5^3/_8 \times 8^1/_2$. 0-486-66014-1

LINEAR ALGEBRA, Georgi E. Shilov. Determinants, linear spaces, matrix algebras, similar topics. For advanced undergraduates, graduates. Silverman translation. 387pp. $5^3/_8 \times 8^1/_2$. 0-486-63518-X

MATHEMATICAL METHODS OF GAME AND ECONOMIC THEORY: Revised Edition, Jean-Pierre Aubin. This text begins with optimization theory and convex analysis, followed by topics in game theory and mathematical economics, and concluding with an introduction to nonlinear analysis and control theory. 1982 edition. 656pp. $6^1/_8 \times 9^1/_4$. 0-486-46265-X

SET THEORY AND LOGIC, Robert R. Stoll. Lucid introduction to unified theory of mathematical concepts. Set theory and logic seen as tools for conceptual understanding of real number system. 496pp. $5^3/_8 \times 8^1/_4$. 0-486-63829-4

TENSOR CALCULUS, J.L. Synge and A. Schild. Widely used introductory text covers spaces and tensors, basic operations in Riemannian space, non-Riemannian spaces, etc. 324pp. $5^3/_8 \times 8^1/_4$. 0-486-63612-7

ORDINARY DIFFERENTIAL EQUATIONS, Morris Tenenbaum and Harry Pollard. Exhaustive survey of ordinary differential equations for undergraduates in mathematics, engineering, science. Thorough analysis of theorems. Diagrams. Bibliography. Index. 818pp. $5^3/_8 \times 8^1/_2$. 0-486-64940-7

INTEGRAL EQUATIONS, F. G. Tricomi. Authoritative, well-written treatment of extremely useful mathematical tool with wide applications. Volterra Equations, Fredholm Equations, much more. Advanced undergraduate to graduate level. Exercises. Bibliography. 238pp. $5^3/_8 \times 8^1/_2$. 0-486-64828-1

FOURIER SERIES, Georgi P. Tolstov. Translated by Richard A. Silverman. A valuable addition to the literature on the subject, moving clearly from subject to subject and theorem to theorem. 107 problems, answers. 336pp. $5^3/_8 \times 8^1/_2$. 0-486-63317-9

INTRODUCTION TO MATHEMATICAL THINKING, Friedrich Waismann. Examinations of arithmetic, geometry, and theory of integers; rational and natural numbers; complete induction; limit and point of accumulation; remarkable curves; complex and hypercomplex numbers, more. 1959 ed. 27 figures. xii+260pp. $5^3/_8 \times 8^1/_2$. 0-486-42804-8

THE RADON TRANSFORM AND SOME OF ITS APPLICATIONS, Stanley R. Deans. Of value to mathematicians, physicists, and engineers, this excellent introduction covers both theory and applications, including a rich array of examples and literature. Revised and updated by the author. 1993 edition. 304pp. $6^1/_8 \times 9^1/_4$. 0-486-46241-2

CALCULUS OF VARIATIONS, Robert Weinstock. Basic introduction covering isoperimetric problems, theory of elasticity, quantum mechanics, electrostatics, etc. Exercises throughout. 326pp. $5^3/_8 \times 8^1/_2$. 0-486-63069-2

THE CONTINUUM: A CRITICAL EXAMINATION OF THE FOUNDATION OF ANALYSIS, Hermann Weyl. Classic of 20th-century foundational research deals with the conceptual problem posed by the continuum. 156pp. $5^3/_8 \times 8^1/_2$. 0-486-67982-9

CHALLENGING MATHEMATICAL PROBLEMS WITH ELEMENTARY SOLUTIONS, A. M. Yaglom and I. M. Yaglom. Over 170 challenging problems on probability theory, combinatorial analysis, points and lines, topology, convex polygons, many other topics. Solutions. Total of 445pp. $5^3/_8 \times 8^1/_2$. Two-vol. set.
Vol. I: 0-486-65536-9 Vol. II: 0-486-65537-7

INTRODUCTION TO PARTIAL DIFFERENTIAL EQUATIONS WITH APPLICATIONS, E. C. Zachmanoglou and Dale W. Thoe. Essentials of partial differential equations applied to common problems in engineering and the physical sciences. Problems and answers. 416pp. $5^3/_8 \times 8^1/_2$. 0-486-65251-3

STOCHASTIC PROCESSES AND FILTERING THEORY, Andrew H. Jazwinski. This unified treatment presents material previously available only in journals, and in terms accessible to engineering students. Although theory is emphasized, it discusses numerous practical applications as well. 1970 edition. 400pp. $5^3/_8 \times 8^1/_2$. 0-486-46274-9

Math—Decision Theory, Statistics, Probability

INTRODUCTION TO PROBABILITY, John E. Freund. Featured topics include permutations and factorials, probabilities and odds, frequency interpretation, mathematical expectation, decision-making, postulates of probability, rule of elimination, much more. Exercises with some solutions. Summary. 1973 edition. 247pp. $5^3/_8 \times 8^1/_2$. 0-486-67549-1

STATISTICAL AND INDUCTIVE PROBABILITIES, Hugues Leblanc. This treatment addresses a decades-old dispute among probability theorists, asserting that both statistical and inductive probabilities may be treated as sentence-theoretic measurements, and that the latter qualify as estimates of the former. 1962 edition. 160pp. $5^3/_8 \times 8^1/_2$. 0-486-44980-7

APPLIED MULTIVARIATE ANALYSIS: Using Bayesian and Frequentist Methods of Inference, Second Edition, S. James Press. This two-part treatment deals with foundations as well as models and applications. Topics include continuous multivariate distributions; regression and analysis of variance; factor analysis and latent structure analysis; and structuring multivariate populations. 1982 edition. 692pp. $5^3/_8 \times 8^1/_2$. 0-486-44236-5

LINEAR PROGRAMMING AND ECONOMIC ANALYSIS, Robert Dorfman, Paul A. Samuelson and Robert M. Solow. First comprehensive treatment of linear programming in standard economic analysis. Game theory, modern welfare economics, Leontief input-output, more. 525pp. $5^3/_8 \times 8^1/_2$. 0-486-65491-5

PROBABILITY: AN INTRODUCTION, Samuel Goldberg. Excellent basic text covers set theory, probability theory for finite sample spaces, binomial theorem, much more. 360 problems. Bibliographies. 322pp. $5^3/_8 \times 8^1/_2$. 0-486-65252-1

GAMES AND DECISIONS: INTRODUCTION AND CRITICAL SURVEY, R. Duncan Luce and Howard Raiffa. Superb nontechnical introduction to game theory, primarily applied to social sciences. Utility theory, zero-sum games, n-person games, decision-making, much more. Bibliography. 509pp. $5^3/_8 \times 8^1/_2$. 0-486-65943-7

INTRODUCTION TO THE THEORY OF GAMES, J. C. C. McKinsey. This comprehensive overview of the mathematical theory of games illustrates applications to situations involving conflicts of interest, including economic, social, political, and military contexts. Appropriate for advanced undergraduate and graduate courses; advanced calculus a prerequisite. 1952 ed. x+372pp. $5^3/_8 \times 8^1/_2$. 0-486-42811-7

FIFTY CHALLENGING PROBLEMS IN PROBABILITY WITH SOLUTIONS, Frederick Mosteller. Remarkable puzzlers, graded in difficulty, illustrate elementary and advanced aspects of probability. Detailed solutions. 88pp. $5^3/_8 \times 8^1/_2$. 0-486-65355-2

PROBABILITY THEORY: A CONCISE COURSE, Y. A. Rozanov. Highly readable, self-contained introduction covers combination of events, dependent events, Bernoulli trials, etc. 148pp. $5^3/_8 \times 8^1/_4$. 0-486-63544-9

THE STATISTICAL ANALYSIS OF EXPERIMENTAL DATA, John Mandel. First half of book presents fundamental mathematical definitions, concepts and facts while remaining half deals with statistics primarily as an interpretive tool. Well-written text, numerous worked examples with step-by-step presentation. Includes 116 tables. 448pp. $5^3/_8 \times 8^1/_2$. 0-486-64666-1

Math—Geometry and Topology

ELEMENTARY CONCEPTS OF TOPOLOGY, Paul Alexandroff. Elegant, intuitive approach to topology from set-theoretic topology to Betti groups; how concepts of topology are useful in math and physics. 25 figures. 57pp. 5³/₈ x 8¹/₂. 0-486-60747-X

A LONG WAY FROM EUCLID, Constance Reid. Lively guide by a prominent historian focuses on the role of Euclid's Elements in subsequent mathematical developments. Elementary algebra and plane geometry are sole prerequisites. 80 drawings. 1963 edition. 304pp. 5³/₈ x 8¹/₂. 0-486-43613-6

EXPERIMENTS IN TOPOLOGY, Stephen Barr. Classic, lively explanation of one of the byways of mathematics. Klein bottles, Moebius strips, projective planes, map coloring, problem of the Koenigsberg bridges, much more, described with clarity and wit. 43 figures. 210pp. 5³/₈ x 8¹/₂. 0-486-25933-1

THE GEOMETRY OF RENÉ DESCARTES, René Descartes. The great work founded analytical geometry. Original French text, Descartes's own diagrams, together with definitive Smith-Latham translation. 244pp. 5³/₈ x 8¹/₂. 0-486-60068-8

EUCLIDEAN GEOMETRY AND TRANSFORMATIONS, Clayton W. Dodge. This introduction to Euclidean geometry emphasizes transformations, particularly isometries and similarities. Suitable for undergraduate courses, it includes numerous examples, many with detailed answers. 1972 ed. viii+296pp. 6¹/₈ x 9¹/₄. 0-486-43476-1

EXCURSIONS IN GEOMETRY, C. Stanley Ogilvy. A straightedge, compass, and a little thought are all that's needed to discover the intellectual excitement of geometry. Harmonic division and Apollonian circles, inversive geometry, hexlet, Golden Section, more. 132 illustrations. 192pp. 5³/₈ x 8¹/₂. 0-486-26530-7

THE THIRTEEN BOOKS OF EUCLID'S ELEMENTS, translated with introduction and commentary by Sir Thomas L. Heath. Definitive edition. Textual and linguistic notes, mathematical analysis. 2,500 years of critical commentary. Unabridged. 1,414pp. 5³/₈ x 8¹/₂. Three-vol. set.
Vol. I: 0-486-60088-2 Vol. II: 0-486-60089-0 Vol. III: 0-486-60090-4

SPACE AND GEOMETRY: IN THE LIGHT OF PHYSIOLOGICAL, PSYCHOLOGICAL AND PHYSICAL INQUIRY, Ernst Mach. Three essays by an eminent philosopher and scientist explore the nature, origin, and development of our concepts of space, with a distinctness and precision suitable for undergraduate students and other readers. 1906 ed. vi+148pp. 5³/₈ x 8¹/₂. 0-486-43909-7

GEOMETRY OF COMPLEX NUMBERS, Hans Schwerdtfeger. Illuminating, widely praised book on analytic geometry of circles, the Moebius transformation, and two-dimensional non-Euclidean geometries. 200pp. 5⁵/₈ x 8¹/₄. 0-486-63830-8

DIFFERENTIAL GEOMETRY, Heinrich W. Guggenheimer. Local differential geometry as an application of advanced calculus and linear algebra. Curvature, transformation groups, surfaces, more. Exercises. 62 figures. 378pp. 5³/₈ x 8¹/₂. 0-486-63433-7

History of Math

THE WORKS OF ARCHIMEDES, Archimedes (T. L. Heath, ed.). Topics include the famous problems of the ratio of the areas of a cylinder and an inscribed sphere; the measurement of a circle; the properties of conoids, spheroids, and spirals; and the quadrature of the parabola. Informative introduction. clxxxvi+326pp. 5³/₈ x 8¹/₂. 0-486-42084-1

A SHORT ACCOUNT OF THE HISTORY OF MATHEMATICS, W. W. Rouse Ball. One of clearest, most authoritative surveys from the Egyptians and Phoenicians through 19th-century figures such as Grassman, Galois, Riemann. Fourth edition. 522pp. 5³/₈ x 8¹/₂. 0-486-20630-0

THE HISTORY OF THE CALCULUS AND ITS CONCEPTUAL DEVELOPMENT, Carl B. Boyer. Origins in antiquity, medieval contributions, work of Newton, Leibniz, rigorous formulation. Treatment is verbal. 346pp. 5³/₈ x 8¹/₂. 0-486-60509-4

THE HISTORICAL ROOTS OF ELEMENTARY MATHEMATICS, Lucas N. H. Bunt, Phillip S. Jones, and Jack D. Bedient. Fundamental underpinnings of modern arithmetic, algebra, geometry and number systems derived from ancient civilizations. 320pp. 5³/₈ x 8¹/₂. 0-486-25563-8

THE HISTORY OF THE CALCULUS AND ITS CONCEPTUAL DEVELOPMENT, Carl B. Boyer. Fluent description of the development of both the integral and differential calculus—its early beginnings in antiquity, medieval contributions, and a consideration of Newton and Leibniz. 368pp. 5³/₈ x 8¹/₂. 0-486-60509-4

GAMES, GODS & GAMBLING: A HISTORY OF PROBABILITY AND STATISTICAL IDEAS, F. N. David. Episodes from the lives of Galileo, Fermat, Pascal, and others illustrate this fascinating account of the roots of mathematics. Features thought-provoking references to classics, archaeology, biography, poetry. 1962 edition. 304pp. 5³/₈ x 8¹/₂. (Available in U.S. only.) 0-486-40023-9

OF MEN AND NUMBERS: THE STORY OF THE GREAT MATHEMATICIANS, Jane Muir. Fascinating accounts of the lives and accomplishments of history's greatest mathematical minds—Pythagoras, Descartes, Euler, Pascal, Cantor, many more. Anecdotal, illuminating. 30 diagrams. Bibliography. 256pp. 5³/₈ x 8¹/₂. 0-486-28973-7

HISTORY OF MATHEMATICS, David E. Smith. Nontechnical survey from ancient Greece and Orient to late 19th century; evolution of arithmetic, geometry, trigonometry, calculating devices, algebra, the calculus. 362 illustrations. 1,355pp. 5³/₈ x 8¹/₂. Two-vol. set. Vol. I: 0-486-20429-4 Vol. II: 0-486-20430-8

A CONCISE HISTORY OF MATHEMATICS, Dirk J. Struik. The best brief history of mathematics. Stresses origins and covers every major figure from ancient Near East to 19th century. 41 illustrations. 195pp. 5³/₈ x 8¹/₂. 0-486-60255-9

Physics

OPTICAL RESONANCE AND TWO-LEVEL ATOMS, L. Allen and J. H. Eberly. Clear, comprehensive introduction to basic principles behind all quantum optical resonance phenomena. 53 illustrations. Preface. Index. 256pp. $5^3/_8$ x $8^1/_2$. 0-486-65533-4

QUANTUM THEORY, David Bohm. This advanced undergraduate-level text presents the quantum theory in terms of qualitative and imaginative concepts, followed by specific applications worked out in mathematical detail. Preface. Index. 655pp. $5^3/_8$ x $8^1/_2$. 0-486-65969-0

ATOMIC PHYSICS (8th EDITION), Max Born. Nobel laureate's lucid treatment of kinetic theory of gases, elementary particles, nuclear atom, wave-corpuscles, atomic structure and spectral lines, much more. Over 40 appendices, bibliography. 495pp. $5^3/_8$ x $8^1/_2$. 0-486-65984-4

A SOPHISTICATE'S PRIMER OF RELATIVITY, P. W. Bridgman. Geared toward readers already acquainted with special relativity, this book transcends the view of theory as a working tool to answer natural questions: What is a frame of reference? What is a "law of nature"? What is the role of the "observer"? Extensive treatment, written in terms accessible to those without a scientific background. 1983 ed. xlviii+172pp. $5^3/_8$ x $8^1/_2$. 0-486-42549-5

AN INTRODUCTION TO HAMILTONIAN OPTICS, H. A. Buchdahl. Detailed account of the Hamiltonian treatment of aberration theory in geometrical optics. Many classes of optical systems defined in terms of the symmetries they possess. Problems with detailed solutions. 1970 edition. xv + 360pp. $5^3/_8$ x $8^1/_2$. 0-486-67597-1

PRIMER OF QUANTUM MECHANICS, Marvin Chester. Introductory text examines the classical quantum bead on a track: its state and representations; operator eigenvalues; harmonic oscillator and bound bead in a symmetric force field; and bead in a spherical shell. Other topics include spin, matrices, and the structure of quantum mechanics; the simplest atom; indistinguishable particles; and stationary-state perturbation theory. 1992 ed. xiv+314pp. $6^1/_8$ x $9^1/_4$. 0-486-42878-8

LECTURES ON QUANTUM MECHANICS, Paul A. M. Dirac. Four concise, brilliant lectures on mathematical methods in quantum mechanics from Nobel Prize-winning quantum pioneer build on idea of visualizing quantum theory through the use of classical mechanics. 96pp. $5^3/_8$ x $8^1/_2$. 0-486-41713-1

THIRTY YEARS THAT SHOOK PHYSICS: THE STORY OF QUANTUM THEORY, George Gamow. Lucid, accessible introduction to influential theory of energy and matter. Careful explanations of Dirac's anti-particles, Bohr's model of the atom, much more. 12 plates. Numerous drawings. 240pp. $5^3/_8$ x $8^1/_2$. 0-486-24895-X

ELECTRONIC STRUCTURE AND THE PROPERTIES OF SOLIDS: THE PHYSICS OF THE CHEMICAL BOND, Walter A. Harrison. Innovative text offers basic understanding of the electronic structure of covalent and ionic solids, simple metals, transition metals and their compounds. Problems. 1980 edition. 582pp. $6^1/_8$ x $9^1/_4$. 0-486-66021-4